Contents

BRE Building Elements

Building services

Performance, diagnosis, maintenance, repair and the avoidance of defects

H W Harrison, ISO, Dip Arch, RIBA

P M Trotman

BRE
Garston
Watford
WD2 7JR

Prices for all available
BRE publications can be
obtained from:
CRC Ltd
151 Rosebery Avenue
London EC1R 4GB
Tel: 020 7505 6622
Fax: 020 7505 6606
email:
crc@construct.emap.co.uk

BR 404
ISBN 1 86081 424 7

Published by
Construction Research
Communications Ltd
by permission of
Building Research
Establishment Ltd

Requests to copy any part of
this publication should be
made to:
CRC Ltd
PO Box 202
Watford WD25 9ZW

BRE material is also published quarterly on CD

Each CD contains BRE material published in the current
year, including reports, specialist reports, and the
Professional Development publications: Digests,
Good Building Guides, Good Repair Guides and
Information Papers.

The CD collection gives you the opportunity to build a
comprehensive library of BRE material at a fraction of
the cost of printed copies.

As a subscriber you also benefit from a 20% discount on
other BRE titles.

For more information contact:
CRC Customer Services on 020 7505 6622

Construction Research Communications

CRC supplies a wide range of building and construction
related information products from BRE and other highly
respected organisations.

Contact:
post: CRC Ltd
          151 Rosebery Avenue
          London EC1R 4GB

fax: 020 7505 6606
phone: 020 7505 6622
email: crc@construct.emap.co.uk
website: www.constructionplus.co.uk

Preface

This book is about building services: the gamut of fuelled, piped, ducted, wired and mechanical facilities which extend over the whole age range of the UK's building stock. In essence, it is the quality of heating, artificial lighting and other services, such as refuse disposal, which makes an otherwise bare carcass habitable – even efficient and enjoyable. However, things do go wrong from time to time – the boiler ceases to function, the television aerial corrodes, the pipes in hard water areas fill with calcium deposits, and, sadly, there is the occasional disaster and tragedy from carbon monoxide poisoning or electrical fault.

Sir John Egan, in his report, *Rethinking construction*[1], has drawn attention to the need, amongst other things, to improve productivity, reduce construction times and cut accidents in the UK construction industry, and he also called for a 20% cut in the number of defects. This book may be seen as a contribution to some of the aims identified in the report, and is drawn from the collective experience of the Building Research Station and its successors, the Building Research Establishment and BRE Ltd, over the years since its foundation in 1921.

Information on faults in buildings of all types shows that a substantial number relate to building services, and some of those are rather elementary in nature. An analysis of the information available to BRE is given in Chapter 0.

Readership

Building services is addressed primarily to building surveyors and other professionals performing similar functions – such as architects and builders – who maintain, repair, extend and renew the national building stock. During the course of routine surveys of existing buildings by surveyors and architects, there will be a need to identify items relating to the services: deficiencies in performance resulting from outdated installations, or breakdown through wear and tear, which demand attention. Bearing in mind the increasing complexity of building services, and of the old adage that a little knowledge can be a dangerous thing, there is still a need for sufficient information to be available to enable building owners to be advised on when to call in specialist consultants to rectify, enhance or replace existing installations. *Building services* is certainly not aimed at the mechanical and electrical engineer nor indeed at the building services engineer, though it will perhaps find application in the education field.

The descriptions and advice given in the pages which follow concentrate on practical details. But there also needs to be sufficient discussion of principles to impart understanding of the reason for certain practices, and some of this information of a general nature is given in Chapter 1.

Scope of the book

All kinds of building services, including space heating and cooling, ventilation systems, piped services of water and gas, refuse disposal, wired services for electricity, telephone and television, and mechanical handling systems such as lifts and escalators are, in principle, covered in the book. It will immediately be apparent that in a production of this limited size the coverage on any particular topic can only be brief. The text therefore concentrates on those aspects with which BRE has been most heavily involved, whether in laboratory research, site investigation or the development of legislation and Standards.

Although lightning protection might possibly be considered as a building service, this topic has already been dealt with in *Roofs and roofing*, Chapter 1.6, and *Walls, windows and doors*, Chapter 1.8; the risk of lightning strikes and detailed discussion of the necessary provision against strikes is not included in this book, though a case study drawing attention to the extent of damage which can occur to a chimney is included in Chapter 2.1. However, since these other volumes were written BRE has issued a new Digest which deals with lightning protection[2].

Building services does not deal with industrial plant engineering, nor in any detail with the design and installation (and faults) of the large scale installations necessary to service larger buildings. Nor can it deal with specialised buildings such as cold stores. There is an almost incredible variety of practices, with corresponding potential for error and breakdown. The saving grace is, of course, the degree of specialisation, competence and professional skills of the larger firms of consultants and contractors in the building services industry who are more usually involved in these larger schemes. It is in the smaller schemes that corners are sometimes cut.

In principle, all types of buildings are included. However, it is inevitable that the nature of installations becomes very sophisticated in some building types such as factories and health buildings, and these installations do not make it easy to provide simple guidance for use of non-specialists on site. The topics differ somewhat in this respect from those which have been covered in the other books in the BRE Building Elements series. However, even the relatively simple systems used in the majority of domestic construction provide adequate potential for improvement.

Both good and bad features of building services are described, and sources of further information and advice are offered. The drawings are not working drawings but merely show either those aspects to which the particular attention of readers needs to be drawn or provide typical details to support text. The discussion is deliberately neutral on matters of style and aesthetics and is wary of suggesting that there is ever a unique optimum solution.

As with the other books in this series, the text concentrates on those aspects of building services which, in the experience of BRE, lead to the greatest number of problems or greatest potential expense, if carried out unsatisfactorily. It follows that these problems will be picked up most frequently by maintenance surveyors and others specifying and carrying out remedial work on building services. Occasionally there is information relating to an item, perhaps a fault, which is infrequently encountered, and about which it may in consequence be difficult to locate information. Although most of the information relates to older

buildings, material concerning observations by BRE investigators of new buildings under construction in the period from 1985 to 1995 is also included.

The case studies provided in some of the chapters are selected from the files of the BRE Advisory Service, the Building Research Energy Conservation Support Unit, and the former BRE Defects Prevention Unit, and represent the most frequent kinds of problems on which BRE has been consulted.

An attempt has been made within the chapters to follow the standard order of section headings adopted for the other books in the series. These standard headings are repeated only where there is a need to refer the reader to earlier statements or where there is something relevant to add to what has gone before.

In the United Kingdom, there are three different sets of building regulations: the Building Regulations 1991 which apply to England and Wales; the Building Standards (Scotland) Regulations 1990; and the Building Regulations (Northern Ireland) 1994. There are many common provisions between the three sets, but there are also major differences. The book has been written against the background of the building regulations for England and Wales, since, although there has been an active Advisory Service for Scotland and Northern Ireland, the highest proportion of site inspections has been carried out in England and Wales. The fact that the majority of references to building regulations are to those for England and Wales should not make the book less applicable to Scotland and Northern Ireland.

In addition to the building regulations, there is also other legislation such as the Electricity at Work Regulations 1989, the Electricity Supply Regulations 1989, the Regulations of the Institution of Electrical Engineers, now published by BSI as Requirements for electrical installations, BS 7671[3], the Gas Safety (Installation and Use) Regulations 1998, and the Water

Byelaws[4] which were succeeded by the Water Regulations in 1999[5].

Although practically all building services are encompassed in the Construction (Design and Management) Regulations 1994, the ramifications for each of the services covered in this book are considerable. It is not practical to spell them out in this book, beyond noting that there must be a Health and Safety Plan and File for buildings constructed after this date which should include information on how to manage health and safety issues after the installation is completed and throughout its life until demolition[6].

Some important definitions

The broad term 'services' usually includes those provisions for meeting the internal environmental requirements that – like heating, lighting and ventilating systems – depend on the consumption of energy and materials. The most common requirements today are for consumable water for life support and for sanitation, for consumable energy (for heating, lighting, ventilation and other purposes), and for means of transportation and telecommunication. All services, of course, require further space for their accommodation within the building, and most, but not all, also require support and enclosure.

For the purposes of this book the word 'chimney' means a structure consisting of a wall or walls containing a flue or flues. This definition includes any part of the fabric of a building or a part separate from it; that is to say, either masonry carcass or metal sheath. The 'flue' is the continuous void which actually carries the products of combustion from the appliance to the terminal. The term 'duct' means an enclosed void which carries one or more pipes from one part of the building to another part. Chimney can also apply to the structure which encloses a vertical ventilation duct. The term duct can also apply to the (usually) sheet metal enclosed void which carries fresh air into the building, or vitiated or exhausted air out of the building.

Since *Building services* is mainly about the problems that occur in building services, two words, 'fault' and 'defect', need precise definition. Fault describes a departure from good practice in design or execution of design; it is used for any departure from requirements specified in building regulations, British Standards and Codes of practice, and the published recommendations of authoritative organisations. A defect – a shortfall in performance – is the product of a fault, but while such a consequence cannot always be predicted with certainty, all faults have the potential for leading to defects. The word 'failure' has occasionally been used to signify the more serious defects (and catastrophes!). The word fault as used here is not synonymous with electrical fault as defined in BS 7671, and defects and unsafe conditions take on a special significance in gas utilisation as controlled by the Gas Safety (Installation and Use) Regulations.

A general requirement for 'safety' arises because many of the means adopted to satisfy the primary user requirements create potential or actual hazards. BRE has been greatly concerned with safety over the years. The most important aspects in the past history of building have been structural collapse and fire. Hazards to health probably come next, though they tend to be more insidious, and less easily recognised and defined. Other aspects include explosion (closely related to fire),

and a variety of possible contributory causes of human accidents such as falls. Safety means the reduction of these hazards and risks of accident to tolerable levels since absolute safety is virtually unattainable. A number of accident rates are quoted in this book for various building services. As noted in the companion book *Roofs and roofing*, it is a matter for the collective judgement of society, operating through building regulations and British Standards, whether these accident rates are acceptable, for it could be very expensive to uprate all Standards to provide for better protection.

Where the term 'investigator' has been used, it covers a variety of roles including a member of BRE's Advisory Service, a BRE researcher or a consultant working under contract to BRE.

Particular terms used in connection with energy, central heating and air conditioning will be found listed in later chapters.

So far as water terms are concerned, there has been a significant change in usage since the 1980s. The term 'potable' water to describe water of a quality suitable for drinking is now no longer popular, though it is still contained in current Standards: the terms 'drinking water' or 'wholesome water' are preferred. 'Grey water' is defined as waste water not containing faecal matter or urine, and 'black water' is defined as waste water which contains faecal matter or urine.

Acknowledgements

Photographs which do not bear an attribution have been provided from our own collections or from the BRE Photographic Archive, a unique collection dating from the early 1920s.

We are particularly grateful to Peter Mapp (Peter Mapp and Associates) who drafted Chapter 5.4, and to Geoff Dunstan (GDK Associates), who provided information and comments on parts of Chapter 2.

To the following BRE colleagues, and former colleagues, who have suggested material for this book or commented on drafts, or both, we offer our thanks:
E Bartlett, A K R Bromley, A J Butler, D J Butler, A Buxton, Sandy Cayless, R Cox, Maggie Davidson, P J Fardell, J Griggs, P Guy, the late Dr J Hall, M Lyons, H P Morgan, Penny Morgan, B Musannif, Dr M D A E S Perera, Prof G J Raw, Dr R Rayment, R E H Read, C Scivyer, J Seller, M Shouler, N Smithies, R K Stephen, Dr P Warren, D Warriner, Dr Corinne Williams and B Young, all of the Building Research Establishment Ltd.

Robert Rayment contributed much of Chapter 2.2, in addition to supplying information for other chapters, and Alan Buxton made a substantial contribution to Chapters 5.1 and 5.2. We have also drawn upon some notes prepared originally by Dr Rowland Mainstone when a revision to *Principles of modern building* was under consideration.

H W H
P M T
July 2000

Some less common technical abbreviations

AAV	air admittance valve	HCFC	hydrochlorofluorocarbon
ABS	acrylonitrile butadiene styrene	HDPE	high density polyethylene
ach	air changes per hour	HVAC	heating, ventilation and air conditioning
AFILS	audio frequency induction loop system	ISP	indoor surface pollution
ATM	air treatment module	LAN	local area network
BDF	building distribution frame	LSZH	low smoke zero halogen
BEMS	building energy management system	MDF	medium density fibreboard
BMS	building management system	MICC	mineral insulated copper covered
CFC	chlorofluorocarbons	MUPVC	modified unplasticised polyvinylchloride
CHP	combined heat and power	MV	mechanical ventilation
CMP	communications metallic plenum	MVHR	mechanical ventilation with heat recovery
COP	coefficient of performance	PCM	phase change material
DHW	domestic hot water	PCV	prescribed concentration value
EN	Euronorm (European Standard)	PIR	passive infrared
ESD	electrostatic discharge	PME	protective multiple earthing
ESFR	early suppression fast response	PP	polypropylene
ETS	environmental tobacco smoke	prEN	Pre-Euronorm (draft European Standard)
		PSALI	permanent supplementary artificial lighting of interiors
		PSV	passive stack ventilation
		PV	photovoltaic
		PVC-U	unplasticised polyvinylchloride

RCBO	residual current breaker with overcurrent protection
RCD	residual current device
RF	radio frequency
RH	relative humidity
SBS	sick building syndrome
SELV	safety extra low voltage
SHEV	smoke and heat exhaust ventilation
SHL	specific heat loss
SR	sound reinforcement
SSV	single sided ventilation
STP	shielded twisted pair
SWA	steel wire armoured
TRS	tough rubber sheathed
TRV	thermostatic radiator valve
TST	total solar transmittance
UF	urea-formaldehyde
UPS	uninterruptible power supply
VAS	voice alarm system
VAV	variable air volume
VOC	volatile organic compound
VRF	variable refrigerant flow
VRV	variable refrigerant volume
μm	micrometre (1/1000 mm)

Chapter 0 **Introduction**

It is taken for granted that modern buildings perform well. Comfortable internal temperatures in all weathers with controllable draught-free ventilation, the absence of dampness, extensive facilities for preparing and cooking food, and communication, lighting and sanitation are all assumed to work properly. To satisfy these expectations at economic cost, designers and builders can choose from a wide range of materials, components and techniques. But the pace at which new practices and products have been introduced and the breadth of choice offered, coupled with the changes in expectations of users, mean that it is often impossible to rely on well tested and traditional methods of selection. Even simple substitutions of new materials for old, if made unthinkingly or without adequate understanding, can lead to serious failures.

Most building services do perform well (Figure 0.1). Having said that, however, there is evidence that avoidable defects in building services occur too often. This evidence is contained in BRE surveys of housing of both new construction and rehabilitation; from the United Kingdom house condition surveys[7–10] (undertaken every five years); from the past commissions of the BRE Advisory Service, especially for building types other than housing; from the records of the Building Research Energy Conservation Support Unit in relation to the use of energy in heating and lighting; and from the Construction Quality Forum database organised by BRE on behalf of the industry.

Records of failures and faults in buildings

BRE Advisory Service and BRECSU records

Since the setting up of the Building Research Energy Conservation Unit (BRECSU) in 1981, most of the enquiries on energy conservation aspects of heating and lighting directed to BRE have been dealt with by this unit. BRE Advisory Service records directly concerned with building services show that enquiries related mainly to problems with heating and condensation, a proportion related to piped hot and cold water, but there has been very little on electrical services.

BRE Defects Database records

Some information on faults which occur in building services in housing is available in the BRE Defects Prevention Unit Quality in Housing database[11]. The database records non-compliance with requirements whatever their origin, whether building regulations, British Standards and Codes of practice, industry standards or other authoritative requirements. It also records actual inspections by BRE investigators or by consultants working under BRE supervision gathered over the years 1980–90. The investigations covered 139 sites, with actual numbers of dwellings ranging from 1 to 60 or more per site; altogether more than 2,500 dwellings were investigated.

The inspections of the sites were thorough, ranging from just under one man-day for individual dwellings up to six or more man-days for larger sites. Investigation teams included building surveyors, architects, various kinds of scientists, engineers and quantity surveyors. The inspection reports record types of faults, with numbers of incidents being estimated and classified as universal (that is to say the fault occurred on all dwellings inspected on a particular site), frequent (more than half of all dwellings), occasional (a few only), or unique (only one on the site). The totals of incidents of faults cannot be deduced with absolute accuracy; nevertheless, the numbers of fault types and incidents are believed to represent a reasonably true picture.

Figure 0.1
Building services are now expected to reach a high degree of sophistication, and to perform impeccably with minimum demands on the environment

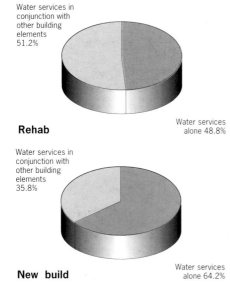

Rehab

Water services in conjunction with other building elements 51.2%

Water services alone 48.8%

New build

Water services in conjunction with other building elements 35.8%

Water services alone 64.2%

Figure 0.2
Distribution of faults identified in BRE DPU site investigations of rehabilitated and new-build housing: water services according to whether the fault occurred within the element or at its junction with another element, 1990

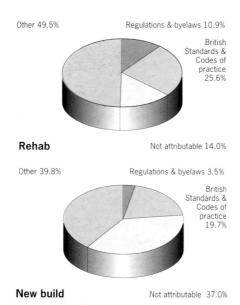

Rehab

Other 49.5%

Regulations & byelaws 10.9%

British Standards & Codes of practice 25.6%

Not attributable 14.0%

New build

Other 39.8%

Regulations & byelaws 3.5%

British Standards & Codes of practice 19.7%

Not attributable 37.0%

Figure 0.3
Distribution of faults identified in BRE DPU site investigations of rehabilitated and new-build housing: water services according to authority contravened, 1990

Data have been analysed broadly into two categories related to date of construction of dwellings: those being built new at the time of inspection and those being refurbished. Only types of faults which occur relatively frequently have been listed.

Water services
This category includes all cold and hot water services including hot water for space heating and above ground drainage. For rehabilitated housing, 129 different items were recorded, and for new-build, 111 items. As already noted, these represented a far greater number of actual faults, though the average number of fault types recorded per site were few. Where a fault occurred at all, it was more than likely to occur on all dwellings in that site.

With respect to element involved (Figure 0.2) for rehabilitated housing, there were almost equal numbers of items relating to water services alone compared with those occurring at the junctions of this service with other elements such as walls, floors etc. For new-build the figures were two-thirds and one-third respectively. Common examples of the former were uninsulated pipes being too close together or touching, and of the latter, pipes not properly fixed to supports.

For authority contravened (Figure 0.3) for rehabilitated housing, just under one quarter of all items related to infringements of codes, Standards and regulations while for new-build it was well over one-third. Common examples within this category included waste pipes not within required limits on length and slope, and open flued boilers with insufficient ventilation.

With performance (Figure 0.4) for rehabilitated housing, safety and habitability related items were more common than in new-build, whereas in new-build the incidence of faults in thermal insulation and heating, and in safety and habitability, was more equal, as one would expect. Common examples in the former category were radiators inappropriately sized, inadequately controlled, or unsuitably positioned.

With origin of faults (Figure 0.5), that is to say who or what was responsible, there was not a great deal to choose between design and site for either category, but materials and components showed quite a large difference. So far this has defied explanation.

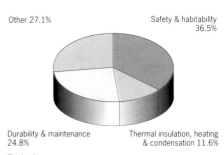

Rehab

Other 27.1%

Safety & habitability 36.5%

Durability & maintenance 24.8%

Thermal insulation, heating & condensation 11.6%

New build

Other 19.1%

Safety & habitability 22.5%

Durability & maintenance 36.4%

Thermal insulation, heating & condensation 22.0%

Figure 0.4
Distribution of faults identified in BRE DPU site investigations of rehabilitated and new-build housing: water services according to performance, 1990

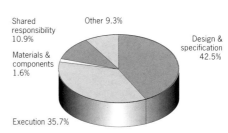

Rehab

Shared responsibility 10.9%

Other 9.3%

Materials & components 1.6%

Design & specification 42.5%

Execution 35.7%

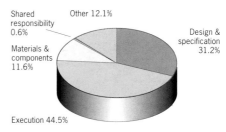

New build

Shared responsibility 0.6%

Other 12.1%

Materials & components 11.6%

Design & specification 31.2%

Execution 44.5%

Figure 0.5
Distribution of faults identified in BRE DPU site investigations of rehabilitated and new-build housing: water services according to responsibility for the fault, 1990

Gas services
This category includes all gas services, both mains and LPG. For rehabilitated housing, 40 different items were recorded, and for new-build, 58 items. Numbers of fault types for gas services were therefore

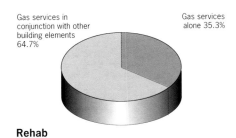

Rehab

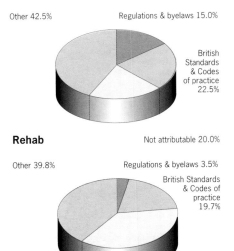

New build

Figure 0.6
Distribution of faults identified in BRE DPU site investigations of rehabilitated and new-build housing: gas services according to whether the fault occurred within the element or at its junction with another element, 1990

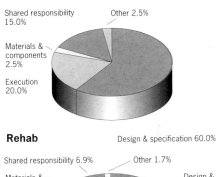

Rehab

New build

Figure 0.7
Distribution of faults identified in BRE DPU site investigations of rehabilitated and new-build housing: gas services according to authority contravened, 1990

far fewer than were observed for other building elements, and so could be less representative of faults actually occurring on the schemes examined. Nevertheless, it can be seen that no gas-related faults were recorded on roughly one-third of all sites inspected (indeed, not all sites were served by gas). The items identified were later verified with design and site representatives.

With respect to element involved (Figure 0.6) for rehabilitated housing, about two-thirds of items related to gas services alone, and one-third at the junctions with other elements such as walls, floors etc; for new-build the figures did not show such a marked difference. Common examples were inappropriate siting or insecure fixings of pipes to walls, and gas meter boxes with no lintels over.

For authority contravened (Figure 0.7) there was rather better observance of Standards, Codes and regulations in new-build than there was in rehab, as one might expect with reuse of old installations. The most frequently encountered examples included no ventilation of gas pipes running in voids. Other, which is the largest single category in both cases, includes such matters as particular client requirements for carcassing, and meter and meter box positioning.

With performance (Figure 0.8) for rehabilitated housing, safety and habitability related items were far more common than in new-build; in fact twice as common. Examples frequently encountered in this category were inadequate (or blocked) supplies of combustion air for open flued appliances in older properties.

With origin of faults (Figure 0.9), that is to say who or what was responsible, design and specification items were more often encountered in rehabilitation work than in new-build. This was almost certainly due to out-of-date and inefficient appliances still in use or new but inappropriate items being specified for replacement purposes.

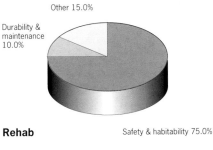

Rehab

New build

Figure 0.8
Distribution of faults identified in BRE DPU site investigations of rehabilitated and new-build housing: gas services according to performance, 1990

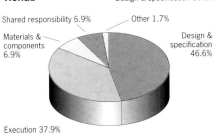

Rehab

New build

Figure 0.9
Distribution of faults identified in BRE DPU site investigations of rehabilitated and new-build housing: gas services according to responsibility for the fault, 1990

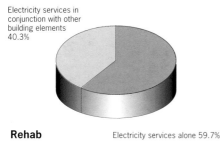

Rehab   Electricity services alone 59.7%

Electricity services in conjunction with other building elements 40.3%

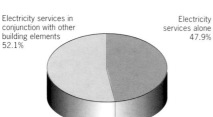

New build

Electricity services in conjunction with other building elements 52.1%

Electricity services alone 47.9%

Figure 0.10
Distribution of faults identified in BRE DPU site investigations of rehabilitated and new-build housing: electricity services according to whether the fault occurred within the element or at its junction with another element, 1990

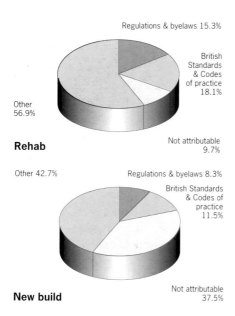

Rehab

Regulations & byelaws 15.3%

British Standards & Codes of practice 18.1%

Not attributable 9.7%

Other 56.9%

Other 42.7%

New build

Regulations & byelaws 8.3%

British Standards & Codes of practice 11.5%

Not attributable 37.5%

Figure 0.11
Distribution of faults identified in BRE DPU site investigations of rehabilitated and new-build housing: electricity services according to authority contravened, 1990

Electricity services
This category includes all consumer units, wiring and terminals. For rehabilitated housing, 72 different items were recorded, and for new-build, 96 items. Numbers of fault types recorded for electricity services were fewer than were obtained for other building elements, and could therefore be less representative of faults actually occurring. The items identified were nevertheless verified with design and site representatives.

With respect to element involved (Figure 0.10) for new-build housing, there were almost equal numbers of items relating to electricity services alone and at junctions with other elements such as walls, floors etc; for rehabilitated the figures were 60% and 40% respectively. Examples related mainly to methods of routeing, clipping or fastening wires, or to the way connections within the system were actually made.

For authority contravened (Figure 0.11) for rehabilitated housing, about one-third of all items related to infringements of Standards, Codes and regulations (in practice this meant BS 7671); for new-build it was around 20%, which might be expected from the increased stringency of inspections. A common example within this category was failure to derate power cables when passing them under or through thermal insulation at ceiling level in single storey accommodation, or through insulated external walls such as in timber frame construction.

With performance (Figure 0.12) for rehabilitated housing, safety and habitability related items formed around three-quarters of the totals. These chiefly concerned the lack of equipotential bonding, wall light switches close to wet conditions etc. With new-build, safety and habitability items were less numerous, though still of concern.

With origin of faults (Figure 0.13), there was not a great deal to choose between design and site for either category, and the same was true for the very few items related to materials and components. These figures bear out observations made in

previous books in this series – it is what people do with components and materials that makes a greater contribution to the incidence of faults than any inherent properties of the components and materials themselves.

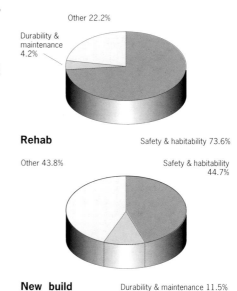

Other 22.2%

Durability & maintenance 4.2%

Rehab   Safety & habitability 73.6%

Other 43.8%   Safety & habitability 44.7%

New build   Durability & maintenance 11.5%

Figure 0.12
Distribution of faults identified in BRE DPU site investigations of rehabilitated and new-build housing: electricity services according to performance, 1990

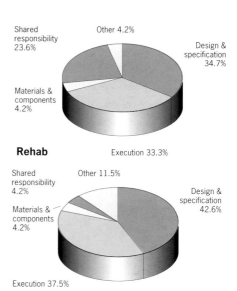

Shared responsibility 23.6%   Other 4.2%

Design & specification 34.7%

Materials & components 4.2%

Rehab   Execution 33.3%

Shared responsibility 4.2%   Other 11.5%

Design & specification 42.6%

Materials & components 4.2%

Execution 37.5%

New build

Figure 0.13
Distribution of faults identified in BRE DPU site investigations of rehabilitated and new-build housing: electricity services according to responsibility for the fault, 1990

House condition surveys

The 1996 English, 1996 Scottish, 1993 Welsh and 1996 Northern Irish House Condition Surveys[7–10] (the most recent available) provide a limited amount of information on, for example, space and water heating provision, adequacy of thermal insulation for installations, gas and electrical systems, and smoke detectors. Data from the surveys are given in appropriate parts of this book.

The surveys are based on structured sampling of the complete national stocks of dwellings. There are 20,321,000 dwellings in England, 2,123,000 dwellings in Scotland, and 573,370 dwellings in Northern Ireland. Up-to-date figures for Wales were not available at the time of writing.

Non-domestic buildings

It is estimated that there are over 1.6 million non-domestic buildings in the UK[12]. Apart from confidential information available to local authorities for local taxation purposes, there are no widely available figures for floor areas. Nevertheless they must represent at least as great, and probably a greater, floor area on a national basis as domestic buildings.

Systematic condition surveys of non-domestic buildings have not been carried out by BRE, nor are they gathered on a national basis. However, the records of the BRE Advisory Service noted above, together with information available through the Construction Quality Forum (CQF), go some way to providing useful information on the occurrence of services faults in non-domestic buildings.

In fact, the main purpose of the CQF database is to draw attention to the most significant defects and their causes. Figure 0.14 gives the breakdown of the contents of the entire database, and Figure 0.15 the breakdown of the separately held building services database in April 1997. The database contains nearly twice as many entries dealing with non-residential construction as with domestic.

Figure 0.14 indicates that building services faults do not loom especially large in the overall picture given by the CQF information.

Figure 0.15 shows that nearly half the 1,231 entries in the CQF database on building services are concerned with mechanical and electrical services; mechanical heating, cooling and refrigeration systems account for one-fifth of all items.

Figure 0.16 gives a further breakdown of the 263 items in the mechanical heating, refrigeration and cooling categories of the CQF building services database.

Records of other organisations

The Building Services Research and Information Association (BSRIA) maintains a database of past defects and failures experienced by the building services industry. Dissemination of the information contained in this database has taken place through published detail sheets, special reports and seminars, and it has also contributed to the CQF database.

Amongst the more recent items identified by BSRIA has been a growth in power failures in buildings built before the 1970s. These have been caused, for example, by high harmonic currents, corrosion of pre-insulated pipework, and legionella from drip trays under apparatus.

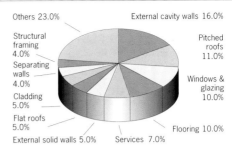

Figure 0.14
Occurrence of defects by element, CQF database, 1997

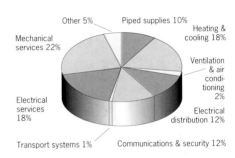

Figure 0.15
Occurrence of defects in building services, CQF database, 1998

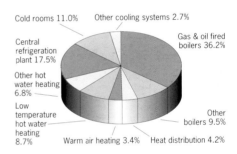

Figure 0.16
Occurrence of defects in mechanical heating refrigeration and cooling categories, CQF database, 1998

BRE publications on building services

Principles of modern building

When the structure of *Principles of modern building*, Volume 1, was being revised[13], no separate volume on building services was envisaged. However, some particular aspects of building services were covered in the chapters dealing with the building as a whole – corrosion of metals in pipework for example. Other aspects were dealt with in outline as parts of the individual elements making up the rest of the series – chimneys and flues being a case in point.

Although the last revision to Volume 1 was published as long ago as 1959, and since the book deals with principles, in some respects it is less out of date than might be imagined.

Those general principles described in *Principles of modern building* which have stood the test of time form the basis of the introductory sentences of some of the following chapters. Descriptions of performance requirements for, and agents affecting, building services are also taken in the main from other BRE texts.

BRE Digests and other papers

The various BRE publications, such as Digests, Information Papers, Good Building Guides and Good Repair Guides, which are listed in the references and further reading lists, together provide a fairly comprehensive coverage of building services. Since, though, they have been published over a time span of many years, they can have received very little cross-referencing between them. All these publications have been drawn upon to a considerable extent in this book.

BRE Reports

Apart from the series of publications referred to above, there are a number of BRE books and reports which examine particular aspects of building services. They include: *BRE housing design handbook*[14], *Assessing traditional housing for rehabilitation*[15], and *Energy efficiency in the work place: a guide for managers and staff*[16].

Changes in construction practice over the years

A historical note

Building services is about those aspects of modern buildings that distinguishes them from those of previous centuries. The fabric of roofs, floors, walls, windows and doors has altered relatively little over the last few centuries compared with the rapid, not to say spectacular, changes seen in methods of servicing buildings. Heating has progressed from open fires to radiators or warm air blowers; lighting from candles, via gas mantles and tungsten lamps, to fluorescent tubes; ventilation from unshielded apertures to sophisticated cooling and cleaning mechanisms; sanitation from the 'three-holer' and chamber pot to the waterborne; and water for household use from the village hand pump to piped supplies to practically all dwellings in the UK.

For most of the past history of buildings, if structural requirements were met and there was some basic provision for lighting and ventilation, and for heating in cold weather, no more was asked. Today the internal environmental requirements are incomparably more rigorous.

Internal thermal comfort is expected in all but the most extreme weathers, and with much less dependence on heavy clothing than previously. Coupled with this are requirements for adequate draught-free ventilation and, in some cases, for humidity control. High standards of glare-free lighting are called for, whether natural or artificial. There are also various requirements relating to the acoustic environment, mostly concerned with the reduction of disturbance from unwanted noise. These are essentially requirements for individual spaces, and may vary considerably from one space to another within a building.

Finally, there are usually some less easily defined requirements chiefly concerned with the visual character of environments; also a requirement that there shall be no visible surface dampness. Arguably, though, these have more to do with the carcass of the building than its services.

From a very small, or even negligible, proportion of total costs of owning and running buildings in the eighteenth century, the proportion absorbed by engineering services has risen inexorably. The most spectacular growth has been in buildings such as hospitals. The average costs of services in the early 1970s were put at 30–45% of the total costs-in-use of the average building[17]. Now there are many buildings where they form the greater part of the running costs.

This book is not the place to present a general history of the development of building services; but for those readers seeking a detailed history, it can be found in *Building services engineering*[18]. Nevertheless, some historical information is included about old practices and installations which may have survived in the present building stock.

Heating

One of the most comprehensive developments in the field of building services has been in heating practices. The earliest buildings were heated, if indeed they were heated at all, by open fres of wood or charcoal placed in the centres of rooms. Sometimes fires were contained within wrought iron braziers, but, more often than not, simply heaped upon flat stones which also served to provide a limited means of cooking food. Although most areas of the UK have long since progressed to more sophisticated forms of heating, it is only within the last two hundred years that some remote rural areas, such as the Highlands of Scotland, finally dispensed with the central open-hearthed fire. In 1827 it was estimated that four-fifths of the Highland population still lived in 'black houses'; that is to say houses, largely without windows, where the smoke from peat fires built on flat stones in the centres of rooms found its way through the thatch without benefit of chimney, or with, at the most, a rudimentary chimney. The underside of the exposed roof might have been lined with jute sacking; where the roof had not been maintained properly, in wet weather, tar would inevitably drip from the underside of the thatch.

During Roman times, heating of larger establishments and villas was often by hypocausts – large voids under suspended ground floors where closely spaced masonry piers carried flagged or tiled decks, forming a plenum for the transmission of flue gases for room heating. The Romans also occasionally used hearths set against external walls, with flues carried up within the walls.

After the Romans left, taking with them the knowledge, skills and resources which went into their heating systems, heating practices reverted once more to the central open hearth. The larger multi-storeyed castles of the eleventh and twelfth centuries could not be heated by a single central fire, and this seems to have been the time when fireplaces, flued within the immense outer wall thicknesses, came into wider use.

The separate masonry chimney was further developed during the thirteenth and fourteenth centuries for use in the better classes of building, and became more necessary during Elizabethan times when sea-coal began to replace wood and peat as the main fuel. Decoration of chimneys externally reached its high point in Elizabethan times with moulded bricks being used to form the characteristic octagonal shapes, and diaper and 'barley-sugar' detailing.

Towards the end of the fourteenth century, the fireplace recess was considerably deepened, and decorative mantel pieces began to come into fashion. In Elizabethan times the large mansions of the day were well provided with these decorative fireplaces, with overmantles in stone, marble or wood, often embellished with heraldic devices of the owner. During this period, too, hearths and chimneys were also inserted into surviving older dwellings.

Some of the flues in these early chimneys were very large indeed, reaching 1 m or more in diameter. They needed to be large for the 'climbing boys' to gain access for

sweeping. Perhaps one of the more significant developments relating to chimneys took place in 1842 when the practice of employing small boys to sweep chimneys was banned by legislation. Of course, when the large spaces were no longer needed for cleaning purposes, the 9 in x 9 in flue of later years became the norm.

The fireplaces, or 'chimney pieces', in these large mansions were sometimes equipped with hoods set over the fire which drew the smoke into the flues. Some even contained dampers to control the supply of air, and a few had underfloor air supply to the grate. It was not until 1796, however, when Sir Benjamin

Thompson, Count Rumford, published his principles of fireplace design[19] (Figure 0.17), that the problem of smoky fres and chimneys was solved, and the way was smoothed for the massive increase in domestic coal consumption. As he said: *'Those who will take the trouble to consider the nature and properties of elastic fluids, of air smoke and vapour, and to examine the laws of their motions, and the necessary consequences of their being rarefied by heat, would perceive that it would be as much a miracle if smoke should not rise in a chimney, all hindrances to its ascent being removed, as that water should refuse to run in a syphon or to descend in a river'.*

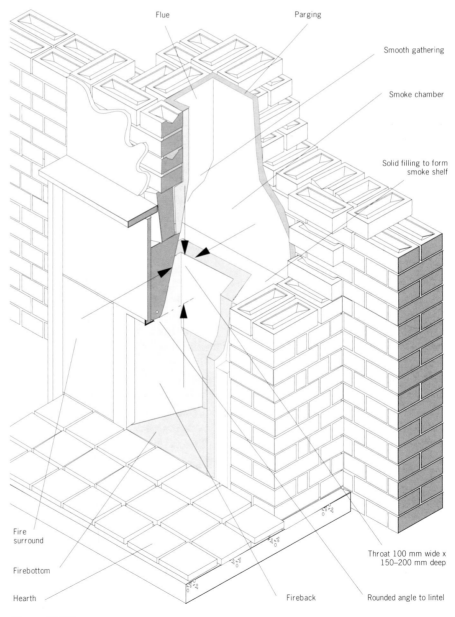

Flue

Parging

Smooth gathering

Smoke chamber

Solid filling to form smoke shelf

Fire surround

Firebottom

Hearth

Fireback

Throat 100 mm wide x 150–200 mm deep

Rounded angle to lintel

Figure 0.17
A fireplace illustrating Count Rumford's principles

However, in spite of this observation, chimneys of his day did sometimes have 'hindrances', which allowed smoke to escape into habitable rooms, and he therefore set about defining remedial measures.

The rules that Count Rumford prepared (in summary) were:

- throat restricted to 4 inches (100 mm) perpendicularly over the fire
- width of fireback to be one-third of the front opening
- fireback to be vertical and finish 6 inches (150 mm) above the lip of the lintel
- a horizontal smoke shelf
- firebrick rather than cast iron for the carcass of the fireplace

Experiments during the 1920s showed that these rules were broadly correct. Efficiencies, that is to say the proportion of useful heat passing into the room, were mostly less than 20–25%, but sometimes a little greater. Although efficiencies are often no better for newly installed open fires, they are paltry by the standards achieved in well designed central heating installations.

Open fires burning house coal were used almost universally up to the 1950s, adding their share to the

Figure 0.18

A cast iron wood burning stove, with clear glazed door and stainless steel flue, installed in 1998 in an old fireplace in a farmhouse in rural Pembrokeshire

smogs and 'pea-soupers' of the Industrial Revolution and its aftermath. The majority of houses built until then incorporated chimneys and fireplaces, but these became virtually unusable when the clean air legislation was enacted in 1956. Heating levels were spartan by present standards, as a quotation from the technical appendices to the *Housing manual 1949*[20] illustrates: *'Where there is only one solid fuel appliance on the ground floor, its most important duty is to maintain the living room at an adequate temperature. In small houses (not more than about 950 sq ft floor area), it may, if it is a stove or a continuous burning open fire with back boiler, also be used to provide heat to one or more additional rooms. This may be done by means of radiators or by convected warm air. If the heat is to be supplied to an adjacent room on the ground floor, eg a dining room or dining kitchen, a small radiator is to be preferred. Convected warm air may be used to warm a bedroom, or, in favourable conditions, two bedrooms. In general, however, moderate warmth in bedrooms, except in very cold weather, will result from air movement and conduction, in any well insulated house that is adequately heated on the ground floor'.* (For later developments see Chapter 2.1.)

Wood burning closed stoves of cast iron began to be used in Germany and Scandinavia in the fifteenth and sixteenth centuries, but did not come into widespread use in the UK until the Industrial Revolution when they tended to be fuelled by coal. Wood burning stoves, though, continue to be popular where there is a continuing fuel supply (Figure 0.18).

Coal or coke fuelled stoves, shaped cylindrically (the so-called tortoises), were used in very large numbers for space heating in premises of all kinds in Victorian times. They were used in the hutted military encampments dating from the Crimean war, having a basic pattern still in use in the 1939–45 war and even later, though by this time burning coke. Many school classrooms were built during the Victorian era and furnished with stoves which were often totally inadequate, even when glowing red,

to heat the large volumes of air beneath the high ceilings; much of the convected heat produced migrated to ceiling level where it was extracted by large ventilators in the ridge.

Anthracite fuelled stoves were also developed, for example by Pithers; these had a different pattern because of the different combustion characteristics of coal and anthracite. The diameter of the spun cast iron flue pipes was either 4 inches (100 mm) or 6 inches (150 mm) depending on stove capacity.

Oil fired domestic boilers date from around 1920, gradually replacing solid fuel. Since that time gas has also significantly increased its market share.

So far as other building types are concerned, such as hotels and churches, there have been changes similar in scope to those taking place in the domestic field. A few warm air coal-and-coke-fired heating installations were introduced into the UK in the first quarter of the nineteenth century, mainly in churches and other non-domestic buildings. With these appliances the air was ducted to a plenum from furnaces remote from the rooms being heated, then fed in through grilles at floor level and exhausted by vents at high level. Many large Victorian hospitals and factories were heated by coal or coke in 'Lancashire' boilers to produce steam; the products of combustion were flued via massive brick chimneys (Figure 0.19), most of which, by now, have been demolished.

Single walled, sheet steel chimneys were the first replacements for brick chimneys, later augmented by twin walled chimneys. More recent chimneys are fabricated with stainless steel. Twin walled chimneys reduce heat losses and are designed to offer some protection to the building fabric. They also allow separate specification of the corrosion resistance of the inner and outer sheaths. While brick chimneys are stable in high winds without the use of stays, slender steel chimneys are not: when first introduced, collapses were not uncommon.

Electricity for heating buildings first began to be used on a significant scale in the final years of the nineteenth century, mainly in wall panel radiators. Enclosed tubular heaters were introduced from the 1920s, but it was only after the 1939–45 war that electric storage radiators, or rather convectors, began to be developed[18].

Cooling

In the days before artificial refrigeration was available, many country houses had 'ice houses' constructed for storing perishable foods, the ice being gathered from ornamental ponds and stored in the ice houses from winter to the following summer. The immense thickness of the earth roofs ensured sufficient thermal insulation to prevent too much heat gain[21]. Mechanical refrigeration became available from around the end of the nineteenth century, though many users still preferred ice which had been produced naturally.

Although domestic refrigerators were available from the 1930s, they did not come into general use until the 1950s. At that time, the Model Byelaws still contained the requirement for a ventilated larder or pantry for storing perishable foods, and many dwellings built after the 1950s have been provided with this facility.

Air conditioning, or cooling of ventilation air, did not come into widespread use until the 1950s.

Ventilation

A brief note on early ventilation practices via windows was given in *Walls, windows and doors*[22]. Ventilation was often linked with the availability of air infiltrating into a building through suspended floors (see *Floors and flooring*[23]).

Rooms heated by central fires burning peat or wood without flues, and ventilated, if at all, by tiny windows or ill-fitting doors, would have presented a very smoky atmosphere to the occupants. The need for ventilation, in today's terms, would have been met by providing sufficient air to feed the fire and to

ensuring that flueing provision was adequate. On the other hand, some houses would, presumably, have been extremely leaky, and therefore may not have had a ventilation problem. The amount of air required for a large fire not having the benefit of a chimney would have been considerable; flue gases rising without constraint would have entrained more air than if they had been contained in a chimney.

Of course flues in rooms with fireplaces, whether fres were burning or not, provided a means of driving ventilation if there was a source of air other than the chimney; experiments during the 1920s recorded air change rates from open fires of the order of 300 m^3/hour. Ventilation through windows, whether acting as input or exhaust, is invariably erratic due to the vagaries of wind, to fortuitous openings in the remainder of the building's surfaces, and to differences between internal and external temperatures. Cold draughts are inevitable in these circumstances, even though the introduction of double hung vertically sliding sash windows around the seventeenth century allowed a fine degree of control over the size of the opening. The later introduction of hoppers helped in diverting draughts upwards so that hot and cold air was mixed above head level.

Statutory minimum requirements for ventilation in habitable rooms goes back at least to the 1870s and 1880s. Indeed, the local byelaws founded on the Model Byelaws and made under the authority of the Public Health Acts were concerned only with ventilation of habitable rooms. The requirement was for every habitable room to be *'provided with a window or windows which shall be so constructed that a total area not less than one-twentieth of the floor area of the room may be opened to the external air, and some part of them so required to open shall be not less than five feet nine inches above the floor. This paragraph shall not apply to any room used or adapted to be used for the lawful detention of any person, or to any room for which adequate ventilation is provided by mechanical or other means'.*

Presumably, in those days the powers that be were not unduly concerned if prisoners suffered!

Attempts to introduce improved means of controlling domestic ventilation seem to have centred during the nineteenth century on ducts provided near fireplaces and flues, using the effect of the heated column of air within the stack to recirculate air in the adjacent room[18]. These developments were not widespread.

Also during the nineteenth century attempts were made to improve ventilation in schools, hospitals and factories. Conditions in most early Victorian workplaces must have been abominable, with lack of ventilation made worse by fumes from primitive industrial processes and heat from the candles used for lighting. Most buildings had pitched roofs (though there were a few with flat roofs), so where improved ventilation was needed, many of these roofs had turrets inserted into the ridges with the slots

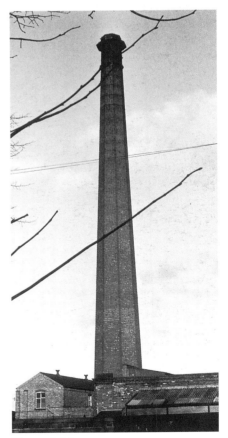

Figure 0.19
The massive brick chimney of Anker Mill, a late Victorian woollen mill in Warwickshire

shielded from the rain. Cowls were also used to some extent, rotating in the wind so that the suction on the lee side aperture would assist in exhausting the vitiated air. Airbricks, chimney pots and vent terminals were made in a great variety of designs (Figure 0.20), although many were primarily concerned with aesthetics rather than with function.

Fans were first used in the early nineteenth century in such places as the Palace of Westminster and hospitals. Further development was encouraged by the need to provide the means of moving very large quantities of air in deep coal mines. The first small sized fans for siting in window panes were introduced just before the 1939–45 war.

Lighting
At night, primitive houses would be lit only by the fire or by splinters of resinous wood called fir-candles. Animal fat tallow or wax candles did not come into general use until the mid-nineteenth century. Even the electric lighting of Victorian towns came late to the country areas; for example, it was estimated that by 1951 only one in six crofts in the Highlands of Scotland had electricity[24].

Water supply and drainage
Lead was used commonly for mains water supplies to dwellings up until the 1930s, and sporadically even later; some dwellings still have lead supply pipes in use. Following the realisation that lead was injurious to health, iron supply pipes gradually began to replace the lead, this process starting in the late 1930s. In Scotland, for example, the number of dwellings where lead was detectable in the water supply fell from just over 1 in 10 in 1991 to around 1 in 14 in 1996[9]. There is still some way to go to achieve complete elimination of lead, both in the internal pipework of dwellings and in the supply pipes from the mains.

The waste water preventer (WWP) or siphon flush for WCs came into general use in the 1860s, using a high level cistern which gave a good pan clearance, but which was rather profligate in the use of water compared with more modern designs. The old cast iron, well bottomed or 'Burlington' type cistern (Figure 0.21), in which a heavy cast iron bell was lifted and allowed to drop so as to displace water over the outlet or upstand pipe to initiate the siphon, was also very noisy. It has long been superseded, and only few examples are now likely to have survived.

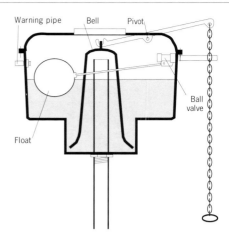

Figure 0.21
Cast iron well-bottomed or Burlington WC cistern

Summary of main changes in common practice since the 1950s

Building management systems (BMS) centralise and automate the monitoring and control of building services The term building energy management system (BEMS) is usually limited to a BMS that manages only energy consuming services. Until the early 1970s, this technology was reserved for large highly technical building complexes or buildings with widely distributed plant (Figure 0.22). Advances in microelectronics since that time have now made it feasible to consider the cost effective application of this technology to many more types of building.

Space heating
Technical Appendix F of the *Housing manual 1949* recommended that facilities should be provided in all new houses for the maintenance of a temperature of 65 °F (18.5 °C) in the living room, and background heating ranging from 45 °F (7 °C) to 50 °F (10 °C) in the remainder of the dwelling.

The appliances which were brought into use after the 1939–45 war differed from older designs in a number of respects. For instance, their efficiency was greater which meant that the amount of heat available from the appliance in terms of space heating, water heating and

Figure 0.20
Air bricks, chimney pots and vent terminals feature in this drawing from an advertisement for fired clay products from Stanley Brothers, Nuneaton, dating from late Victorian times

Figure 0.22
The building management system in this fairly typical example of a large office building in the Government estate was monitored by BRE in the early 1980s

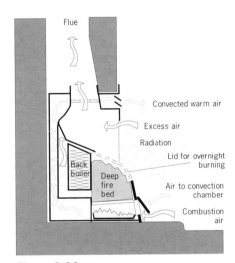

Figure 0.23
A so-called 'closable open fire'. The design of this high efficiency fire with back boiler dates from 1949. The additional efficiency over and above that of a simple open fire stems largely from the convected air circulating in a separate airway round the back boiler and fire bed, and from the facility for fine control of the combustion air supply to the fire bed

cooking, was higher in relation to the heat content of the fuel consumed. The appliances operated either on smokeless solid fuel, or on house coal. Many of the appliances, including those fitted with boilers, burned overnight at a reduced rate without attention. One of the disadvantages of the old fashioned open fire with front fret was the difficulty of regulating the air supply to the underside of the fire and, therefore, of controlling the rate of burning of the fuel in the appliance and, hence, the heat output. In the improved appliances, the rate of air supply and therefore of heat output could be accurately controlled by means of a damper[20] (Figure 0.23).

From central or programmable heating being a luxury and a comparative rarity in dwellings, there has been a very rapid growth in the last half of the twentieth century. Taking England as an example, from a very small number in the interwar period, the proportions of dwellings having central heating grew to 73% in 1986[25], 84% in 1991[26] and 87.6% in 1996[7].

Improvements made to the energy efficiency of the housing stock between 1970 and 1985 resulted in a 6% reduction in the energy consumption of the average household. This was accompanied by substantial improvements to space heating and internal temperature conditions as households moved from single room heating to whole house heating. The main driving forces with respect to new buildings were the changes in the thermal insulation requirements of the various national building regulations. Building regulations have contained some provision for thermal insulation since they were first established. Changes in the Building Regulations (England and Wales) are shown in Table 0.1.

Space heating standards for non-domestic buildings built before the mid-twentieth century were largely influenced by the recommendations of the Institution of Heating and Ventilating Engineers (now the Chartered Institution of Building Services Engineers – CIBSE), although legislation was later

Table 0.1 U values required by the Building Regulations (England and Wales)			
	Roofs	**Walls**	**Floors**
1965	1.42	1.70	1.42[†]
1976	0.60	1.00	1.00[†]
1982	0.35	0.60	0.60[†]
1990	0.25	0.45	0.45[‡]

† Applies to exposed floors only

‡ Applies to all floors including those in contact with the ground

introduced which covered particular buildings, notably shops and allied premises. Typical values of that time included, for example, 62 °F (17 °C) for school classrooms, 55 °F (13 °C) for factories, and 60 °F (15 °C) for shops.

The 1965 Regulations for thermal insulation were pitched at a level that could be met by a standard brick–brick or brick–block cavity wall and 20 mm of glass fibre quilt in the roof. Those levels remained in force until concern about energy conservation grew in the early 1970s and following the international oil crisis. Regulations were formally revised in 1976 including the specification of better U values and a provision for limiting the total area of windows. Further improvements were introduced in 1982 and 1990 with significant increases in insulation on each occasion. The 1990 revision is notable because it introduced ground floor insulation for the first time in the UK. It also allowed much more flexibility than the previous Regulations, making it possible for builders to compensate for reduced insulation in one element by increasing it in another.

Construction methods have evolved to take account of the changing Regulations. The 1976 revision could be met by the adoption of lightweight concrete

blocks for the inner leaves of cavity walls. The 1982 revision increased the use of insulation within the cavity itself but also led to the development of high performance insulating blocks which allowed the cavity to be kept clear.

By the late 1980s it was possible to observe that heating in British homes had undergone a quiet revolution over the previous 20 years, being much better heated but still using about the same amount of energy, or even less, on average. This was achieved through the installation of central heating and better insulation[27].

Boilers for supplying space heating in large buildings were originally of wrought iron, with welded steel coming into use around the time of the 1914–18 war[18].

Perhaps one of the most significant developments related to the use of forced warm air heating; many schools built in the 1950s and 1960s had this kind of heating installed. One example is the Andrews Weatherfoil system developed in conjunction with the Hertfordshire County Council architects and BRE, and incorporated in a number of schools in the county. The installations were at first solid fuel fired, but later ones were oil fired. The heat exchanger and fan unit casings were situated wherever

required in the plan, and the distribution system was largely prefabricated and built into the carcassing of the building as it was erected[28].

Although air conditioning has been widely used in office building in the UK since the early 1950s, there has been a recent change in attitude towards its use since it became clear that many of the advantages which it offered could be obtained by other means significantly lower in energy consumption (see Chapter 3.3).

Establishment of BRECSU in 1981, sponsored by the then Department of Energy, saw the beginnings of a new concerted effort by Government to reduce the nation's fuel bill. The Best Practice Programme run by BRECSU has been a major benchmarking activity, providing advice on achievable performance. Application has been by publications and seminars, and practical demonstrations on site. Other outputs have included:

● the BRE Domestic Energy Model, (BREDEM) was launched in 1985 and since then has formed the basis of revisions to building regulations, and the foundation of National Home Energy Rating (NHER) and the Standard Assessment Procedure (SAP).

● BRECSU took over responsibility for all Energy Efficiency Office (EEO) demonstration projects in building in 1986, and since then has transformed the take-up of energy efficiency measures in buildings. To take just a single example, in 1989 BRECSU demonstrated annual energy savings of 28–40% in new factories in Wales, producing potential savings of £30 million per year.

Also around this time, 1989, there was a concerted effort to target energy saving in schools, and a number of initiatives were taken in conjunction with the then Department of Education and Science (Figure 0.24).

Enquiries received by BRECSU under the Best Practice Programme for information and publications have been compared for the years

Figure 0.24
A school benefiting from energy conservation measures as part of the BRECSU energy conservation programme in the mid-1980s

1994 and 1997 for the different sectors of the industry; that is to say the category into which the building type fell where it was clearly identifiable. About half the requests in both years were for information on more than one sector and not readily allocatable to a specific sector or building type; these have been omitted from the charts which are shown in Figure 0.25.

It is clear that enquiries became more focused in 1997, as can be seen by the large reduction in the Miscellaneous category. This category covers requests for all publications produced by BRECSU, no matter what building type, together with general items such as new technology and higher education. The reduction was largely offset by a very large increase in requests for information on energy management of buildings, but there has also been a growth of interest in the commercial sector, presumably where the largest potential financial savings are to be found. The other categories show little change, with the domestic sector forming nearly half of all enquiries in both years.

Domestic hot water systems

Small domestic hot water boilers, the so-called 'pot boilers', were widely used in the UK in the 1930s, fuelled by coke or later by anthracite. They had largely gone out of use by the late 1950s.

Provision was sometimes made for heating water alongside open fires in the cast iron ranges of the late nineteenth century (Figure 0.26), although they were very inefficient by recent standards. From the early twentieth century, cast iron or copper back boilers were frequently installed into the backs of open fires. The products of combustion passed under and alongside as large a heating surface as possible. Special L-shaped boilers were available to increase the area of boiler surface in the horizontal plane (the 'boot boiler') as well as the more common rectangular box shape. Back boilers were normally equipped with dampers to divert the flue gases when more hot water was needed. Open fires with back boilers

were still being widely installed in houses until the early 1960s.

A significant change in domestic hot water systems came in November 1985 when unvented systems were permitted. Prior to that time, the Water Byelaws 1966 in Britain prohibited the connection of water apparatus in excess of 2 gallons directly to a service pipe, with exceptions for instantaneous water heaters and other water heaters of limited storage capacity. Similar provisions applied in Northern Ireland where water installations were controlled by Water Regulations. In 1986, a revision of the Water Byelaws for England and Wales allowed the installation of unvented domestic hot water systems. The Building Regulations 1985 had introduced a new requirement, G3, and its associated Approved Document which was concerned with the safety of unvented hot water supply systems with a storage capacity exceeding 15 litres. Now, of course, the Water Supply (Water Fittings) Regulations 1999 have replaced the Water Byelaws in England and Wales.

In the housing drives of the period immediately after the ending of the 1939–45 war, much use was made of pre-assembled plumbing units, sometimes referred to as heart units or modular plumbing units, which were lifted into position by crane. There is no doubt that these units gave a saving in site man-hours. They were also claimed to give better quality control over the finished product than was possible with site assembly. However, in order to reduce the total volume of the building occupied by services, some of the designs were crammed into very small spaces, which has made the replacement of individual pipes and fittings very difficult.

In the late twentieth century similar but larger units, called pods, had been employed on a large scale in fast track building programmes for office, hotel and health buildings. It is to be hoped that these units will prove easier to maintain or replace than their predecessors of two or three decades earlier.

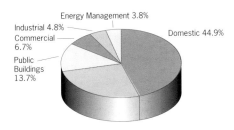

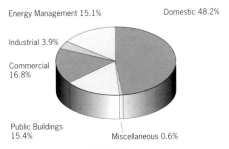

9,164 enquiries in 1994

9,388 enquiries in 1997

Figure 0.25
Analysis of enquiries for publications received by BRECSU according to sector or building type

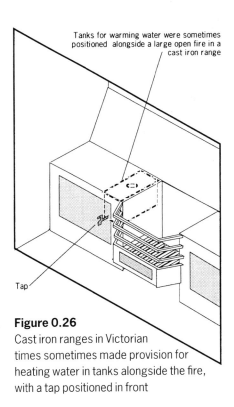

Figure 0.26
Cast iron ranges in Victorian times sometimes made provision for heating water in tanks alongside the fire, with a tap positioned in front

So far as domestic clothes washing arrangements are concerned, there have been considerable changes. Wash boilers were still being installed in houses in the early 1950s. In many mining districts these were coal fired, built into the masonry of the scullery. In one group of houses in the West Midlands inspected by BRE in the 1960s, where the 'coppers' were still in occasional use, the flue was shunted into that from the adjacent main living room fireplace. In another example in Norfolk in the 1970s, BRE investigators found the coal fired copper still in use. Gas or electric boilers were used elsewhere. Wash boilers, or coppers, were not always made of copper. They might have been made of cast iron, or galvanised or tinned sheet steel, with covers of wood or steel. If inadequately protected, rust from the metal could have stained the clothes being laundered. Now, of course, the automatic washing machine is king.

The most recent developments have included boilers with low-water capacity, constructed of steel rather than cast iron, and also condensing designs. The 1986 Model Water Byelaws permitted the use of hot water storage systems without a vent for the first time in the UK; the Building Regulations 1985 provided the safety requirements for these systems.

With regard to other building types, a similar situation to domestic buildings can be seen. Solid fuel largely gave way to gas and oil over this period as the run down of the UK coal industry progressed, and some non-domestic installations even used off peak electricity.

Reducing pollution and improving the environment

These are issues that have become prominent in the last quarter of the twentieth century[29]. They recognise:
- the impact of building on global atmospheric pollution through the 'greenhouse effect' (use of energy for buildings is responsible for half the UK's annual production of greenhouse gases, two-thirds of this from housing), acid rain and ozone depletion. Buildings are also responsible for a significant proportion of emissions of oxides of sulphur and nitrogen, and of chlorofluorocarbons (CFCs)
- the influence of buildings on the health, comfort and safety of building occupants through the effects of indoor pollution. Since the highest concentrations of most airborne pollutants are found in indoor environments, and the adult populations of Europe and America, for example, spend 90% of their time indoors, indoor air pollutants have great potential to affect health

- the impact of buildings on the local outdoor environment and depletion of resources
- the impact of climate change on buildings (eg wind loading and indoor cooling requirements). The changes in climate predicted by some environmentalists could have considerable implications for the location and design of buildings.

Following the ratification of the 1992 Rio Climate Change Convention[†], the UK undertook to reduce its emission of carbon dioxide and other greenhouse gases to 1990 levels by the year 2000. The construction industry is still a major area for action since the construction and use of buildings account for about 50% of all UK greenhouse gas emissions. In 1991, the UK building services sector was estimated to have consumed 808 PJ of energy, emitting 89 million tonnes of carbon dioxide. British industry consumed an estimated 330 PJ of energy for building services, emitting 27 million tonnes of carbon dioxide.

Ventilation and air conditioning

Natural ventilation of habitable rooms in dwellings has remained much the same since the early 1950s for the majority of homes in the UK. Mechanical extraction from toilets in multi-storey blocks dates from about this time. The high incidence, though, of condensation in dwellings, after reduction of fortuitous ventilation by weatherstripping windows and doors in attempts to conserve energy, have led to requirements being introduced into the building regulations for air extraction in kitchens and bathrooms. Although building regulations are not retrospective, there may be good reasons for upgrading existing ventilation provisions to the new standards (Figure 0.27).

There has also been a considerable tightening up of requirements for the provision of air to combustion

Figure 0.27
A new extractor fan fitted into the wall of a kitchen. The windows had been replaced with new weatherstripped opening lights – the old airbrick with hit-or-miss slider was assumed no longer to be adequate to provide ventilation

† At the 1992 Earth Summit in Rio de Janeiro, many governments committed themselves to limiting emissions of the main greenhouse gas – carbon dioxide – by signing the Climate Change Convention. Buildings-related emissions, as a by-product of energy consumption, contribute substantially to the release of carbon dioxide.

appliances, and the control of extract fans situated elsewhere in the rooms where they might influence the volume of air available to the appliances.

By far the greatest change in practice follows increasing concern with the airtightness of buildings. Two main factors are at play:
- the commitment on the part of the UK government to reduce carbon dioxide and CFC emissions as part of the Rio Agreement
- the preferences of the occupants of buildings for natural ventilation

These considerations apply not only to new buildings, but also, to an increasing extent, to older buildings and their refurbishment; some of the implications are spelled out in Chapter 1.

In relation to refurbishment, action has been taken over recent years by occupants, community action groups and others to conserve heat in homes by fixing simple draughtproofing around external doors and windows. Some house occupiers may seal room vents, fit replacement windows, or install secondary glazing further to reduce unwanted cold draughts and heating costs.

These draughtproofing measures can improve the living conditions for occupiers **provided** there are still supplies of fresh air to rooms and their occupants. If full draughtproofing measures are taken without consideration for the ventilation requirements of a particular room, fuel burning appliances may not function properly, airborne contaminants may not clear, and condensation and subsequent mould growth may become a problem.

The 1980s and 1990s have seen changes in the previously accepted conventions relating to building services. In former years great efforts were made to hide them within the carcass. As they have increased in size and complexity, that process has become more difficult. It is now more acceptable that building services installations are seen in full view: as part of the outward expression of the functional requirements of the building (Figure 0.28).

Artificial lighting

The supply of electricity to homes in the UK is now virtually completed, including those in remote areas. When refurbishment does include wholesale renewal of the electrical system, consideration will probably automatically be given to fitting energy saving lighting; for example, high efficiency fluorescent units in place of tungsten filament bulbs (Figure 0.29). This practice alone can make a considerable contribution to the national targets for energy conservation.

Water supply and drainage

Perhaps the most significant innovation with regard to the means of supplying water to properties has been the introduction of copper and polyethylene pipes in substitution for lead and cast iron, and plastics and vitreous clay drainage and wastes in substitution for cast iron and asbestos cement.

A consequence of the introduction of polyethylene mains has been to remove a convenient means of earthing the electricity supply: the metal incoming pipe fixed into damp ground and formerly thought to be adequate for providing an effective earth. It is arguable that this practice would have ceased for other reasons too, in particular the introduction of protective multiple earthing (PME) referred to in Chapter 5.1. It is now no longer permissible to use the incoming water main as a primary earth, and this should be considered when replacing metal water mains.

Internal distribution of water in thin-wall copper tubing instead of steel or lead came into general use during the years between the two world wars. Thin-wall stainless steel tubing was introduced during the late 1960s during a copper shortage, and some systems in this material may have survived. So far as materials for drainage are concerned, soil and waste pipes of PVC were introduced in the 1960s, largely replacing the cast iron, lead, or sometimes copper or asbestos cement used formerly.

Figure 0.28
Services installations, the Lloyds Building, London

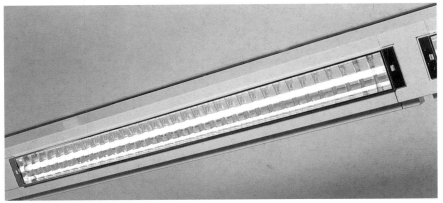

Figure 0.29
High efficiency luminaires can make a substantial contribution to energy saving, with little or no sacrifice of utility

Figure 0.30
Ice formation from a persistent drip
beneath a warning pipe. Large quantities of
water can be wasted if maintenance is
neglected

Water conservation has become increasingly important in the UK with the growing population, with the number and capacity of water-consuming appliances and activities has increased, and with climate changes that have reduced average rainfall in many regions. There is growing public awareness of the scarcity of water in some areas of the UK and of its economic value. The benefits of conserving water are being recognised. They include: maintaining the availability of the water supply during the drought periods; reduced pumping and treatment costs (with associated energy savings); reduction of waste water; protection of the environment; and the possible reduction in costs for the consumer. It is confidently expected that this trend will accelerate in future.

Measures have long been in place to avoid wasting water within properties. Warning pipes (commonly misnamed as overflow pipes) fitted to storage cisterns are required to discharge as a way of alerting occupiers to a problem in the water system and the need for prompt repairs. Unsightly limescale deposits and algal growths on brickwork and concrete, and, in winter conditions, ice formations on the outsides of buildings, give an indication of how much water is wasted by lack of adequate maintenance of the installation (Figure 0.30).

Chapter 1

Building physics (services)

This first chapter deals in simplified form with some of the basic underlying scientific and engineering principles which tend to affect the internal environment of the building as a whole. It also concentrates on those uses of energy which seem to be of most consequence to the occupants and which have a direct effect on their comfort, both thermal and visual.

As noted in the introductory chapter, the oldest buildings in the UK building stock have relatively little in the way of building services. In the days when these buildings were built, their occupants had to make do with the crudity of what was currently available. The application of scientific principles to the design of buildings, although receiving some impetus with the technological developments of the Industrial Revolution, did not take off until the 1920s and did not really form a significant part of the education of architects until the middle years of the twentieth century.

The highest available knowledge of pure science and the most effective methods of research are needed in building as in any other field of research. Building research as a whole, however, is concerned with the principles of an exceptionally wide range of science. Results of past scientific research are not at present fully utilised in building because there is no suitable bridge between the research worker and the architect or designer.' (The Department of Scientific and Industrial Research, 1919).

Although the principles underlying design for user satisfaction might not have changed significantly during the interwar years, the 1920s and 1930s, the application of scientific principles to the design and construction of buildings received a boost when the first volume of the book *Principles of modern building* was published in 1938 (see Chapter 1.1).

There is not the slightest doubt that since that time the increasing sophistication and consequent demands of users for improved standards in all kinds of buildings are driving an accelerating rate of change in the technological development of building services. In consequence, building services plant is getting more and more sophisticated (Figure 1.1), and services are taking an ever-increasing share of the total costs of a building project.

Figure 1.1
Part of the plant room for a small office building. The rate of change of technological development is increasing, and plant rooms can become congested

Chapter 1.1   **The building as a whole**

At the time that the Building Research Station (BRS) was first established, relatively few houses had bathrooms and water closets, relying instead on outside 'privies'. Central heating was rare, even in non-domestic buildings. Washday for many households began by lighting a gas or coal fire under the 'copper', stirring the clothes with a 'dolly', and 'mangling' them by hand to a semi-dry state. Standards since then have risen out of all recognition.

When the first edition of *Principles of modern building*[13] was published in 1938, it was possible to discuss in physical terms the functions and performances of the types of construction that were then being built, to indicate ways of predicting some of the performances, and to distinguish between good and bad practices. In the revised and expanded editions of 1959–61 this approach was further developed and extended to floors and roofs as well as to walls. But the focus on the building element – the wall, floor, or roof – remained.

This concentration on building elements had the merit that it directly reflected the main interests of the designer at the detailed design stage. On the other hand it offered little general guidance on, for instance, the design of a complete spatial enclosure, the performance of which was of more interest to the user; and, for this reason, it could lead to overemphasis on some features and the relative neglect of others of equal or greater importance. There was a need to think more in terms of the whole system, at least when contemplating any major departure from already proven practice.

Part 1 of the third edition of *Principles of modern building* was entitled 'The building as a whole'. The text dealt with a number of important aspects of the performance of the whole building such as stability, ventilation, thermal and sound insulation, fire protection and daylighting, but there was little examination of the role of building services and the part they played in establishing comfortable conditions for the occupants. Since the 1960s, BRE has put much effort into examining the influence of one services subsystem upon another; for example, the inter-relationship of different forms of heating systems with thermal capacity, thermal insulation and ventilation provision, and the effects of extraneous air leakage. The performance of the whole building ought to be viewed as a complex interaction of all its parts and all its subsystems, and what is in balance for one set of circumstances may not be the same for another. Although the carcass of the building can provide the occupants with some protection from extremes of climate, both winter and summer, it is the servicing subsystems which now provide the fine tuning and correction of any imbalances in comfort levels (Figure 1.2).

The UK Climate Change Impacts Review Group have published estimates of the changes to the British climate that are expected to result from global warming over the

Figure 1.2
Simple protection from the weather, which might have sufficed in years gone by, is no longer enough. When this substantial dwelling was built in the nineteenth century, the many chimneys now surviving indicate that the main rooms were heated by open fires; even so, some rooms were unheated. Used since the early 1920s as offices, the servicing systems have needed to change out of all recognition

next 60 years. Climate change has particular relevance to buildings because they last a long time. Buildings now being designed or extensively refurbished are expected to last well beyond the time when significant climate change is expected, and many aspects of buildings are sensitive to climate. Current design procedures rely on historical climate data for the assessment of risk, but, if climate is to change, risk needs to be reassessed. Given the high degree of uncertainty about future climate, the first requirement is to determine the extent to which aspects of buildings are sensitive to climate change. The results form a basis for deciding where changes to design conditions are required. For building services, the greatest changes are likely to be in heating and cooling requirements. In the short term, the most important effects are likely to derive from initiatives to limit global warming by improving energy efficiency, rather than from the direct effects of climate change[30].

BREEAM – the BRE Environmental Assessment Method for buildings – was launched in July 1990. It remains the main working method worldwide for assessing environmental performance, and indeed has become a *de facto* standard for environmental performance[29].

Integration of building services into the overall design process

Although it might be thought that services would always be fully integrated into the carcasses of buildings from the start of the design process, with coordination proceeding through to the site assembly, it is only within recent years that significant progress has been made. When buildings were simple, and had few services other than provision for space heating, a hand pump in the kitchen, and bell wires to summon the servants, coordination of the installations was relatively unimportant, and could be accommodated piecemeal. Some were left exposed in any case.

BRS became more heavily involved in the rationalisation and integration of building services in the late 1950s following the discovery that it was common on site inspections to find that no such planning had taken place. Each specialised contractor would do his own thing. The first one on site had a clear run, and all the others following on had to fit their pipe runs and cables around what was already there. A mechanical and electrical engineer of a brand new hospital under construction in 1960 was asked by a BRE investigator if he could show a drawing of the services at one pinch point in the structure, an underground duct joining two buildings. The engineer produced 24 separate drawings, each showing different installations passing through the same location! No one had thought to coordinate them, and a veritable cat's cradle had resulted on site. The legacy lingers on in many of the buildings of those days, with a few honourable exceptions.

This lamentable state of affairs prompted BRE to include the integration of building services in the series of books aimed at the educational field, and entitled *Designing for production*[31].

In these publications, building services are categorised into three types:
- large self-contained elements with few connections to other services, such as lift installations
- utility services, such as sanitary accommodation, usually grouped into fairly well defined areas of the building, with connections with hot and cold water systems, and to drainage
- environmental services, which by their nature extend on a significant scale throughout a building; these include heating, ventilation, lighting and communications subsystems (Figure 1.3)

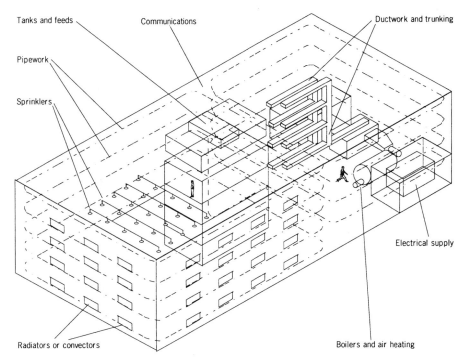

Tanks and feeds

Communications

Ductwork and trunking

Pipework

Sprinklers

Electrical supply

Radiators or convectors

Boilers and air heating

Figure 1.3

A schematic illustration adapted from *Designing for production*[31] showing some of the environmental services for a large building

Although cables are relatively small, and can be routed through buildings, the same is certainly not true of air handling trunking. One of the most common design faults with regard to the installation of services is inadequate allocation of space; this fact is likely to constrain, if not to govern, the adaptation or provision of building services in the future, let alone the ease with which buildings can be completed in the first place[32]. Building services now form a major proportion of the total costs-in-use of most non-domestic buildings.

Some servicing systems tend to create noise and provisions for their distribution may provide pathways for the transmission of sound between rooms. There may sometimes be requirements for background noise in some circumstances; for example, to enable conversations to be conducted unobtrusively.

Service installations are potential causes of accidents. Reduction of the risks calls partly for suitable shielding of, for instance, surfaces that are too hot to touch, but reliability and ease of operation will also contribute. Some of the more important circumstances are examined in later chapters.

Chapter 1.2

Energy, heat transfer and thermal comfort

In most European climates a heating system is necessary to provide thermal comfort in colder weather. A cooling or air conditioning system may or may not be necessary to provide comfort in hot weather.

These processes consume considerable amounts of energy. For example, in 1990, energy used in dwellings accounted for about 30% of all energy consumed in the United Kingdom, and a similar proportion of energy related emissions of carbon dioxide released to the atmosphere. The average household devoted 6% of its total annual expenditure to meeting its energy needs.

In 1993, in the non-domestic field alone, it was estimated that the UK's public and commercial buildings consumed 781 PJ of delivered energy resulting in the emission of approximately 83 million tonnes of carbon dioxide. Of this energy consumption, 57% was used for space heating, 9% for hot water heating, 10% for lighting, 7% for cooking, 5% for air conditioning, 4% for refrigeration of perishables and 8% for miscellaneous power purposes.

The requirements for thermal comfort might be achieved in more than one way, except that some characteristics (eg temperature gradients, and the balance between air temperature, radiant temperature and air movement) may assume greater or lesser relative importance for some situations or building types.

In addition, since heating is a major energy consumer and contributor to a building's running costs, the patterns of occupation are of great importance. The extent to which heating output can be varied to match the requirements of intermittent occupation depends, however, on the dynamic thermal response characteristics of the spatial enclosure system – that is to say, on whether it has large thermal storage capacities which warm up relatively slowly (eg insulation on the outside, Figure 1.4a), or less capacity which can warm up rapidly in response to a change in internal air temperature (eg insulation on the inside, Figure 1.4b).

In the modern domestic environment, heating in living rooms has always been considered to be a basic requirement[33]; an effective and economic heating system is therefore always required. Heating bedrooms, especially children's, was also desirable.

BRE investigators have noted that the operation of heating systems with multiple components such as a timer, thermostat and immersion heater switch may confuse some occupants, and so lead to dissatisfaction. On the other hand, it did not take much technical competence to light a candle or stoke an open fire, or even to press an on–off switch. (All installations, therefore, should have clear instructions and be easy to operate.)

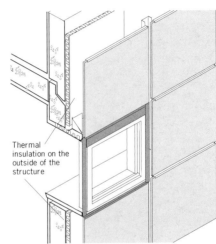

Thermal insulation on the outside of the structure

1.4a

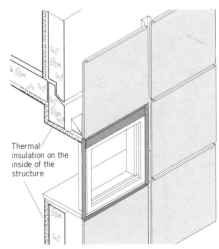

Thermal insulation on the inside of the structure

1.4b

Figure 1.4
Thermal storage capacities. The internal spaces of buildings which have thermal insulation on the outside (1.4a) warm up relatively slowly. Those with thermal insulation on the inside can warm more quickly (1.4b)

Required and achieved temperatures

As far back as the 1950s, the requirements for public sector housing[20] were:

● that the living room temperature should not fall below 65 °F (approximately 18.5 °C).
● that background heating throughout the rest of the house should be 45–50 °F (approximately 7–10 °C)
● that advantage could be taken of heat from the ground floor living areas penetrating the upper floor spaces, thereby reducing the amount of direct heat needed in the bedrooms

Although average living room temperatures did not seem to alter much during the next 30 years, temperatures in the remainder of the house did. During February and March 1978, spot readings of wet and dry bulb temperatures were made in each room of 1,000 homes nationwide in the most extensive survey of domestic dwelling temperatures to have been taken in the UK. On the same occasion, an interview was conducted with an adult member of the household to obtain information on thermal comfort conditions and heating patterns. Mean living room temperature was recorded as 18.3 °C, somewhat lower than the ideal quoted above, mean kitchen temperature 16.7 °C, and the mean temperature of the warmest bedroom 15.2 °C. The average dwelling temperature was 15.8 °C. The major sources of variance in temperature were the type of heating system and its controls, the method and patterns of operation, the age of the dwelling, household income, the time of day and geographic location[34].

Since 1978, average household temperatures have risen in line with the increase in the use of central heating, and accompanied by a decrease in the amounts of clothing worn indoors.

By the mid-1990s, standards had again improved, though only marginally. These standards were:

● comfort level: 21 °C in the living room, 18 °C in other rooms, and 16 °C as a minimum for the whole dwelling
● minimum level of heating required to prevent ill-health: 18 °C in the living room, 16 °C in all other rooms, and 12 °C as a minimum for the whole dwelling

In well designed and well managed buildings, comfort and energy efficiency can go together. Occupants should enjoy reasonable comfort under automatic control, but should also be able to alleviate discomfort manually when necessary. BRE studies show that improved controls for temperature, lighting and ventilation will lead to energy savings[35].

Energy used in heating

In 1974, oil accounted for about half of UK energy consumption, and the sharp rise in oil prices at this time had a considerable impact on the direction of BRE research as well as on the heating practices of the nation. Buildings were then estimated to consume 40–50% of all energy used. A BRE study indicated that energy savings in buildings could reduce consumption by at least 15% of total UK primary energy. A report prepared by a BRE working party assessed the proportion of national energy used in buildings, examined how it might be reduced without impairing environmental standards, assessed the total energy savings nationally, and evaluated the cost effectiveness of various alternative energy saving measures. Among its conclusions was that domestic buildings used about twice the energy consumed by all other types: housing was therefore an obvious priority for energy saving measures.

It was calculated that most of the energy used in houses was for heat, and of this about two-thirds was for space heating. Actual figures are, on average, 18 500 kWh per year for space heating, 720 kWh for lighting, and around 1000 kWh by white goods.

One-fifth of the energy consumed at the point of final use (ie in housing) was electricity but, because of the losses incurred in the conversion of primary energy into electricity at power stations, the use of electricity actually accounted in primary terms for about half of all energy used. Possible energy saving measures either depended on improvements to the fabric of the building or on energy efficient use of fuels. BRE's research programme therefore changed to reflect greater emphasis on research into energy saving by these two routes.

During the 1980s, despite the rapid growth of central heating, improvements in heating and comfort standards were not accompanied by inordinately large increases in energy consumption. The modest rate of increase of energy consumption in the housing stock (about 1.1% per year) closely matched the rate of increase in the size of the housing stock (1% per year), so that, on a per household basis, energy consumption changed very little. Having said that, there is still considerable potential for the application of energy efficiency measures in the future.

Continuing concern for the environment and for the need to conserve finite energy resources prompted two organisations to announce, in 1990, energy labelling schemes for housing. BRECSU gave technical support to the development of these schemes, both of which were based on the BREDEM. Similarly, BREAAM was applied to the assessment of no fewer than 70 new office building designs during its first two years of operation. The assessments took account of a wide variety of environmental concerns, including:

● greenhouse gas emissions
● ozone depletion
● reduction of rain forests
● desirable reuse of an existing site
● poor quality of indoor air
● risk of Legionnaires' disease
● risks from hazardous materials

One factory monitored by BRECSU using BREAAM achieved a 35% reduction in space heating energy consumption between 1986 and 1991, by introducing gas fired warm air and radiant heating. Good use was also made of a building energy management system (BEMS) and sub-metering[36].

Significant and continuing energy savings can be achieved by following good housekeeping practices such as switching off lights whenever a room is left empty; keeping windows and outside doors closed when the heating is on, and switching off electrical equipment when it is not being used[37].

Energy definitions and units of measurement

Primary energy is the energy content of fuel entering the energy economy – effectively at power stations, oil refineries, solid fuel depots etc. It also includes energy used during the mining and extraction of the fuel itself, losses in distribution networks and fuels delivered directly to consumers.

Delivered energy is the energy content of fuel delivered to the user. For space heating, it is the energy inputted to the heating system.

Useful energy represents the energy required to perform a specific function at the point of consumption. For space heating it is measured as the energy output of the heating system.

Energy units

The joule is the basic international unit of energy, abbreviated to J. Because of the joule's small size in relation to the annual energy requirements say of a typical household, the gigajoule (GJ) is used. (The petajoule, PJ, or 10^{15} joules, was used earlier in the text.) Other familiar units of energy are the therm, used for gas, and the kilowatt hour which is used for electricity.

$1 \, GJ = 10^9$ joules
$\qquad = 278$ kilowatt hours
$\qquad = 9.48$ therms

The watt is the basic unit of power or rate of flow of energy. It equals 1 J/s.

Heat losses

When a building is heated, the indoor temperature is higher than the outdoor temperature and heat is lost by conduction through the fabric of the building. The fabric heat loss rate is quantified in terms of heat flow in watts per degree of temperature difference between indoors and outdoors (W/K). Heat is also lost through the replacement of heated air by fresh air drawn from outdoors. The amount of air involved is the product of the internal volume of the building and the rate at which air is replaced. This rate must include both adventitious ventilation, or infiltration, as well as deliberate ventilation by opening windows, using extract fans etc. The ventilation heat loss rate, like fabric heat loss rate, is quantified in W/K. Together, fabric and ventilation heat loss rates make up the specific heat loss rate; this is a good measure of the thermal performance of the building and gives one basis for comparing its performance with other buildings of similar size.

Heat losses and draughts due to uncontrolled air infiltration can be minimised by ensuring that external openings are weatherstripped (Figure 1.5) and, if possible, by making the building envelope airtight. Draught lobbies can be provided to frequently used external doors which also shelter them from prevailing winds. Local radiant heat loss and downdraughts from large areas of glass can be minimised by using double or triple glazing, insulated blinds and low-level heating.

Thermal insulation of the building fabric has been described in *Roofs and roofing*[21], Chapter 1.4, *Walls, windows and doors*[22], Chapter 1.4, and *Floors and flooring*[23], Chapter 1.3.

Infrared thermography is now established as a valuable tool for measuring the thermal performance of whole buildings and for assessing, in situ, the resulting surface temperatures of a wide range of building products and components[38] (Figure 1.6).

Figure 1.5
Draughtstripping to control unwanted air infiltration. This example is factory applied. Where remedial draughtstripping is applied to existing windows, it is often the corners that are badly done

Figure 1.6
Infrared thermography image of part of a three-storey office building. This technique is used to identify thermal bridges or areas of reduced thermal insulation, and therefore higher surface temperatures where heat is being lost to the outside of the building

Effects of different forms of glazing on energy consumption
Experiments on the comparative energy consumption of houses with single, double and heat reflecting glazing were carried out in the BRE 'matched house pairs with simulated occupancy'. The pairs of semi-detached houses were heated intermittently with gas fired boilers and radiators controlled by thermostatic radiator valves. The main glazing areas faced northwest and southeast and average winter weather conditions were close to that normal for Cambridgeshire. The main results showed a saving in space heating consumption of 9% (±2%) for double glazing and 10% (±2%) for heat reflecting glazing compared with single glazing, using the same PVC-U frames. The glass area was 7% of the external wall area. Theoretical savings for this glazing were 7% and 9% respectively for double and heat reflecting glazing. Extrapolating to a 12% glazing area gave savings of 12% and 16%. This substantially supported the Building Regulations Approved Document L2/3 in which glazed areas may be doubled for double glazing and tripled for double glazing with a heat reflective film [39].

Draughts

Draughts are a common cause of complaint in many situations in buildings where the design of the ventilation and heating system has been deficient. The complaints relate to nuisance or discomfort felt by occupants. But there can also be real problems when draughts are of such a high velocity as to disturb papers in offices.

Apart from downdraught from single glazed windows, which ought to be avoided by the provision of heat emitters underneath the window, multiple glazing or double windows, the main problems seem to centre on the use of low-level opening lights in windows in general and the lack of fine degrees of control in particular.

A possible solution to this problem may be to provide for fast moving air to enter a room at high level, allowing it to mix with stiller air already in the room. The air then slows down before reaching the lower, occupied region of the room. BRE studies show:

- that standard windows and high level windows of the same area give similar whole room ventilation rates
- that the distribution of air within the room is also similar for both standard and high level windows
- that high level windows reduce the risk of draughts at working level
- that comfort conditions can be maintained for most of the time with single-sided ventilation (ie on one wall)

Improving thermal insulation in existing buildings

Many existing dwellings were built before even basic thermal insulation provisions were introduced and many more before the standards of 1976 and 1982 were established in the Building Regulations. Therefore there are considerable opportunities for improving energy efficiency in existing dwellings. Those opportunities depend upon how individual buildings are constructed both for the practical difficulty of carrying out the improvements and their cost effectiveness.

Roofs

It is usually easy to insulate pitched roofs with accessible lofts, giving highly cost effective results. Many improvements have already been carried out, often supported by the Government's Home Insulation Scheme. Flat roofs are more difficult to insulate as the space between ceiling and roof deck is often inaccessible. However, it is possible to install insulation above the roof deck using warm deck or inverted construction (Figure 1.7). This is likely to be more cost effective when the weatherproofing must anyway be renewed. Further information is available in *Roofs and roofing*, Chapter 1.4, from which Figure 1.7 is taken.

External walls

The practicality of insulating external walls depends particularly on whether they have cavities which can be filled with insulating material. Cavity insulation can yield a large and cost effective improvement to

energy efficiency. However, it should be undertaken only by a competent installer working to the relevant British Standards, or covered by a British Board of Agrément or other third party certificate.

The solid brick walls of older homes (built before about 1935) may also be insulated but usually at higher cost than for cavity walls. Insulation applied externally is particularly appropriate when the existing wall is in poor repair or an external render needs renewing. In these cases the marginal cost of insulating at the same time as re-rendering may be low enough to make the insulation cost effective, at the same time greatly reducing the risk of condensation which is often found in solid walled dwellings. Insulation may also be applied internally either using composite insulation boards or by placing insulation behind a plasterboard dry-lining. A vapour control layer is needed on the warm side of the insulation to prevent moisture from inside reaching the cold surface behind the insulation.

Windows account for a substantial proportion of total heat loss in a typical dwelling. Double glazing will usually reduce the loss through windows by about half; low-

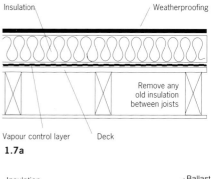

1.7a

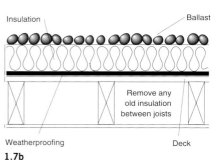

1.7b

Figure 1.7
Warm deck and inverted flat roof construction

emissivity glass and inert gas fillings give further reductions. Although replacement windows are not usually cost effective purely on the basis of energy savings, it makes sense to specify double glazing if windows are being replaced for other reasons.

In flats, windows can form a large proportion of the external surface area and, consequently, can be the dominant factor determining space heating needs. In these cases, double glazing may result in much lower demands being placed on the heating system. Further savings can be achieved, when the heating system has to be replaced, by installing smaller capacity units. Further information is available in *Walls, windows and doors*, Chapter 1.4.

Draughtproofing the windows and external doors of a UK dwelling can be an effective and relatively cheap means of improving comfort and reducing heat loss by natural ventilation, provided the profiles are suitable. Double hung sash windows are some of the most awkward to draughtproof (usually with wiping seals), and care is needed to maintain ease of opening. In most cases, draughtproofing is unlikely to cause any significant deterioration in the quality of indoor air in the dwelling. There are however a number of simple checks which should be made before installation to ensure that the ventilation requirements of the dwelling and its occupants will be satisfied after completion; this includes the availability and ease of control of opening lights and the existence of any non room-sealed combustion appliances.

Draughtstripping of windows and doors in non-domestic buildings can also make a significant reduction to the heat losses of a building. Prediction models have been used to determine the effectiveness of draughtstripping windows on the energy consumption of an office building, and to compare these with the rates actually achieved. Predicted infiltration rates compare well with measured values, and energy calculations have shown that draughtstripping windows reduced the seasonal gas consumption for heating an office building by 21%[40].

Some windows and doors will remain which are all but impossible to draughtstrip, particularly those in heritage buildings (Figure 1.8).

There is further discussion on the air permeability of windows in *Walls, windows and doors*, Chapter 4.1.

Floors

It may also be feasible to improve the thermal insulation of some types of suspended ground floors. Further information is available in *Floors and flooring*, Chapter 1.3.

Improvements to heating systems

Other improvements to energy efficiency in existing dwellings include the insulation of hot water storage cylinders. There is still a residue of uninsulated installations, in spite of publicity (Figure 1.9). Of course, where installed in an airing cupboard, it might be argued that the heat lost from an uninsulated cylinder is actually performing a useful function. The perceived wisdom, however, is that it is preferable to insulate, as enough heat is emitted from adjacent pipework.

There is also considerable potential for improving the efficiency of existing heating systems. This has already occurred on a large scale through the replacement of open fires by central heating systems. Existing gas central heating boilers may need replacement and improvements in efficiency can be obtained simply by replacing them with modern equipment. Condensing boilers offer further opportunities for savings although they may be more expensive. Typically these operate at an annual average efficiency of 85% compared with about 70% for a standard type, depending on type and model. Since some of the boilers to be replaced will have been installed in the early 1970s or before, they have even lower efficiencies and it is possible to reduce fuel consumption by up to one-third. This improvement is greatest where demand for heat is highest (ie in large, badly insulated properties) and is often the most cost effective measure available. Clearly,

Figure 1.8
It would hardly be worthwhile, even if practical, to attempt to draughtproof the few tiny side hung casements of this fine old half timbered façade. Two side hung casements can be seen open on the top storey

Figure 1.9
Uninsulated hot water storage cylinder in a dwelling about to undergo refurbishment in the late 1980s

BRE Domestic Energy Model (BREDEM)

BREDEM, simply described, is a model for the calculation of the annual energy requirements of domestic buildings, and for the estimation of saving resulting from energy conservation measures. The model lays down the factors to be included in the calculation: transmission and ventilation losses, efficiency and responsiveness of the heating system, the user's choice of temperatures and heating periods, internal and solar gains, and external temperature. The level of sophistication is tailored to meet specific applications, resulting in a number of versions of the model which differ according to such factors as the level of information available, the means of calculation to be used, and the form of output required. This framework allows new information to be readily incorporated. The model can also be used to deduce internal environmental conditions for a given energy input.

Traditionally, calculation methods have concentrated on heat losses and ignored heat gains. This meant that they were suitable for estimating the maximum loads to be met by heating systems but were poor at estimating annual energy needs. BREDEM has been developed to make realistic estimates of annual needs simply and conveniently. It is based on experience gained from measurements made in a large number of occupied dwellings and the results of research into many aspects of dwelling design which relate to energy use. There is also a worksheet version of BREDEM which can be operated using either a hand-held calculator or a personal computer. It includes standardised values for internal temperatures, solar and internal gains, hot water usage and lighting and appliance usage. It provides a robust basis for assessing the energy efficiency of dwellings and the benefits deriving from energy efficiency measures.

the best opportunity for installing a condensing boiler is when the existing boiler needs replacement: at that time only the marginal cost of the condensing boiler over the standard type needs to be considered. Replacement of a boiler with many years of useful service ahead of it is less likely to be economic. There is further discussion of this topic in Chapter 2.2.

Calculation methods

In the early 1990s, the average UK household spent over £500 per year on heating, lighting and power for domestic appliances but there was a huge variation between individual household expenditures. Field trials, in which the energy use in occupied dwellings is measured, show that the physical characteristics of the dwelling and the lifestyles of the occupants are about equally important in determining energy consumption. It is clear, therefore, that realistic estimates of domestic energy consumption can only be made if both these factors are considered together. BREDEM (see feature box) was developed, with support from the Departments of Energy and Environment, to make these estimates and has become the most widely applied energy calculation procedure for housing in the United Kingdom[41].

Heating system efficiency

The efficiency of a heating system is the ratio of the heat it produces to the energy it consumes; it is usually expressed as a percentage. It can vary over a wide range and therefore have a strong effect on the amount of fuel consumed. The benefits of good insulation can be offset by an inefficient heating system while, conversely, a hard-to-heat building can benefit greatly from the installation of a highly efficient system. A calculation of running costs allows investment in an improved system to be compared with investment in better insulation to produce the most cost effective combination of measures to be selected.

The controls fitted to a heating system are also important because they affect both the efficiency of the system itself and the extent to which it matches the requirements of the occupants.

Methods for calculating building heat losses

U value
The rate at which heat is lost through an element of a building expressed in $W/(M^2K)$.

Fabric heat loss (FHL) rate
The rate at which heat is lost through all the enclosing elements of a building (eg the walls, roof and windows). It can be calculated by adding up the products of the areas and U values of each individual element:

$$FHL = \sum_{i=1}^{N} A_i U_i \; W/K$$

Ventilation heat loss (VHL) rate
The rate at which heat is lost through the replacement of air in the building by fresh air drawn from outdoors. It may be calculated from the volume of air replaced and its specific heat and density:

$$VHL = n \, V \sigma \rho \; W/K$$

where:
n is the number of air changes per hour
V is the enclosed volume of the building (m^3)
σ is the specific heat of air in J/kg K
ρ is the density of air in kg/m^3

Using typical values for σ and ρ:

$$VHL = 0.33 \, n \, V \; W/K$$

Specific heat loss (SHL)
The rate for a building is the sum of the fabric and ventilation loss rates:

$$SHL = FHL + VHL \; W/K$$

$$= \sum_{i=1}^{N} A_i U_i + 0.33n \, V$$

These are expressed in terms of a temperature difference of 1 kelvin (K) between indoors and outdoors. **Design heat loss (DHL)** is an alternative expression to SHL based on the difference between indoor and outdoor temperature under design load conditions. It is normally expressed in kilowatts:

$$DHL = SHL \times (T_i - T_e) \; kW$$

where T_i and T_e are indoor and outdoor design temperatures respectively.

Storage of heat

Loads on heating and cooling plant in buildings can often be reduced, and at the same time occupant comfort improved, by storing excess heat (or excess coolness) within the building fabric. A potentially attractive way of storing this extraneous energy until needed is to exploit the latent heat of fusion of materials with a melting point around room temperature. Such materials – referred to as phase change materials or PCMs – allow the thermal storage capacity of a building to be increased substantially without an undue increase in building mass or volume. A number of phase change materials with a melting point around room temperature have been specially developed for the purpose, particularly during the last 30 years, in relation to passive solar design. However, although the use of PCMs (including ice) for bulk storage in air conditioning plant is now well established, they are still rarely found within the occupied space of a building[42].

Thermal storage using vessels containing eutectic salts have been installed in at least one major office building. These salts change from solid to liquid at certain temperatures, drawing in heat in the process, and releasing latent energy when they revert to the solid state.

Energy rating for dwellings

An energy rating for a dwelling aims to provide a single expression of its energy efficiency. The rating gives households a means of comparing one dwelling with another, or the same dwelling before and after refurbishment. In particular, the use of energy ratings should enable householders to take energy efficiency into account on a more rational basis when buying or renting. For architects, ratings can be used as a design tool for optimising energy efficiency in new dwellings.

Two home energy rating schemes were launched in 1990: National Home Energy Rating (NHER) and Starpoint. Both schemes rely on a computer program to calculate the annual energy requirements of the dwelling from a description of its construction and its heating system. Both also base their respective scales for energy efficiency on estimated annual energy costs, assuming a standard pattern of heating and occupancy. The Government has developed a Standard Assessment Procedure (SAP) for calculating energy ratings to enable comparisons to be made between the two methods. Both schemes have agreed to incorporate the SAP in their calculations.

The SAP, originally developed by BRE for the Energy Efficiency Office, provides a common basis for the energy rating of dwellings. All other approved energy labelling schemes are now obliged to include SAP ratings, and the ratings are referenced by the relevant parts of the Building Regulations.

The procedure is based on calculated annual energy costs for space and water heating, and assumes a standard occupancy pattern derived from the measured floor area of the dwelling and a standard heating pattern. The rating is normalised for floor areas so that the size of the dwelling does not strongly affect the results which are expressed on a scale of 1–100: the higher the number the better the standard. The method of calculating the rating is set out in the form of a worksheet, accompanied by a series of tables. A calculation may be carried out by completing, in sequence, the numbered boxes in the worksheet. The tables include data to support the assumptions of standard occupancy and heating patterns[43].

Improvements in the use of energy are necessarily gradual. There is a long way to go before all dwellings will meet even the average, let alone the best of current new-build standards, and for a large number of dwellings this is not even a realistic long term proposition. For example, it may be noted that the mean NHER rating for dwellings in Scotland in 1996 ranged from 3.9 for pre-1919 dwellings to 5.1 for post-1982 dwellings compared with a target of around 7 on a 10 point scale. The mean SAP rating was 43[9].

The importance of good control

For economical operation avoiding wasteful use of energy, good control is necessary. In principle this can be either manual or automatic or some combination of the two. The trend is increasingly towards requiring automatic controls, leaving the user only to select the programme in terms of certain measurable characteristics of output in relation to the external environment.

Where controls are not automatic, they should be easy to operate and should not encourage the user inadvertently to do the wrong thing.

Reliability of operation is also highly desirable, particularly if the advantages of efficient automatic controls are to be realised. It is, however, difficult to specify meaningful criteria of reliability except in terms of tolerances on actual performance over specified periods. Even this performance is bound to depend, to some extent, on the manner of use and the way in which each system is maintained.

The latest control and energy management systems promise to improve individual comfort and reduce energy consumption. However, fully automatic control is only part of the answer: the user interfaces, both for the individual and for building and organisational management, also need to be understood. BRE studies have revealed that:
- control systems need to be matched closely to the way in which buildings are actually used and managed, particularly in multi-tenanted buildings. Otherwise systems tend to be left on 'just in case', causing considerable energy waste, particularly with air conditioning
- individual occupants require systems not only to provide comfortable conditions but also to respond rapidly to alleviate discomfort. Air conditioned buildings fall short in the latter respect, which may help to explain why occupants often seem less satisfied with them than might be expected

Figure 1.10
Solar panels installed on the roof of the Integer House on the BRE site at Watford. The design of this house and its services ensures that as small a contribution as possible is made to the problem of global warming. However, very large collecting surfaces are needed to make a significant contribution to the energy requirements of even a small dwelling

● with suitable management, modern controls can deliver high levels of comfort and energy efficiency. However, they and the systems they control are often too complex for the average user, and need to be designed for better and easier manageability[44]

The greenhouse effect

As already noted, about half of all emissions of the major greenhouse gas, carbon dioxide, can be attributed to the energy used by buildings, and about 60% of this is attributable to dwellings. It has been shown to be possible that carbon dioxide emissions from dwellings can be reduced by 35% using proven technologies. About two-thirds of these savings are due to improved insulation standards while the remaining third are due to improved appliance efficiencies.

New designs of heating plant are much 'greener'. Given similar standards of heating, a dwelling with an NHER of 1 emits four times as much carbon dioxide as one with an NHER of 7.

Safety

Installations that use electricity or solid or gaseous fuels create potential fire hazards. Some, chiefly those using gaseous fuels, also create explosion hazards. These hazards must be minimised by appropriate attention both to the design of appliances and to provisions for storing and distributing fuel and power. The risks to health and safety are dealt with in subsequent chapters.

Summer overheating

Internal conditions affecting comfort vary with time in naturally ventilated buildings. However, it should be possible to maintain a reasonably comfortable range of conditions, without resorting to the provision of artificial cooling, by manipulating such measures as shade, thermal mass, air movement, lighting controls, and low-energy lighting and IT equipment.

Cross-ventilation through openable windows should generally provide sufficient flow rates in conventional shallow-plan buildings to avoid overheating. However, there will be cases where natural ventilation alone will not be satisfactory (see Chapter 3).

Ventilation can be used during the night to cool the building structure and so limit the temperature rise during the following day. The effectiveness of night-time ventilation for cooling depends on the difference between the ambient temperature and the internal temperature as well as on the thermal characteristics of the building.

Chapter 1.3   **Condensation**

Incidence of condensation

There are a number of factors which determine the production of condensation (the water in air being deposited on the colder surface of a material).

- The relative humidity (RH) of the air reaches 100% (saturation point – the point at which the air can hold no more water in suspension). Any extra water (or moisture) in the air must then be deposited (condense) on the surface

- The surface must be impervious to water, otherwise the excess water in the air would be soaked up by

the material. Moisture in the air could, theoretically, condense on an already saturated surface but would probably be indistinguishable from the water already in or on the surface

- The point at which air reaches saturation (ie 100% RH), also called the dewpoint, depends on the temperature of the air. The higher the air temperature, the more water the air can hold in suspension before it, the air, becomes saturated. (This is exemplified by hot, humid summer days; cold, dry winter days.)

While condensation in itself may not give rise to more than a temporary nuisance, the mould growth which often accompanies persistent condensation is more likely to give serious concern.

The two most common situations in which condensation occurs on the vertical external envelope are:
- single glazed windows
- walls with high thermal capacity where the surface temperature is unable to follow rapid changes to the air temperature, and can often be below the dewpoint

The causes of condensation are not always easy to identify. In many cases, condensation will be confused with other forms of dampness such as rising damp and rain penetration. There is a much fuller description of condensation and its remedial treatment in roofs in *Roofs and roofing*, in floors in *Floors and flooring*, and on walls and windows in *Walls, windows and doors*.

The problems of condensation loom largest in the domestic field, though the underlying physics is the same for all building types. The occurrence of condensation is a function of four factors: heating, ventilation, insulation and occupant activity. It is quite possible that correction or enhancement of any one of these factors will not necessarily reduce the problem. Such a case might be in a poorly heated dwelling in which the occupants routinely dry their washing indoors. All four factors need to be in balance or under control to eliminate condensation (and, therefore, mould growth) –

Figure 1.11
Kitchens are among the worst locations for the production of condensation, usually from cooking, particularly in older properties where both heating and ventilation provision are inadequate. Persistent condensation leads to mould growth. In this particular case it seems as though additional factors, such as thermal bridging at ceiling level (seen as a differential pattern of mould growth) and poor air circulation (which has led to the crescent shaped pattern of mould on the ceiling at the far corner), may be partly to blame

Condensation on the walls of a 1920s terrace house

BRE investigators were presented with a problem of mould growing on the inside faces of the external walls in bedrooms and bathroom. The external walls were 225 mm solid brick with sand:cement external rendering.

They first looked at the lifestyle of the occupants, one adult and two children. The house was left empty in the daytime during weekdays because the adult was at work and the children were at school. Heating of the house was by radiant gas fire in the living room, bottled gas heater in the hall, and paraffin heater in the bathroom. The heating was off during the day while the occupants were out. Windows were opened occasionally first thing in the morning.

Normal levels of moisture were produced in the kitchen and bathroom for a family of three, although additional moisture was produced by the ancillary heaters. The house felt cold and damp during the day, but warm and steamy in the evening.

The main cause of the mould growth was condensation. The solitary gas fire was not sufficient to keep the whole house warm – the reason for the occupants using ancillary heaters in other rooms of the house. These heaters were not adequately vented to the outside air. It was the classic situation of the occupants arriving home to a cold house, turning on the heating and generating a considerable amount of moisture by cooking and washing, though in this case the problem was made worse by the use of moisture-producing ancillary room heaters.

The remedy was clear enough: providing a heating system sufficient to heat all the rooms without introducing water vapour into the air. If bedrooms and bathroom were to be heated only occasionally, ancillary heaters which do not produce water vapour (eg electric storage heaters) would largely obviate the problem. The occupants were advised that a continuous, low background heat during the daytime would keep the house warm and condensation free at little extra cost than heating the cold house quickly on their return from school and work.

reasonable heating and ventilation provision, the absence of thermal bridging in the building's materials and components, and the occupants avoiding excessive production of moisture[45].

In 1988, BRE investigators carried out postal and interview surveys in bedsitting room and one-bedroom homes. The principal aim of these surveys was to examine problems relating to condensation. Half the homes studied had enough condensation to cause pools of water on the window sills, and 1 in 6 had mould growth that caused stains to plaster or woodwork. Problems reported by occupants strongly correlated with observations made by interviewers. In these small homes, condensation problems were related to location in the UK (being worse in warmer areas), age of respondent (retired people having fewest problems), household size, insulation standards, home heating (particularly the use of bottled gas) and air movement within the home, but not ventilation habits[46]. Condensation and mould growth certainly used to be widespread problems in all housing sectors, but especially so in tenanted accommodation. In many cases it was difficult to identify the underlying cause; this was often complicated by social issues.

The national house condition surveys contain some information on the incidence of condensation[†].

In England approximately 1 in 5 dwellings suffer from dampness, which in two-thirds of the cases is due to penetrating or rising damp, and the remainder to condensation[7]. The situation has hardly changed since 1991[26].

In Scotland there has also been no substantive change in the incidence of condensation between 1991 and 1996 as assessed by the surveyors for the Scottish house condition investigations. Almost 1 in 3 households reported some level of condensation, with steamed up windows being the most common

complaint, though a substantial number (approximately 1 in 6 households) reported other more serious problems including mould growth on walls or carpets. The worst affected houses, as might be expected, were those with no central heating system or which used methods of heating other than by gas or electricity.

Furthermore, in the one-third of dwellings in Scotland which suffered from condensation over the period up to 1982 the risk was not related to age. In dwellings built after that date the proportion at risk fell to 9%[9].

Condensation on walls

The most common occurrence of surface condensation is on solid 225 mm thick masonry walls. Although these walls do not generally suffer from interstitial condensation (ie within the structure of the wall), under certain prevailing temperature and internal environment conditions, condensation can be a considerable problem. Mould growth is invariably a symptom of persistent surface condensation.

Walls, windows and doors described investigations, carried out by the BRE Advisory Service during the period 1970–74, in which there were about twice as many cases of surface condensation on external walls as there were of interstitial occurring in the walls. By 1987–89, cases of

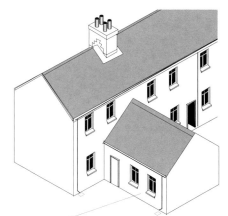

Half brick walls, often found in 'byelaw housing', are highly susceptible to mould growth internally

Figure 1.12
Projecting back rooms in 'byelaw housing' often suffer from condensation and mould

† Comparable figures for Wales and Northern Ireland were not available for this book.

condensation investigated were fewer than in former years, but in the following years the case load of the Advisory Service rose again.

Reduced wall thicknesses (and hence reduced insulation) may exist in porches, previously bricked-up openings and where fireplaces have been removed. Half brick solid walls, often found in projecting back rooms of 'byelaw housing', are highly susceptible to surface condensation and mould growth (Figure 1.12).

● Condensation can be a problem at window and door reveals where there may be locations with reduced wall thickness. Corroding metal corner beads around window or door openings may indicate that condensation has been a problem

● Impervious external wall coatings may prevent or restrict drying out of water vapour from the wall and lead to dampness and an increased risk of condensation

● Changes to heating, ventilation and occupancy patterns on the other hand can often lead to condensation and mould where previously no problems were encountered. There is further discussion of condensation on walls and its treatment in *Walls, windows and doors*, Chapter 1.4.

Condensation on windows and doors

Of all dwellings in England[7]:
● 30% have double glazing on all windows
● 30% have double glazing on some windows
● 40% have no double glazing

It is older dwellings which are less likely to have double glazing (more than two-thirds of those dwellings dating from before 1919 have no double glazing), most likely because of the retention of desirable architectural features such as wooden sash windows with slender glazing bars which will not take sealed double glazing units. Further information is given in *Walls, windows and doors*, Chapter 4.1.

In Scotland, with a colder climate, nearly two-thirds of dwellings in 1996 had full double glazing, compared with just over one-third in 1991[9].

Condensation on floors

The insulation value of many kinds of floor can be degraded by thermal bridges where high thermal transmission materials penetrate layers of low thermal transmission material; this may occur at thresholds or sleeper walls. Thermal losses due to thermal bridges are often ignored in calculations, especially where thin sections are

Figure 1.13
Older windows, such as this standard steel frame, will not take double glazing. Signs of condensation, which has occurred on both the frame and the single glazing, are evident on the sill

Figure 1.14
The sarking was still waterproof, but not vapour permeable, so it was concluded that condensation and not rain penetration had caused the deterioration in these roof timbers

involved, but these and other materials such as concrete floor beams become more important as thermal insulation standards increase. Some thermal bridges are also important because they produce inside surface temperatures below the dewpoint of the air, leading to selective condensation on parts of the flooring.

The two most common situations in which condensation occurs on floors are:
● where they are adjacent to exterior perimeter walls and heat is lost to the outside via a thermal bridge
● where they have high thermal capacity and their temperatures are unable to follow rapid changes to air temperatures which can often be below the dewpoint. This particularly affects floors in warehouses or in concourses open to the external air when cold weather is followed by a warm front

Further information is given in *Floors and flooring*, Chapter 1.3.

Condensation in roofs
Pitched roofs
The 1996 English House Condition Survey noted that, of the 14 million or so houses in England with pitched roofs, 208,000 (say 1 in 50) displayed signs of condensation in the roof space. Interestingly, the age of the property does make some difference to the incidence of condensation. The ratio of dwellings with condensation in the roof (Figure 1.14) ranges from around 1 in 50 of those built before 1944 to about 1 in 85 for those built after 1980. Cases of the latter period were caused, for the most part, by blockage of the eaves ventilation in refurbished cold deck roofs insulated to higher standards. Unintended air movement within the roof void may also carry water vapour to areas where condensation can cause problems. Similar conditions also exist in other parts of the UK.

Flat roofs
Cold deck flat roofs present a high risk of condensation, whatever the form of heating in the building, and these roofs should be converted to warm deck wherever possible. In existing roofs, especially flat roofs where the diagnosis of the cause of dampness is often complicated, it may be worth calculating the risk of condensation occurring. The calculation can be done using the procedure outlined in BS 5250[47]. There is more information on condensation in roofs and its treatment in *Roofs and roofing*, Chapter 1.11.

Mould
Most of the problems which result from surface condensation are caused by moulds (Figure 1.15). As well as being unsightly, mould growth is thought to cause respiratory problems in susceptible individuals. It is the reason for many complaints by building occupants, and therefore it should be prevented from occurring in all parts of a building. There is a fuller description of mould growth and its treatment in *Walls, windows and doors*, Chapter 1.4.

Evidence from the 1991 English House Condition Survey showed that, in houses built in the previous 10 years, mould growth occurred in 5% of houses where extract fans were present and in 11% where there were no fans. Mould growth occurred in 17% of all dwellings of all ages, whether there was a fan or not.

Remedies
There is much to be said for a careful investigation of problems relating to condensation. Only rarely is the diagnosis straightforward since there are other sources of dampness which can easily lead to confusion.

The following procedure is recommended by BRE[45]:
● diagnose the sources of dampness
● look at the factors
● identify the causes
● select the remedies
● apply the remedies
● follow up to assess the success of the remedies

Figure 1.15
Mould has grown on the external solid wall of this toilet, in spite of the presence of an airbrick

Most porous building materials will retain some moisture, though the actual moisture content will vary widely. The presence of moisture that is not apparent on the surface of a material should not normally be a cause for concern. However, excessive dampness which shows on the material or component should always be investigated. Possible causes which need to be eliminated include leaks of all kinds, retained construction water, and hygroscopic salts on walls resulting from previously occurring rising damp or inundation or from animal contamination.

When it has been established that the cause of excessive dampness is condensation, the next step is to study the occupancy conditions of the building, in particular what the heating and ventilation regimes are which might lead to the generation of condensation. Causes of condensation could include the conditions being:

- too cold: inadequate heating, heating too expensive to run, underused heating, inadequate thermal insulation, or too much ventilation
- too wet: high moisture-emitting appliances, poor layout of the dwelling and poor household management
- too little ventilation

The remedies to these problems include increasing heating, increasing thermal insulation, reducing fortuitous ventilation (air leakage), increasing controllable ventilation, and reducing moisture content of fabric by improvements in domestic management.

Mild cases of mould growth will often yield to simple changes in the heating and ventilation regime in the dwelling or to cosmetic treatments during redecoration, perhaps with fungicidal paint. In more severe cases of mould growth, fungicidal treatments may be little more than a useful holding operation if major rehabilitation is not possible for some time. More severe cases will usually require improvements to thermal insulation, greater heat inputs and adjustment of ventilation arrangements (either natural or mechanical).

BRE investigators have consistently found it necessary to follow up site investigations of condensation to determine whether recommended measures have actually been carried out, and, what is perhaps more important, whether the occupants understand the part they need to play in its control.

Chapter 1.4   **Artificial lighting**

Historical note

Until electricity started becoming the main form of energy for providing artificial lighting, a variety of methods was used – from wax and tallow candles, and oil lamps, through to gas lamps fed from public, piped gas supplies, globed paraffin lamps and pressure paraffin (Tilley) lamps.

Coal gas was first produced in the seventeenth century, but it was not until late in the next century that it was occasionally used to light buildings from a fixed installation, and public supply gas works were starting to be built just before 1820. The first coal gas lighting installations had jets, and the mantle was invented in the middle of the nineteenth century, though not made in significant quantities until 40 years later[18]. Gas carcassing for lighting purposes still exists in many old buildings, though the systems would have become disused when lighting by electricity became practical.

Major buildings in London were beginning to be lit by electricity in the last two decades of the nineteenth century. At first, arc lamps were employed, then filaments of various materials with progressive improvements. The first British Standard for the filament lamp was issued in 1920. The fluorescent lamp was invented in the UK and first marketed during the 1939-45 war[18].

From 1950 onwards studies were pursued at BRS of artificial lighting; research related, in particular, to layout in office design. Work on permanent supplementary artificial lighting of interiors (PSALI) provided the ground rules for producing lighting levels appropriate to specific tasks. In the early 1950s studies began of natural and artificial lighting in hospitals, greatly influencing ward design; the new concepts were first applied in wards built in 1954 and 1957 (Figure 1.16). Measurements made at full scale in these wards confirmed the validity of the laboratory predictions.

Current position

Artificial light is required for part of the day for all activities except sleep, and may be required for the whole day where access to daylight is poor or unavailable. The precise requirements differ slightly according to whether all light is artificial or the artificial light is only supplementary. In the latter case, balance between the two kinds becomes important in terms of both intensity and colour.

The properties which need to be considered include the method of distribution, relative illumination of task and background, avoidance of glare, definition of form, the role of occupancy sensing, avoidance of flicker and superfluous heat.

Lighting is frequently the largest single item of energy expenditure in UK offices, accounting for about 35% of energy costs. Hence the importance of making improvements in this area as well as considering the contribution lighting makes to the

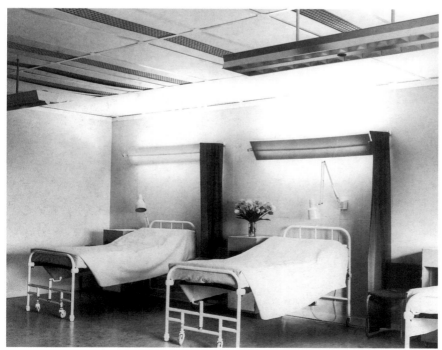

Figure 1.16
Experimental indirect lighting in a BRE mock-up of part of a hospital ward in 1957. Luminaires and task lighting are also installed but not in use

example, lighting can account for around 50% of the electricity used and lighting costs can even exceed those for heating. The application of current best practice can reduce costs significantly; practical cost effective examples have demonstrated savings typically in the range of 30% to 50%[48].

In many situations, provided appropriate lighting and heating controls are fitted, building energy use varies only slightly with changes in glazing area. Window sizes should not be dictated by shallow optima in energy use, but instead be chosen to produce a comfortable and visually appealing interior[49].

Lighting equipment
Domestic
Lampholders on wall brackets in accessible positions may be preferred to the traditional pendant in the centre of the room. In addition to offering potentially more flexible lighting arrangements, there may be a safety benefit, as many elderly people may be unable to reach pendant fittings to replace faulty bulbs. In family housing this factor is less important than in housing occupied by one or two people. Using energy saving compact fluorescent lighting with appropriate sockets and fixings can save considerable amounts of electricity.

Figure 1.17
The Art Library at the Victoria and Albert Museum. Task lighting is provided over each desk, while the candelabra and perimeter luminaires give a subdued background light during the hours of darkness

provision of a safe, comfortable and productive work environment.

In housing, the requirements are normally met by providing wall or ceiling lighting points, controlled from one or more positions. Automated response to external daylighting levels is rare. The costs of running artificial lighting form a relatively small proportion of total energy consumption, though it is still worthwhile to consider installing low-energy luminaires or bulbs.

In the non-domestic field, automatic response, either of full or supplementary, artificial lighting is more common. The pattern of running costs may also differ from the domestic situation. In offices, for

Figure 1.18
Large floodlights coupled with subdued perimeter lighting in this swimming pool. It is the flood lights which cause most of the reflections on the pool surface

Figure 1.19
A typical multi-person office with uniform lighting controlled from a manual switch panel at the entrance to the room. This discourages the occupants from taking advantage of the available daylight

Unfortunately, misunderstandings abound amongst the occupants of domestic buildings on the characteristics of these. A common misconception encountered by BRE investigators on site is that long-life tungsten bulbs save energy.

Non-domestic
The design of luminaires is covered in various parts of BS 4533[50] and BS EN 60598-1[51], with special requirements for particular types of buildings; for example, swimming pools in Section 102.18, and hospitals and health care buildings in Sections 102.55 and 103.2 of BS EN 60598-1.

Lighting calculations
There are a number of computer programs available to perform lighting calculations. The underlying algorithms are described in *Interior lighting calculations: a guide to computer programs*[52].

Lighting controls
Good design of artificial lighting aims to provide visual satisfaction within the lit space and, where appropriate, an environment that enables tasks to be performed efficiently and comfortably. Although the main benefit and prime reason for interest

in lighting controls has been the saving of electrical energy, providing localised controls to enable occupants to exercise discretion over the lighting of their particular work areas can significantly increase user satisfaction[53].

Lighting controls should help to provide:
● appropriate, safe, comfortable and healthy indoor environments
● suitable lighting for the needs of the situation
● sufficient, understandable and responsive local control
● efficient use of energy, avoiding waste from over-lighting or from lights being on unnecessarily when spaces are unoccupied or when daylight is sufficient
● planning flexibility, allowing workstations or partitioning to be moved without altering fixed wiring, and luminaire outputs to be adjusted to match lighting requirements

The systems themselves should be:
● effective and reliable in operation
● readily usable and easily managed
● limited in their susceptibility to failure
● able to recover rapidly from failure

Energy saving
Electronic lighting controls were first installed largely for their energy saving benefits, and substantial savings have been documented. The 1994 amendments to the Building Regulations for England and Wales required that artificial lighting systems in buildings should use no more fuel and power than is reasonable, and should have reasonable provision for control. Approved Document Part L[54] advocates controls to encourage the maximum use of daylight and to avoid unnecessary lighting when spaces are unoccupied, local switches are placed in accessible positions, and other automatic controls are installed as appropriate. However, automatic switching should not endanger the movements of building occupants.

If sufficient daylight is available to meet lighting requirements for significant parts of the working day, energy savings from suitable switching regimes can be considerable. BRE studies have shown that the probability of switching on artificial lighting on entering a space correlates closely with the daylight availability at the time, but switching off rarely occurs until the last occupant leaves. The control strategy should invite switch-off decisions to be taken when daylight availability improves (time switching) or place switches in

Figure 1.20
Luminaires are individually switched, providing opportunities for saving energy

locations where the occupants' perceptions of daylight adequacy will be more relevant to their needs than for locations at the entrances to spaces (localised switching)[55].

For many types of installation, these provisions will be adequate. However, further savings can be achieved by using automatic sensing of daylight levels (photoelectric daylight linking) or occupancy (occupancy linking). The use of these options depends on the type of occupation. For spaces with negligible daylighting, a combination of time switching and localised switches will cover most situations although care must be taken to ensure that timed-off control does not produce dangerous blackout conditions. For installations with sparse and intermittent occupancy, such as large store rooms and warehouses, localised (distributed) switching will eliminate the need for 'blanket lighting' of the whole space; occupancy detector control will also be a major option.

Substantial energy and cost savings may be available if lighting is appropriately controlled to make full use of available daylight and to eliminate lighting use when rooms are unoccupied[56].

Recent case study work by BRE has shown the performance of lighting control systems generally to fall short of expectations. It has been suggested that this is due to the changing nature of use, in particular deep plan office spaces with extensive use of computer screens, and failure to tailor the control system to match the needs of occupants. Extended building operating hours (eg night working) will change the economics of applying lighting control measures.

Greater consideration will need to be paid to areas such as corridors and meeting spaces where no particular person has responsibility for controlling the lights. Corridor lighting tends to be over-bright for its function, particularly in office buildings, and a person entering a room from a well lit corridor may feel it to be too dim and switch on additional lights.

Occupancy sensing – either of presence or of absence – can play a role in energy conservation in lighting; but a system designed so that lights can be triggered by a person passing an open doorway of an empty room should be avoided.

Other benefits of good control

Controls can yield a variety of benefits other than energy savings, the principal ones include:

● *Occupant satisfaction and convenience*
Individual occupant control has become more common since the 1980s, reinforced by workplace legislation, in particular display screen equipment requirements, and a more general concern for comfort and productivity.

● *Planning flexibility*
Electronic controls may be specified to improve the flexibility of working spaces; for example, by allowing workstations and partitioning to be more easily moved. Increased controllability also allows for arrangements to be shared where the occupancy of workstations is variable or unpredictable, or where requirements may change rapidly between successive occupants.

● *Better management information*
The more sophisticated lighting management systems can provide information on utilisation for monitoring and energy management, and assist in planned maintenance.

● *Simplified design and installation*
Electronic controls can reduce detailed design requirements and installation time. Final connections to luminaires can be made with plug-in flexible cables, while switches and sensors are connected in inexpensive safety extra-low voltage (SELV) wiring; or self-contained 'intelligent' luminaires can be used, with built-in sensors to detect occupancy, ambient light levels or signals from hand-held controllers[55].

Automatic switching off of lighting system in an office block
An installed system enabled all lighting to be switched off from a central point at appropriate times (eg at the end of the working period) while allowing individual occupants to override the control by operating remote switches at each luminaire. Apart from a few faults which were identified and corrected during the commissioning period, the equipment functioned well. Initial monitoring over a four month winter period indicated that savings of 42% in lighting costs were achieved compared to the same period in the previous year (ie before the system was installed). Extrapolation of this figure to a full year, with allowance for greater daylighting in summer, indicated that annual savings in electricity costs of 40% could continue to be expected.

Time switches

Available hardware includes the well tried-and-tested, multi-position, electromechanical time switch and various solid state alternatives that have been developed recently. Time control can represent an important element of a building's energy management system.

For small installations (below, say, 2 kW) the time switch can be used to switch the lighting load but, more often than not, it provides time signals for operating relays which control power to the lighting circuits.

Time switching necessarily implies that the timed signals are transmitted to the controlled elements via some communication channel. This can be the mains supply itself, with the signals merely an interruption of the mains supply. If the time switch is remote (or centralised) and override facilities are required (eg to provide localised switching for occupants), some other communication channel may be necessary to give override facilities and it may be desirable to allow different switching patterns for different parts of the lighting installation. Low-voltage wiring systems can be used for this purpose. Recently, mains-borne signalling using microelectronic coding techniques has become available.

A further alternative using the mains wiring involves using latching relays at the luminaires or luminaire switch panels so that, when a mains-interrupt pulse of short duration (about 1 sec) is generated by a time switch, the relays latch off even when mains supply is restored. Override or local control involves resetting the relay to the on position using conventional switch hardware such as a rocker switch or pull cord[53].

Localised switches

Convenient localised switching can be accomplished merely by providing more mains switch positions than has been normal practice. With due regard for safety requirements, switches should operate on lighting in the immediate vicinity of the luminaire and wherever possible the switching arrangement should reflect the profile of penetration of daylight. So, for an installation consisting of rows of fluorescent lighting units parallel to a window wall, there should at least be as many switches as individual rows. Economies will result from:

● having the switches for each row placed at approximately the same distance from the window wall as the lighting itself
● having a further breakdown into groups of luminaires within rows, with switches close to their area of influence

Figure 1.21
Corridor lighting controlled by passive infrared detector

Alternatives to wall mounted mains switches include low-voltage switching, luminaire mounted pull cords, and even remote ('wireless') switches using infrared or ultrasonic transmitting and receiving devices similar to those commonly used for controlling TV or sound systems.

Photoelectric switches

Two basic forms of automatic photoelectric switching control can be used to link electric lighting to daylight availability.

● The cheaper method is to switch the lights by a photocell, mounted externally or facing out of a window to sense daylight only (open loop control). This is similar to photoelectric street lighting control, except that calibration is required in situ because the daylight level for switching depends on the particular characteristics of the building. For example, a given daylight illuminance at the sensor can result in a variety of internal illuminances depending on window size, furniture layout, decorations etc. The use of an external sensor is not suitable where there are complex shadows across the façade.

● The alternative approach has the sensor mounted in the area to be controlled to sense both daylight and artificial light (closed loop control). In this case, a switch on level and switch off level need to be set so that activation of the electric lighting does not immediately cause switching off. For either type of photoelectric switch, the device should allow a delay to be set so that after a switching action has occurred no further action can take place until after the set delay. This minimises the problem of rapid switching caused, for example, by fast moving clouds.

Photoelectric on–off switching causes sudden and noticeable changes in lighting levels leading to complaints from occupants of distractions. For this reason, this type of switching is recommended only for working areas close to a window wall where the frequency of switching will be least. Successful installations which make use of multi-level switching in multi-lamp luminaires for a large working area exist, but clearly require careful planning, and their complexity will result in higher costs.

Lighting switched on by passive infrared movement detection and switched off by time delay may be a useful method of saving energy where internal spaces are infrequently used (Figure 1.21).

Lighting requirements

People, particularly those who are elderly, need good artificial lighting conditions. Those with restricted mobility may not easily be able to take their work to the light, and enhanced general illumination levels may be desirable. A survey carried out in 1983 found that two-thirds of the homes visited were inadequately lit[57]. Poor lighting could be particularly linked to accidents[58].

Most pipes and cables can be made to be as unobtrusive as possible. Artificial lighting, however, plays a major visual role, even if its source is deliberately concealed; it may largely determine the character of an interior, or even the external appearance, when it is in use. It is desirable therefore to specify or agree upon the character at which to aim – stimulating, assertive, sharply modelled; or neutral, flat, restful etc.

Task lighting

At present, most lighting installations are designed to give a specified average illuminance on a horizontal working plane, although sometimes predictions are made of the level of discomfort glare to check that it is within recommended limits. In many situations no other lighting design work is done. To ensure good visual task conditions (eg in office environments) it is important that other aspects of the lighting are

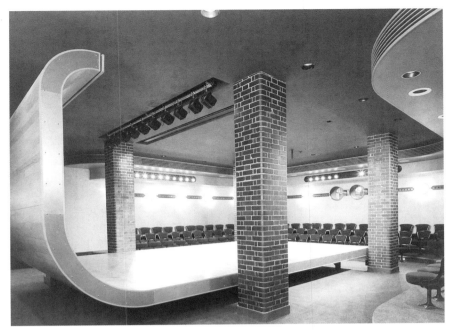

Figure 1.22
In this installation at the Theatre Museum, London, a variety of lighting sources gives good flexibility

Figure 1.23
This installation of luminaires set into the highly serviced ceiling will give more or less uniform illumination

considered. By doing so, more efficient energy use and a more efficient working environment will result with less likelihood of visual distraction or fatigue.

Office lighting is designed primarily to enable visual tasks to be performed by the occupants; it should be designed, therefore, to make the task as visible as possible. Bright reflections in the task area will reduce its visibility and may cause discomfort[59].

The contrast provided by visual task material depends on the intrinsic reflectance properties of the material and on the way it is lit. If the task material is completely matt, incident light is reflected equally in all directions, therefore the direction of incident light is unimportant. However, very few visual tasks are composed of materials which are truly matt and, if the surfaces of these materials produce a specular or mirror-like component of reflection, defocused images of bright sources such as luminaires are seen reflected in them. These bright images produce a veiling effect, reducing the task contrast and hence its visibility. The incident light direction is therefore crucial to the observed task contrast. This is most apparent with very glossy materials.

If daylight from side windows can be used, desks should be positioned at right angles to the windows for high contrast rendering factor values. In an installation using both daylight and artificial lighting, the luminaire rows should therefore be parallel to the window wall. Reflections need to be avoided where computer screens are in use.

Conservation lighting
Where high quality lighting, which does not affect delicate paintings and tapestries detrimentally, is needed, it is often desirable to reduce the amount of daylight, and particularly sunlight, from reaching the items. Techniques such as using fibre optics to direct artificial light produced remotely onto exhibits, thereby ensuring that the risk of damage by heat is reduced, can now replace old tungsten bulbs.

Illumination levels
The following typical levels of standard maintained illuminance are suggested:
- entrance halls, passageways   200 lux
- staircases   100 lux
- kitchens   150–500 lux
- bathrooms   150 lux
- bedrooms   50–100 lux
- libraries   300 lux
- restaurants   50–200 lux
- boiler houses   100–150 lux
- control rooms   300 lux

Supplementary lighting should be provided in local areas where background lighting does not give adequate illumination levels; for example, at mirror positions in bathrooms. In some areas, such as kitchens, it will be necessary to avoid shadows on work surfaces.

Very many more suggested levels for other types of spaces in different working environments are given in the CIBSE *Code for interior lighting*[60].

Dimmer switches allow users to adjust lighting levels to suit a wide range of activities.

The illuminance uniformity criterion used in electric lighting design has an influence on the appearance of the space and the energy consumption of the installation. A subjective study of the acceptability of illuminance differences between two desks was carried out in a simulated office at two room illuminances with light and dark desk tops. The results showed that illuminance ratios between work stations of at least 0.7 are generally acceptable. Although there is a trend of decreasing

Figure 1.24
Photoluminescent markers on this experimental staircase at BRE have proved an acceptable alternative to emergency lighting

acceptability at lower illuminance ratios, there are indications that, under some conditions, lower illuminance ratios may also be acceptable[61].

Photoelectric dimming
Unlike switching, 'top-up' control is unobtrusive and ensures that at all times the sum of daylight and artificial lighting reaches a given minimum (the design level). This is achieved by photoelectrically sensing the total light in the controlled area and automatically adjusting electric light output to top-up to the set level. If daylight alone is adequate, electric lighting is dimmed to extinction. The energy saving potential of top-up control is greater than for photoelectric switching and the mode of control is likely to be more acceptable to occupants. However, although savings will be greater, costs are higher because of the more complex control electronics and special lamp ballasts required.

Photoelectric switching can be applied to a wide range of lamp types; for high intensity sources special ballasts may be required to minimise restrike times.

Photoelectric dimming is most satisfactorily applied to fluorescent lamps, although this is not yet possible with the recently developed smaller diameter tubes. Although dimming of tungsten lamps presents no problem, the light output or power characteristic means that energy savings are unlikely to be achieved cost effectively. (In practice, these lamps are unlikely to be used when energy efficiency is a major concern.) Dimming of some types of high intensity lamp is possible although there is little experience of its use in the UK. Doubts have been raised about the effects on lamp life of photoelectric dimming, so the technique is not recommended for high intensity lamps.

Emergency lighting
Emergency wayfinding systems consist of luminous tracks, emergency exit door marking and exit signs. They come into operation when the main lighting fails or when smoke is detected by the sensors (see Chapter 5.3).

Luminaires for emergency lighting, which may contain direction indications for means of escape, are available with standby batteries for when the mains system fails. The relevant Standards are BS 4533-102.22[62] and BS 5266-1[63].

BRE investigators have concluded that 0.2 lux on the centre line of the floor of an escape route is a reasonable standard, but that 1 lux might be better for stairs[64].

Lighting positioned at low levels is preferred to lighting positioned at high levels since it is less likely to be obscured by smoke.

BRE has also carried out a number of trials of photoluminescent markers sufficient to enable people to negotiate escape in fire conditions. These can be applied to risers in staircases and to skirtings in corridors used for means of escape. There are no current British Standards, though it has been seen that markers are at least as good, if not better, than conventional emergency lighting[65].

User satisfaction criteria
Occupant surveys suggest some general points for selecting lighting controls for usability and acceptability.

● People are good at judging whether they need the lights on. Decisions about whether lights are on or off should be left to building occupants.

● People are not good at switching lights off. Automatic switch-off should be provided, preferably to operate when users are absent or daylight is good. Otherwise dim-out is better for both safety and unobtrusiveness.

● People dislike automatic systems which distract them or do not do what they want. Ideally systems should switch off either when people are absent, or imperceptibly (eg by dimming gradually).

● Local controls should be accessible and intuitively obvious (to avoid embarrassment of users). Switches and controls should be close to hand at the points of decision (eg at desks or at doors). The further a switch is from a door, the less likely that it will be used.

● Building managers also need good, accessible, usable controls. Control consoles should be readily accessible at points of decision: easy to see, easy to use and not tucked away.

● People want a rapid response. Some control systems do not respond instantly, some lamps take several minutes to warm up, and some high intensity discharge lamps will not strike again for up to 10 minutes after being switched off.

● People do not like automated systems which switch lights on (or keep them on) when they could have been off under manual control. Problems like these can undermine confidence in the system concerned, in management and in energy saving programmes generally.

● People do not like being plunged into darkness. This may also be unsafe or embarrass occupants (eg when visitors are present). Where possible lights should be dimmed down, switched off in stages or, at the very least, give a warning of switch-off (eg some systems flash the lights a few minutes before they go off).

● Not everyone will need the specified illuminance level for the same tasks. Visual acuity and perception varies. For example, although the recommended illuminance standard for offices is 500 lux, lower levels are often preferred for work on computer screens.

● Daylight is not always usable, and daylight and sunlight are not always an asset; they can cause thermal discomfort, glare of various kinds and reduce computer screen visibility, particularly when screens catch reflections from windows or have windows behind them.

● Working patterns can be diverse. Switching lights off at lunchtime and at the end of the normal working day has been suggested, but, with working hours frequently being longer and less routine, the lunchtime switch-off has often had to be abandoned and the end-of-day switch-off delayed.

Chapter 2

Space heating and cooling

This chapter deals with all forms of space heating and cooling, including open fires burning solid fuels, boilers and hot water radiator systems, forced warm air systems and air conditioning (Figure 2.1). There are brief sections on floor and ceiling heating, heat pumps and solar heating devices.

It is appropriate, first of all, to deal with topics relevant to all or most forms of heating. These include a comparison of the frequency with which systems will be encountered on site, although the available figures relate to the domestic market rather than to the non-domestic; also the relative efficiencies of the various kinds of fuel.

Heating provision in dwellings

Central or programmable heating of one kind or another is found in 87.6% of all dwellings in England, although 2% of these installations are over 30 years old[7]. Most of the remaining dwellings have fixed room heaters or convectors, although some 106,000 dwellings still have no fixed heating of any kind.

When the age of the dwelling is taken into account, it would seem that it is the pre-1919 dwellings that have the highest numbers with no fixed heating provision; some 44,000 dwellings are in this category, or roughly 1% of the total stock[7].

So far as the type of central heating is concerned, the majority of systems, nearly two-thirds, are powered by gas single-purpose boilers. Gas back boilers account for around 14%, electric storage radiators 9%, oil and solid fuel 3%, gas powered ducted hot air 3%,

with electric floor or ceiling heating and communal systems each taking a 1% share[7].

With regard to secondary heating provision, the most popular form in England is a mains gas fire or convector installed in just over 10 million dwellings[†]. Fixed or portable electric heaters are used in 3 million dwellings, with solid fuelled open fires or stoves in 2.5 million. Portable LPG or paraffin heaters are comparatively rare[7].

In Scotland, full or partial central heating is now installed in 78% of all dwellings with by far the most popular system being full gas central heating – in fact over half of all dwellings have full gas central heating. Full or partial electric central heating is installed in around 18% of dwellings[9].

In Northern Ireland, although some 87% of all dwellings have full or partial central heating, the situation with fuels is quite different, with the most popular being oil. Electric storage is used in around 1 in 10. As elsewhere in the UK, ceiling or floor heating is rare, less than 1 in 500. Around 1 in 400 have communal heating. For those dwellings without central heating, which number around 1 in 8 of the total stock, by far the largest number, nearly 60,000 dwellings altogether, use solid fuel open fires or stoves[10].

Figure 2.1 Space heating and cooling systems can take up a considerable amount of space. Here ducting, about 1 m in diameter, is being installed in a large building with complex servicing systems

† Comparable figures for Wales were not available for this book.

District heating, and combined heat and power (CHP)

District heating, or the closely related community heating, is a method of providing space heating and hot water supply to a number of buildings from a central source. The heat may be derived from any of the ordinary fuels, but the larger boiler plants normally associated with these schemes facilitated the use of low grade fuels which, because of small size and low calorific value, were not suitable for individual house installations.

District heating was intensively developed in the USA during the last quarter of the nineteenth century. The principle has been widely used since then in continental Europe, but there were only isolated schemes in the UK until the 1950s.

By the early 1950s, schemes had been designed for a number of housing estates varying in size from about 100 to 8,000 houses and flats. In some cases industrial and public buildings were included, and, in others, provision was made for supplying process heat services to industrial premises. Supplies to separate establishments were metered, but the meters were not always reliable.

Data have been gathered by BRE investigators on the reliability and costs-in-use of heat distribution networks in district heating schemes throughout the country; these data are based primarily on the experience and records of owners and operators. The performance and maintenance costs of the great majority of schemes compared well with those to be expected of engineering services generally. A few schemes, however, incurred much larger annual repair costs[66].

To give some idea of the overall performance of district heating schemes, peak heating demands and daily heat consumption were recorded by BRE investigators at 15 district heating schemes during the winter of 1975–76. At most sites, the peak space heating demand was less than 60% of the connected space heating load for the complete system, while the daily heat consumption for space heating was about 60% of the figure predicted by conventional design procedures. Annual load factors were in general about 50%, but this somewhat high figure included the effect of the heat losses from the distribution mains which commonly exceeded 25% of the annual heat consumption. Boiler efficiencies averaged 80%, irrespective of the fuel used[67].

Buildings on the BRE site used to be heated by means of a small district heating scheme provided from a central boiler house, using low grade fuel. Heat losses from the transmission mains, housed in concrete ducts recessed into the ground, but within which the pipes were poorly insulated, meant that, with rises in the price of fuel, the system became uneconomic.

Damage to the insulation from infiltrating water was also a problem requiring constant maintenance. One beneficial, though expensive, side effect of the large transmission losses was that there was always a snow and frost-free system of pedestrian pathways because of heat losses from the mains! Individual gas fired boilers now ensure that much higher efficiencies are achieved.

In combined heat and power (CHP) schemes the generation of heat in the central plant is combined with the generation of electricity, which enables heat and electricity to be produced with a lower fuel consumption than would be required to produce similar amounts of heat and electricity in separate plants[68].

Case study

Community heating scheme

A city-wide community heating scheme exists in Sheffield, completed in 1991, which is a joint venture between local government and the private sector. It uses up to 30 MW of heat recovered from a municipal refuse incineration plant, and serves up to 3,500 dwellings. Hot water, at temperatures of up to 120 °C and at pressures up to 16 bar, is distributed around the city through an underground network of pre-insulated steel pipes to heat exchangers situated in individual buildings connected to the system.

A key provision in the system is that the distribution pipework needs to be very reliable. Previous client experience with distribution mains in other schemes was that they had proved unacceptably unreliable. In the present scheme, thermal movements in the pipework were carefully accommodated by a suitable layout using curved pipes, and each part of the system was the responsibility of a single manufacturer to ensure compatibility of fittings, and every weld tested on completion. Electronic monitoring detects the presence of leaks.

When electricity is being generated, only a proportion of the input fuel is converted into electricity (typically 30–50%). The remainder of the energy consumed in electricity generation is not at present usable.

In the first three years of its operation, 670 GWh of fossil fuel energy was saved[69].

Figure 2.2 Standby boiler plant at Bernard Road Incineration Plant, Sheffield. The plant needs to be used occasionally to provide heating when the main plant is not in use

In a thermal-electric station providing both heat and electricity, there is a problem in that the heat and electricity demands do not normally coincide. At one time it was suggested that the difficulty might be met in large stations by a combination of condensing and back pressure turbines. Another method which has been used to overcome the lack of balance is the use of heat accumulators; in these, excess heat, produced when the heat demand is low and the electricity demand high, is stored until the conditions reverse and the demand for heat exceeds that for electricity. These accumulators store heat in the form of hot water and need to have very efficient insulation.

Electric storage heaters are designed to take advantage of off-peak tariffs. It is usual to design systems based on storage heaters so that they are able to meet some of the heating needs by using electricity at the off-peak rate. In practice, the proportion of electricity actually consumed at the off-peak rate can vary quite widely according to the way the heating is used. Systems used for heating only part of a house generally require a greater proportion at the on-peak rate.

With electric storage heaters, heat leakage may provide heat when not wanted, therefore utilisation is less than 100%.

As can be seen from the table aside, condensing boilers save energy and help to reduce fuel bills. In non-domestic buildings the annual energy and cash savings can be dramatic, leading to the recovery of the capital investment within a relatively short period of time.

Table 2.1 Efficiencies of domestic heating systems and fuels (as measured on site)

	%
Gas fired boilers	
Older boilers	55–65
Modern boiler	60–70
Condensing boiler	80–85
Oil fired boilers	
Modern boiler	60–70
Condensing boiler	80–85
Solid fuel boilers	
Auto-feed boiler	60–70
Manual feed boiler	55–65
Electric off-peak boiler	100
Central heating hot air systems	
Gas	65–75
Oil	65–75
Electric	100
Gas fires	
Modern	55–65
Wall heater with balanced flue	65–75
Electric fires	100
Solid fuel	
Room heater	55–65
Open fire	30–35
Open fire + throat restrictor	40–45
Open fire + back boiler	50–60
Electric storage heaters[†]	
Old heaters, manually controlled	100
Old heaters, automatic	100
New heaters, manually controlled	100
New heaters, automatic	100
Fan assisted heaters	100
Underfloor heating	85

† The table gives typical annual average efficiencies, full load or 'bench' efficiencies are higher but will not be achieved in normal usage. **It does not take into account primary energy use, for example in relation to electrical energy.**

Chapter 2.1   **Solid fuel, open fires and stoves**

There has been a dramatic reduction in the use of open fires, in parallel with the growth in central heating, and the effects of the Clean Air Act[70]. However, millions of existing dwellings still have chimneys, and where circumstances are appropriate, for example in rural areas, they are still used to burn wood, coal or peat in open fires. These fuels, including anthracite, can be used in closed stoves, again where circumstances are appropriate. Whether or not using open fires, which means burning solid fuels, is desirable from an environmental standpoint is outside the scope of this book. Irrespective of the emotive issue of UK deep-mined coal production and peat extraction, supplies of wood, although in theory renewable, are very restricted. It has been estimated that a three bedroom house heated entirely by wood burning stoves, but depending on the most efficient use of thermal insulation and ventilation, needs 3 ha of woodland of which 0.2 ha is harvested annually[71]. Perhaps, albeit on a small scale, this could be a prime use for 'set-aside' farmland. Coke, of course, virtually disappeared when natural gas replaced town gas, though smokeless fuels of similar characteristics are still available.

Over the years there have been a number of problems with chimneys, as commented on in Chapter 0. Some of these are explored in this chapter, and they may be summarised as follows.

● The functions of a domestic chimney are to vent the products of combustion safely to the outside air, and to induce sufficient airflow through the flue to suit the appliance and its heat output. A chimney may fail to work correctly if the outside air temperature is only marginally different from that inside the flue, especially on start-up

● Many problems causing poor chimney performance are due to the design and construction of the flue as well as to the relation of the flue to the many different appliances which may be connected to it

● The chimney flue is required to have minimum heat loss. In principle, a chimney built within a dwelling rather than on an outside wall has the advantage of creating both a warmer house and a warm flue, although this is less important with a well insulated chimney

● If the airflow (induced by the action of the flue) through an appliance is insufficient, combustion will be incomplete. If in addition a flue is badly constructed fumes may discharge back into the dwelling

● If the chimney terminates below the main ridge line, under certain wind conditions there could be turbulence or positive air pressure which may result in products of combustion 'blowing back' into the dwelling

Figure 2.3 Practically every building in this archive photograph, perhaps even the cathedral, has a masonry chimney

Figure 2.4 An open fireplace at Bolsover Castle, Derbyshire. The castle dates from the seventeenth century

● If the flue temperature is too low, updraught will be poor and condensation occurring in the flue may cause damage to the chimney fabric[72]

● There is a potential risk of carbon monoxide poisoning if the products of combustion are not exhausted efficiently

Many, if not a majority, of dwellings built since the 1960s will not have chimneys. Whether or not it would be possible to install a retrofit lined masonry chimney for a solid fuel fire or stove depends on the design of the dwelling and its situation, though it will be expensive and may conflict with the appearance of the house. Although no useful general guidance can be given, it may be possible to use a proprietary design in metal, or perhaps one of the flueblock packages with prefabricated appliance chambers.

Characteristic details
Space and detailed design requirements
Open fires

The earliest open fireplaces, fuelled by wood, tended to be quite large, since they were usually required to heat very large rooms (Figure 2.4). The construction was of ordinary brick or stone. With the introduction of coal for domestic heating, however, fireplaces became much smaller, until it became possible to manufacture virtually the whole of the rear surround (the fireback) in one piece. Materials were sometimes cast iron, but the majority were fireclay. Firebacks have been made in various sizes in the past, but the standard size, now smaller, ranges from 350–500 mm[73].

There is insufficient evidence on which to assess the probable value of a smoke shelf, but it seems likely that, in some circumstances, it will help to reduce the risk of smoke entering the room due to downdraught. *'The throat, at its least cross-section, should be about 4 inches from front to back and of a width equal to three-quarters of the nominal width of the grate. The entry to the throat should be rounded, of a smooth surface and tapered'*[20].

Closed stoves

The earliest freestanding closed stoves were usually installed in front of a register plate fitted into the fireplace opening, the stove having a back or top flue outlet. Otherwise, unless provision was made for sweeping the chimney flue from outside the building, early stoves had to be removed to sweep the flue; moreover, if a boiler was fitted to the stove, it was not practicable to remove the connecting pipe for this purpose.

The inset-type openable stove which supplies convected warm air to the room in which the appliance is fitted, or through ducts arranged in the chimney structure to a room or rooms above, was brought into use in the early 1950s. Access to the flue adaptor of the stove was achieved either by removing panels at the top and sides of the stove, or by removing a panel at the back of the stove accessible from an adjoining room or from outside. Most of these inset stoves were designed so that the chimney could be swept through the front door of the stove, although, in cases where the stove was accessible only from the back, a handhole for chimney sweeping was provided in the flue adaptor of the stove.

Boilers and cookers can stand within a recess, under a register plate or raft lintel, with access for cleaning either at the front of the chimney stack or from under the register plate. This may be effected with an offset to the fluepipe passing into the chimney, or the gather at the top of a builder's opening may provide sufficient space (BS 8303-1[74]).

Flues, chimneys and terminals
Flues for open fires

The minimum size for flues from open fires is largely governed by the deposition of soot and provision for its removal. In past centuries, as noted earlier, flues for wood burning open fires tended to be very large.

For the last three centuries a nominal 9 in × 9 in brick flue has been found to be both convenient in its relationship to brick sizes and satisfactory in relation to its effectiveness. *'With open fires, kitchen ranges and combination grates, a large amount of excess air is carried up the flue with the gaseous products of combustion, smoke and soot. So much heat also passes*

Figure 2.5 A wood burning closed stove without boiler imported into the UK from Norway and installed in the 1970s

Figure 2.6 Multiple flues can take up a large proportion of the available space in a building. There are only around a dozen flues in this combined stack. The arches were introduced to provide mutual support

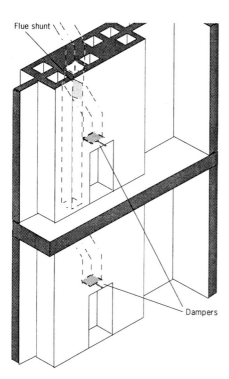

Flue shunt

Dampers

Figure 2.7 Combined flues were sometimes built in blocks of flats in attempts to save space

up the flue that the diluted gases rarely reach the condition in which moisture and acid gases condense on the flue lining, and for this reason the normal brick flue with a parged lining is satisfactory. Flue parging is required for two reasons: to reduce internal roughnesses to which soot would adhere, and to prevent the leakage of smoke through cracks in the joints of the brickwork. The traditional parging mixture embodying cow-dung is now rarely available and a fat lime mortar, to which has been added one part in twelve of Portland cement, is suitable.

'The position of the flue terminal is of importance. The ideal position is above the ridge of the house: any other position is liable to be unsatisfactory. In closely developed areas, difficulties arise in buildings against higher premises and there is no certain cure in such cases other than to provide a flue discharging above the level of any higher adjoining property.'[20].

Flues were provided separately for each fireplace, and in large buildings the stack size containing multiple flues could take up a large proportion of the available space in the building (Figure 2.6). In order to concentrate the flues into a common stack, it was usual practice to gather the flues by corbelling the brickwork. This practice was claimed also to prevent rainwater getting into the fireplace, which is only marginally true, and to improve the performance of the fire, which is a myth. If the angle of the flue was too sharp, the resistance to the upward flow of gases could be increased and smoke would spill into the room.

Flues and chimneys for solid fuelled appliances have always been something of a problem, not only for avoiding smoke and fumes being drawn back into habitable spaces but also with the structural stability of masonry stacks. For example, it is said that in the great storm of 1703, 2,000 chimney stacks were blown down in London alone, and this at a time when there were far fewer dwellings than in recent times. It was reported, in one instance during this storm, that a large stack fell through several floors to the cellar[75]. Such cases still occur.

As already noted, during the interwar period, BRS, as it then was, published an abridged version of a 1796 essay on chimney fireplaces by Count Rumford. Experience gained in BRS investigations of unsatisfactory installations showed that there were sound underlying principles in the essay. By 1937 BRS had investigated over a hundred cases of smoky fireplaces and other forms of unsatisfactory performance, applying Count Rumford's criteria as the basis for recommendations for remedial work. These criteria, listed in the introductory chapter, had proved to be very effective in practice and, through BRS publications, future fireplace design was influenced for the better; faults which were once commonplace became relatively rare.

The deterioration of flues from the operation of domestic boilers was a problem for some years and prompted the erection at Garston in 1948 of a pilot scale laboratory. In this laboratory (and another erected in 1957), investigations were made into, for example, heat transfer, condensation, flue gas velocities and the durability of flue materials. The research identified the dimensional needs that characterised satisfactory flues and led, ultimately, to the inclusion in building regulations of the requirement to provide brickwork flues with suitable linings[76].

Some blocks of flats were built in the second half of the nineteenth century and the early years of the twentieth century with combined or shunt flues (that is to say, serving more than one fireplace) in an attempt to reduce the considerable space taken up by multiple flues rising through many floors. Fireplaces were equipped with a shutter or damper to close off the opening when the fireplace was not in use (Figure 2.7). It is just possible that some of these installations are still in existence, though, it is to be hoped, not now in use.

It is important that flue gases are maintained at high temperatures in order to promote the upward flow of combustion products and to avoid both tar deposition and condensation of water vapour on the walls of the flue. This is especially for appliances burning wood with its higher moisture content than other fuels. This is why insulated chimneys are now preferred. Anyone who has tried to light a solid fuelled appliance after the building has stood empty in winter conditions will know that it can take some time to get a masonry flue up to working temperature, and to persuade the appliance to draw properly.

Flues for slow-combustion appliances

'Slow-combustion stoves, independent boilers and continuous-burning cookers burn their fuel at considerably higher efficiency than open fires. Far less heat passes up the flues (especially when fires are banked) and a much smaller volume of air passes into the flue without going through the fire. The products of combustion are therefore more concentrated, and deposits in the flue are potentially more destructive. The flues for such appliances should therefore be lined with material, such as glazed earthenware pipes, resistant to attack. The flues to these appliances should not be placed on an outside wall; but if this is unavoidable they should be separated from all surfaces externally exposed by 9 in. of solid, or 11 in. of cavity brickwork or equivalent.

In general, a separate flue should be provided for every appliance. If that is not feasible, dampers should be fitted so that the flue or pipe from any appliance not in use can be shut off. Awkward gatherings, sharp bends and pockets where soot could accumulate should be avoided, but where that is not possible a soot-door should be provided'[20].

A flue for a closed stove may sometimes be sized in accordance with the opening in the casing, subject of course to a minimum height of the flue. The manufacturers' literature needs to be consulted. If the appliance is categorised as closed, both British Standards and the building regulations give flue sizes without reference to opening sizes. On the other hand, if the appliance can be used as an open fire, BS 6461-1[77] gives the size of the flue to be used related to the size of opening, although the wording is a little ambiguous.

The British Board of Agrément has published a Method of Assessment and Test for prefabricated chimneys[78].

Fuel consumption

The consumption of coal, including the coal equivalent of all gas and electricity required to provide space heating, hot water and cooking in what were then average-sized well insulated houses, was examined by BRE in 1950. The fuel consumption of an efficient continuously burning open fire or stove supplying heat to the living room, domestic hot water service, and some heat to other rooms by means of radiators or convected warm air but not providing for cooking, was estimated to be approximately 3–3.5 tons of solid fuel per heating season of 33 weeks. The fuel consumption of an improved type of combination grate or freestanding cooker for space heating in the kitchen and living room, and cooking and water heating throughout the year, was estimated to be about 4–5 tons excluding any provision for bedroom heating; to heat a bedroom added about 1 ton per year[20].

With the increase in thermal insulation standards since the 1950s, newer dwellings heated by solid fuel should require considerably less.

Figure 2.8 A collapsed stack. Incredibly, no bricks seem to have fallen to the ground (Photograph by permission of the Lincolnshire Echo)

Figure 2.9 These stacks, probably rebuilt some years before the photograph was taken, are leaning so much that rebuilding was again being contemplated

Main performance requirements and defects

Strength and stability

The exposed parts of chimneys above roof level are very vulnerable to wind pressure, and strong winds usually produce a crop of incidents involving falling masonry chimneys (Figure 2.8 on the previous page). Wrought iron tie rods and straps encircling the chimney and taken back to a firm fixing on the roof construction are occasionally seen. However, BS 6461-1 calls for the use of stainless steel or non-ferrous metal in this situation. The junction needs to be weatherproof.

Although there has been a requirement for chimneys to be lined since the mid-1960s, unlined masonry chimneys, which are still estimated to form the majority of those in use, are susceptible to migration of salts through the walls; the salts are mobilised by condensation from within stacks or driving rain on the outside. The salts can be deposited in the mortar joints of a stack, leading to sulfate attack; this can then cause the mortar joints to expand on one side of the stack more than the other. Consequently the stack begins to lean, aggravated, often, by additional cooling on the side facing the prevailing wind (Figure 2.9 on the previous page). Not only that, BRE investigators on site have seen stacks so jelly-like and insecure from severe sulfate attack that they could be moved by hand pressure alone. The mechanisms of sulfate attack are explained in Chapter 2.2 of *Walls, windows and doors* [22], and techniques of repair in BRE Good Repair Guide GRG 15 [79].

Exclusion and disposal of rain and snow

It is important that adequate provision either has been or is made for dampproofing a masonry chimney stack where it penetrates the roof line (Figure 2.10). Not all old chimneys were provided with a DPC, and some of these chimneys can be sources of continuous problems. In one old farmhouse seen by BRE investigators near the coast in Pembrokeshire, damp walls had existed as far back as the occupants could recollect. After many attempts to waterproof the rubble stone walls, the problem was finally diagnosed as the absence of DPCs in the brickwork stacks. Since the fireplaces were no longer in use, and the building was not listed, the stacks above the roof line were demolished and adequate provision made for ventilating the remaining disused flues.

Heat outputs required

Existing dwellings may have open fireplaces of many different kinds and sizes: from inglenooks providing space for roasting spits, to Victorian and Edwardian cast iron grates (Figure 2.11), and shallow twentieth century firebacks. They do have one property in common, however – practically all will be very inefficient by modern standards. Some old fireplaces did attempt to provide for

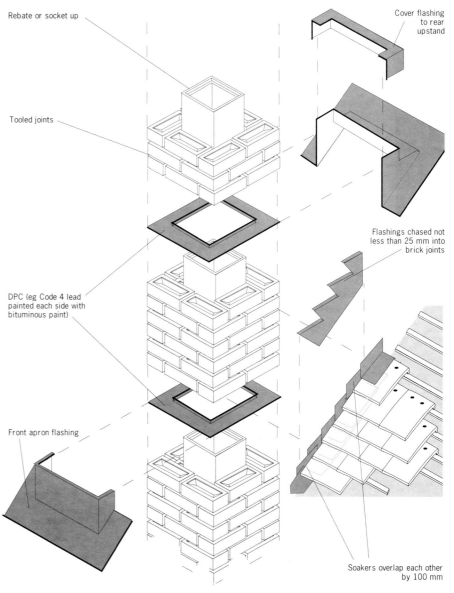

Rebate or socket up

Tooled joints

DPC (eg Code 4 lead painted each side with bituminous paint)

Front apron flashing

Cover flashing to rear upstand

Flashings chased not less than 25 mm into brick joints

Soakers overlap each other by 100 mm

Figure 2.10 A damp proof course should be provided in a stack; two may be needed where the chimney is severely exposed or the roof slope is steep

better control over air supply, with tight fitting frets and hit-or-miss vents (Figure 2.11). Worthwhile improvements in efficiency date from the post 1939–45 war period. Improved solid fuel appliances were classified in 1949 as:

- openable stoves with convection heating, with or without boilers
- open fire, solid smokeless fuel grates without convection heating, with or without boilers
- open fire, solid smokeless fuel grates, with convection heating, with or without boilers
- continuously-burning cookers with boilers
- independent boilers

The overall efficiencies of improved grates and stoves were estimated to be around 40% without a boiler and 50% with a boiler; minimum outputs were set for public sector housing. For example, an open fire, 18 inch wide grate without convection, for use with solid smokeless fuel was expected to give at least 9500 BTUs per hour (say, nearly 3 kW)[20].

In more recent years, the outputs of these appliances have varied very widely – even those classified for domestic use and sometimes ranging up to 30–40 kW. Manufacturers' literature should be consulted.

Figure 2.11 This Edwardian cast iron fireplace has a tight fitting fret to allow better control over the rate of combustion

Choice of solution

From the time of universal availability of domestic coal in the early nineteenth century, it became common practice to incorporate open fireplaces in all the principal reception rooms of dwellings, and even in most bedrooms; it was rare, though, to provide them in all bedrooms. This practice continued for the bulk of new housing throughout the remainder of that century, and even the first half of the twentieth.

However, by the early 1950s it was common practice in new dwellings to provide only one solid fuelled open fire with a back boiler as the sole means of heating the whole household. It was pointed out at the time that this would be insufficient to heat the whole dwelling to the provisions then in force for all public sector housing when outside temperatures fell below freezing for considerable periods[20].

Unwanted side effects
Heat loss

One of the main drawbacks of the open fire was the rate of heat loss up the chimney, although many people maintained at the time that there was a significant benefit in the much increased ventilation rates when the fire was alight. Technical Appendix F of the *Housing Manual 1949*[20] drew attention to the fact that estimated heat losses in a room with an open fire with unrestricted throat would be nearly 11 000 BTUs per hour (say around 3.5 kW) because of the increased flow of warm air up the chimney flue. If the nominal output of the appliance was low there could, in theory, and under certain operating conditions, be a net heat loss to the room. The benefit of the open fire was more apparent than real!

Sound transmission

The ducts in two-storey houses fitted with inset openable stoves, which supplied convected hot air to both floors, were found by BRE investigators to provide pathways for significant sound transmission between downstairs and upstairs.

Migration of soot and tar through chimney walls

A further side effect can frequently be seen where the flue is unlined. Over a period of years, combustion products can migrate through the masonry to the outside surface of the chimney (Figure 2.12). Soot and tar is transported by rainwater soaking into the chimney, and by condensation on cold internal chimney surfaces, evaporating on the outside face. Tar deposits are invariably a problem to remove and treatment is usually confined to disguising the deposits or stains by applying a sand:cement render mix over eml. (*Walls, windows and doors*, Chapter 9.2, suggests suitable coatings and mixes.)

Figure 2.12 Although this stack had been rebuilt some ten years before the photograph was taken, some place (eg underburnt) bricks from the original stack had been reused. Although they may have been whole, they were insufficiently durable for the onerous conditions in the stack. Since they were already contaminated by the products of combustion, the problems recurred, showing up against the cement based paint

Smoky chimneys

Another unwanted side effect was that some chimneys tended to smoke due to unfavourable weather conditions or poor chimney design, or a combination of both (Figure 2.13). One of the main problems, where the design of the opening was satisfactory, seems to have been a chimney pot sited below the ridge of the roof, whereas it should have been sited in an area of negative air pressure above the ridge and away from adjoining higher buildings.

Whatever the type of chimney, and whatever the type of solid fuel and method of combustion, the proper dispersal of the products of combustion is essential, not only for the health of building occupants but also for the prevention of corrosion and deterioration of the chimney fabric. Limiting the height on aesthetic grounds is rarely desirable, and some installations provide for fans to be connected to increase the velocity of the effluent to throw it clear of wind eddies.

The conditions causing smoke to escape before it can rise in the flue include:

● insufficient draught up the flue
● blowing back of smoke when wind blows in certain directions (downdraught)

The reason for insufficient draught may be that the flue is not tall enough. The situation tends to arise on single storey buildings where the flue does not terminate above the ridge of a pitched roof. In the immediate vicinity of a building or group of buildings, wind changes speed and direction rapidly, depending on the form and scale of the building. Severe eddies and vortices occur (eg at corners, eaves and ridges, and at any projections such as dormer windows or tank rooms) thereby creating positive and negative pressures which affect smoke issuing from, or being blown back down, flues. The slope of the roof is also an important factor. A further condition might be that the flue is of incorrect cross-section, or that it needs sweeping.

An open fire will need around 250 m^3 per hour of air for efficient combustion and sufficient volume of air to carry the smoke up the flue, though the air for combustion depends on the calorific value of the fuel and the size of fire. If the room in which the fire is situated is very well draughtproofed, then the fire may be starved of air. The remedy for this may lie in improving the air supply to the room without, of course, causing draughts. Closed stoves usually require a lesser volume of air; something like one-tenth the amount needed for open fires.

Inglenook fireplaces with hoods may present further problems for which ready solutions may not be available. The ratio of front opening area to flue area is important - for dog or basket grates standing in large openings the ratio of cross-sectional open area to that for flue area should be 8:1 for a house and 6:1 for a bungalow (13–16% of the front opening) (Figure 2.14).

Open fires are normally provided with a flue of around 190 mm × 190 mm square section or 210 mm diameter circular section. This approximates to the traditional parged 9 in × 9 in brick flue size used for centuries, and which ought to provide suitable conditions. The pot should not significantly constrict the flue size, and where round pots sit on top of square flues, the transition from one shape to the other should be gradual.

The velocity of gases over the top of the fire should be sufficiently high to carry away all the products of combustion. This is the reason why fireplaces have narrow throats. Remedial work may involve inserting a throat restrictor.

Figure 2.13 A problem which has plagued householders from time to time through the ages. If the fireplace and flue do not conform to the rules first formulated by Count Rumford, and summarised in Chapter 0, downdraught can cause reversal of the smoke flow

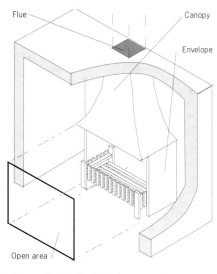

Figure 2.14 Ratio of front opening area to flue area for a basket grate

Figure 2.15 The metal mesh curtain of this individually designed screen is both decorative and, when pulled across, prevents flying sparks

Health and safety

There are around 60 deaths each year in domestic buildings as a result of burns and scalds[80]. Naturally, not all these are due to open fires, but there is a perceived risk of loose clothing catching fire when reaching up to the mantelpiece. The greatest risk to building occupants of course comes from the flames of the unprotected solid or gas fuelled fire or the red heat of the radiant electric fire. All need to be guarded with wire or perforated metal shields which keep loose clothing or tiny fingers well clear of flames and elements.

Sparks from open fires, especially those burning wood, are an obvious hazard (Figure 2.15) and spark guards of finely woven wire mesh have been widely available, probably since the Industrial Revolution. They tend, however, to reduce the amount of radiant heat output into the room, though there will be some convected heat from the screen.

There is some evidence to suggest that using coal for domestic heating may have caused respiratory damage in the past, but this is unlikely now to be a widespread, serious or long term hazard, particularly as open fires are normally associated with high ventilation rates.

Concern has been expressed about the numbers of carbon monoxide related incidents concerned with using solid fuel. Although the total numbers of incidents involving each of the two fuels, solid and gas, are roughly similar, since fewer installations use solid fuel rather than gas, the proportion of incidents of carbon monoxide poisoning is greater with solid fuel.

There have been cases where the products of combustion have found their way into bedrooms via open windows and by inadequately constructed and maintained chimneys. The defects become significant when the chimney is operating under positive pressure conditions such as may be induced by wind effects. On one site examined by BRE, wisps of smoke could be seen emanating from cracks in the plastered chimney breast of a bedroom at first floor level in a 1930s house – the parging in the flue from an open fireplace had failed, and sulfate attack had caused expansion in the masonry and cracking.

It used to be the case that the air required for combustion in a space heating appliance, plus the air required for the efficient operation of the flue, used to enter the room through gaps around normally constructed windows and doors, so that the provision of a special air inlet was not necessary. Then underfloor ducts were designed into rooms to deliver the air needed for combustion and flue operation to a position near the fireplace in an attempt to eliminate draughts; however, BRS experiments of the 1950s showed they were not effective unless their area was large compared with that of the fortuitous gaps.

Now that better fitting and weatherstripped doors and windows are installed in housing, it has become necessary to introduce the air needed for combustion appliances by means of permanent vents drawing the air from the outside of the house, or via a ventilated lobby[77].

Preventing chimney fires

It is important that flues for solid fuel appliances – as indeed for those using other fuels – are well built and kept away from combustible material. BS 6461-1 calls for the external surfaces of flues (eg chimney breasts) not to exceed 70 °C when in use. So far as masonry flues are concerned, there is a deemed-to-satisfy assumption with masonry that is not less than 100 mm thick, and with wythes of the same thickness between flues. These dimensions are exclusive of lining thicknesses.

Combustible material should not have been incorporated into any part of a building's fabric within 200 mm of a flue, and all combustible material except wood trim should have been kept clear of the chimney wall by at least 40 mm (Figure 2.16). Metal

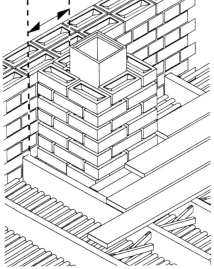

Joists and other timber components at least 200 mm clear of flue

Figure 2.16 Combustible material (other than trim) should be at least 200 mm clear of the flue

Lightning strike on a domestic chimney
One house of a pair of semi-detached four
bedroom houses built in 1898 in the West
Midlands was struck by lightning in a
summer storm. The owners were out at the
time. On their return home they found
broken brick, pot and tile debris scattered
across the front garden. Further
investigation showed that one of two
chimney stacks on the gable wall had been
partially demolished, nearly to ridge level,
and rainwater had already poured into the
bedrooms of the house. Later that same
evening a tarpaulin was temporarily placed
over the holes by a local jobbing builder
working from roof ladders before erecting
scaffolding for rebuilding the stack.

The following morning the damage was
examined by a BRE investigator. The brick
masonry of the surviving chimneys
appeared to be in sound condition, the
buffeting winds had abated and no further
immediate dislodgement was anticipated
from the damaged stack. There was no
evidence of fire. The pressed bricks in the
stack were of a semi-engineering quality.
The strike had thrown brick debris from the
stack to a distance of approximately 6 m,
falling largely on the south western slope of
the roof, breaking some 50 plain tiles in the
process, and the loose masonry and the
broken tiles had then slid down the steeply
pitched roof to be jettisoned from the eaves
up to a further 4 m from the walls of the
dwelling. It was clear that the strike had
actually fractured several bricks, and had
not merely displaced whole bricks.

The houses were in open country, though
not at an unduly high location compared with
the immediate neighbourhood; there were a
number of mature beech trees in the vicinity.
Although the trees were higher than the
dwellings, they had clearly afforded no
preferential path to earth. The investigator
concluded that at least two terminals and
conductors would be required to offer
adequate lightning protection to each of the
dwellings, but that the risk of any being
struck again, with strikes to ground in that
area being around 0.6 per km^2 per annum,
was acceptably low. (The old adage of
lightning not striking the same place twice
needs to be discounted).

inserts, shoes etc should have been
kept at least 50 mm from the flue
wall. In new construction, it is
expected that European Standards
will in future specify the distances to
combustible materials for particular
chimney applications.

Special precautions relating to
chimneys passing through a
thatched roof are mentioned in *Roofs
and roofing*[21], Chapter 2.7.

A space or 'soot box' should be
provided below the point of entry of
an appliance's outlet into the main
flue so that falls of soot or parging
do not block the duct. Condensation
within the flue may also need to be
prevented from soaking into the base
of the flue by means of a tray.

Chimneys and lightning
Since they are nearly always situated
at the highest point of a building,
chimneys also tend to be the most
vulnerable parts of a building to
lightning strikes (see case study
aside). The risk of a lightning strike,
and the protection necessary against
it happening, have been briefly dealt
with in *Roofs and roofing*, Chapter 1.6,
and *Walls, windows and doors*,
Chapter 1.8. Further information is
available in BRE Digest 428[2].

Figure 2.17   The side of the stack (centre of picture) has fallen away in this view of the ruins
of Guisachan House, Inverness-shire, and the ceramic chimney liners (reddish colour) have
been exposed to view. The house had been built in the third quarter of the nineteenth century

Figure 2.18 Eight different pots on these two chimneys provide evidence of past problems. When a pot of the same height as its neighbours was not available, a rather crude *ad hoc* solution was employed. The stacks are flaunched with a mortar fillet instead of flashed, and no DPC is evident

Durability

Appliances

Cast iron, stone and timber surrounds to fireplaces far outlive the firebrick shells which contain the open fire. The life of a shell depends almost entirely on the treatment it receives; large fires, and careless or bad firing practices and poor maintenance, can lead to breakdown in relatively few years. Fortunately, the fact that standard dimensions were introduced some years ago for firebacks means that replacements are still available. The current Standard is BS 1251.

The situation with closed and openable stoves and cookers with boilers is similar in one respect: the longevity of appliances depends on firing practices. Even vitreous enamelled cast iron can be destroyed if temperatures of the castings reach red heat, and BRE has seen this type of damage to examples only three years old. Cast iron firebar gratings tend to distort if good firing practices are not observed. The availability of replacement firebricks for such items depends entirely on the policy of the original manufacturer and whether they remain in business, but 10 to 15 years seems to be common for the continuing availability of spares. This is just around the time when corrosion caused by condensation begins to take its toll on appliances that have not been in continuous use during winter months.

Flues

Unless they have since been relined, the flues of most chimneys built before February 1966 are likely to be of 225 mm × 225 mm brickwork construction, parged, but not otherwise lined. Having said that, however, there were exceptions, and some flues built in former times may have been lined (Figure 2.17).

A chimney used for burning solid fuel may have successfully vented the combustion products (tar acids, ammonia, sulphur compounds and water vapour) to the atmosphere. However, the fuel and the appliance can be changed many times during the lifetime of a dwelling and the flue may eventually become unsuitable.

If water vapour condenses continuously in the flue, the chimney fabric may deteriorate. Sulfates and acids will attack both the parging and the mortar joints, and expansion will occur, possibly leading to the chimney leaning. Sulfates may be dissolved out of brickwork by the condensate which will accelerate the attack on the mortar. Tarry deposits and salts may be carried through the bricks and plaster to damage internal decorations. Great care should be taken when installing gas fires into hearths which formerly were used for open fires; further comment on this is offered in Chapter 4.6.

The chimney pot, of course, is the part of the flue which is subject to some of the most onerous service conditions (Figure 2.18). While it is the furthest part of the flue from the appliance, and therefore the coolest, it is subject to the extremes of the weather as well as to the products of combustion. Glazed earthenware pots have a good track record for durability, though cracking is the greatest problem, particularly where the flaunching has deteriorated.

Maintenance

Access for chimney sweeping is important. Normally this is provided by a double walled door at the base of the flue, but some single thickness doors will also be found.

Relining a defective chimney can be considered as a major maintenance item. Old cement or lime based parging on the walls of masonry flues does not last indefinitely, and the sight of quantities of mortar crumbs in the soot when the chimney is swept is probably also a sign that the chimney needs relining (Figure 2.19). This should in any case be done where an old flue is to be reused for a new appliance.

There are various proprietary systems available for relining a flue, not all of which involve flexible metal liners. Sometimes it is possible to add structural strength to a chimney where only a smaller diameter flue than the existing one is needed, and a specially formulated concrete can be poured round a withdrawable former. The British Board of Agrément maintain a list of third party certified chimney lining installers[81].

Figure 2.19 A flexible stainless steel liner being inserted into an old flue

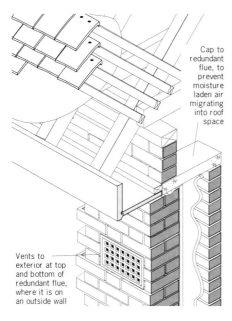

Cap to redundant flue, to prevent moisture laden air migrating into roof space

Vents to exterior at top and bottom of redundant flue, where it is on an outside wall

Figure 2.20 Where stacks are taken down to below roof level, if they are on an external wall, they still need to be ventilated top and bottom

Work on site

Workmanship

Installation of continuous burning combination grates often proved rather complicated. To quote from the technical appendices in the *Housing Manual 1949*[20]: '*In the past, faulty installation has been the cause of many failures in operation. In the case of combination grates, and back-to-back grates, it is essential that all joints around the appliance are made good so that air cannot enter the chimney except by the oven flues. With self-contained oven flues, it is desirable to fill in the space between the brickwork and the oven flue casing with rubble and weak lime mortar, because a large air space reduces the temperature of the flue gases and this leads to inefficient operation*'.

In replacing with improved open fires, care must be taken to ensure that there is no potential for air leakage below and around the front fret, since this would destroy the advantages of the controllable air supply to the fire. Where convection heating is provided, models with double casings simplify the installation of the appliance. In other types, the passage for the convection air is formed by the space between the jacket of the appliance and the surrounding masonry, and these types must be very carefully installed in accordance with the makers' instructions.

Supervision of critical features

In rehabilitation work, particularly where central heating is being or has been installed, chimneys which formerly served open fires are often taken out of service without proper consideration being given to internal ventilation. It is particularly important that existing or new combustion appliances have an adequate and continuous air supply.

Disused stacks taken down to below roof level do not normally need ventilation to keep them dry unless they are on an external wall. But, if disused stacks that are on an external wall or remain exposed above roof level are not protected from ingress of rain, and the flues are not ventilated, internal damp staining is likely (Figure 2.20). If flues, whether on external or internal walls, are ventilated so that water vapour is conducted from rooms to the roof void, copious condensation in the void may occur. And if internal ventilation is inadequate when a room is no longer ventilated by a flue, condensation in habitable areas may also result.

Figure 2.21 The flue and chimney breast below this roof space have been removed, and new steel brackets now provide additional support to the masonry near the foot of the corbel. The raking line of the old flue can be seen left and the brickwork now overhangs some one and a half bricks. The trays catch rainwater drips from the defective flaunching above

Any flues still in service should be checked and maintained to ensure that they remain sound and complete. Any disused flue in a stack containing others still in use should be ventilated at the base and capped at the top with a proprietary ventilating cap that allows a small quantity of air to pass through the disused flue. If the disused flue is completely capped, there is a risk that flue gases may leak into it through the surrounding masonry or the separating wythes and be drawn down into habitable areas.

Where chimneys and chimney breasts are removed and the stack remains in place, additional support may be required (Figure 2.21).

Inspection

Unlined masonry chimneys are notoriously liable to permit the transfer of combustion products to the interior of a building. The extent of any staining within the building should be checked: light staining of plaster may be sealed by using an adhesive backed metal foil. If the staining is confined to the chimney breast, dry-lining with foil backed plaster board fixed to preservative treated timber battens may be used – serious staining will require the removal of all render and finishes, and making good with a 1:3 cement:sand render and finish which should extend at least 300 mm beyond existing staining.

In installations where the appliance provides convected warm air, it is not always easy to prevent the products of combustion passing from the appliance or flue adaptor to the warm air chamber or ducts. These installations need very careful inspection.

The problems to look for are:
◊ disused flues not vented
◊ exposed disused stacks not weatherproofed
◊ flues vented but open to roof spaces
◊ room ventilation inadequate following sealing of flues
◊ corroding chimney stays
◊ crumbs of mortar parging falling down flues
◊ sulfate attack leading to leaning chimneys

Chapter 2.2

Boilers and hot water radiator systems

When boilers producing steam or hot water for space heating purposes were introduced in the middle years of the nineteenth century, coal was the predominant method of fuelling them. Initially, boiler installations were limited to industrial and public buildings, and the very largest of private houses. Installations suitable for small single dwellings began to be introduced in the interwar period (Figure 2.22), although the rapid growth in central heating did not come until later. Coal remained dominant until the middle of the twentieth century; indeed until the introduction of natural gas and the nationalised gas utility laid a comprehensive national grid. Technological advances helped to drive the changes: the widespread adoption of small bore copper piping, the glandless pump, higher customer expectations for improved heating standards, including whole house heating, and, of course, the new sources of oil and gas from the North Sea.

Hot water and steam radiators in cast iron were widely used in Victorian times, and some of these systems are in use still. Quite elaborately decorated columned or finned designs were not unusual.

Full central heating is now normal in new houses and, invariably, a priority when refurbishing existing dwellings without it. Primarily, central heating installations are designed to produce comfortable living conditions, but also to avoid or reduce problems such as condensation, mould growth and, even, hypothermia.

> **Heating misconceptions**
>
> Unfortunate misunderstandings about domestic central heating systems still prevail amongst users. The most common found by BRE investigators on site visits have been:
> - leaving the heating on 24 hours a day to prevent 'all that gas' being consumed every morning
> - turning the thermostat up to maximum to heat the house up more quickly
>
> The most surprising one, though only occasionally encountered, is to insulate radiators to stop them losing precious heat.

Though not mandatory, there is a tacit assumption in Approved Document L[82] of the Building Regulations that most houses will have central heating, and that it must then conform to some minimum standards of control. Also the Standard Assessment Procedure (SAP)[83] rating, now part of the Building Regulations, gives more points for more efficient heating systems. In these cases it may then be possible to relax some other requirement; for example, insulation in wall cavities.

Thermostats for controlling the air temperature have been based on a variety of operating principles – mainly pneumatic or electric – until electronic versions became available in the late 1970s; electronic thermostats did not become popular until the 1980s.

As noted in the introductory chapter, building management systems, largely for non-domestic buildings, did not come into general use until the 1980s.

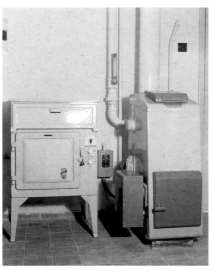

Figure 2.22 A small domestic boiler, dating from just after the 1939–1945 war, used in BRE heating experiments at Abbots Langley in the early 1950s. Alongside is an electric cooker of similar vintage

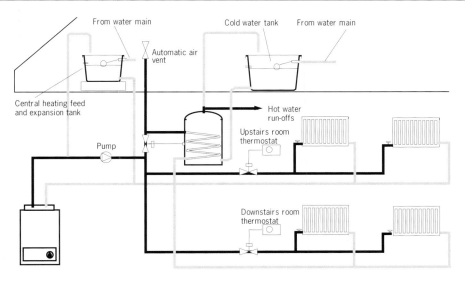

Figure 2.23 A traditional system of domestic gas fired central heating

Characteristic details

Descriptions of systems in general use

Boiler systems can be fuelled by natural gas (currently in the majority of installations), liquified petroleum gas (LPG or bottled gas), oil or solid fuel. Mixed systems are feasible, but they are not usually recommended due to their impracticability and cost. A summary of designs for boilers and heating systems generally is available in BRE Good Repair Guide GRG 26 Part 2[84]. (Descriptions of combination boilers, or 'combi-boilers', that is to say boilers used for the supply of both direct domestic hot water and space heating, are given in Chapter 4.2.)

Because of the complexities of boilers producing steam, and health and safety questions relating to explosion risk, providing space heating by means of steam has largely disappeared from common use in favour of low pressure hot water installations. Another factor is the high temperatures produced in steam fed radiators: control of temperature and the protection of users is much more easily achieved with hot water.

Open vented systems

The traditional, commonly gas fired, central heating installation (Figure 2.23) employs a feed and expansion cistern to allow for changes in water volume, and for filling the boiler and radiator circuits. This system is open to the external air. Also, domestic hot water is stored in an insulated cylinder, fed from a cistern usually located in the roof space. There are several varieties of pumped and gravity fed systems; Figure 2.23 shows a fully pumped system with thermostat control over the upstairs and downstairs radiators as well as over the temperature of the hot water cylinder. This system is well understood by installers and allows considerable flexibility.

Sealed systems

As an alternative to an open vented heating system, a sealed system may be installed. The advantage of these pressurised (sealed) systems is that they can be used where there is no space for a feed-and-expansion tank or insufficient head room to provide the necessary pressure. Pressure may be supplied from the mains or from a pneumatic device (eg a pump applied to the pressure expansion vessel).

A sealed system requires:
● a boiler which is approved for this purpose
● special safety controls
● an expansion vessel which will accept the changing volume of water as the temperature changes

Air in the pipework system is generally vented through automatic release valves. The heating circuit must not be permanently connected to the mains water supply. For filling or repressurisation purposes, a temporary hose with adequate backflow protection must be used.

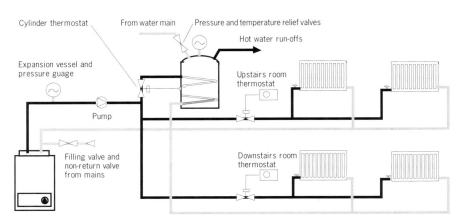

Figure 2.24 A sealed (unvented) heating system

Figure 2.25 This large 1970s open flued boiler takes combustion air from the room in which it is situated, and therefore needs a permanently open vent to the external air. It is too heavy for wall mounting

Descriptions of boilers in common use

Apart from the solid fuelled type, boilers may be divided into those that take their air supply for combustion purposes from outside the habitable space (called room sealed units), and which are the preferred type, and those that take their combustion air from within the habitable space (called open flued types). Solid fuelled boilers are exclusively of the open flued kind. Electric boilers do not need a combustion air supply.

Boiler efficiencies in the 1990s increased significantly, in part due to the EU Boiler Efficiency Directive.

Solid fuel boilers

The smaller domestic solid fuelled boilers of course were almost invariably refuelled by hand, though some had magazines of limited capacity; automatic chain grate stokers were introduced for larger models of boilers from the middle of the nineteenth century, and screw fed stokers from the first quarter of the twentieth. As already noted, these boilers were all open flued; that is to say they took their combustion air supply from the habitable space. Flues suitable for solid fuel boilers are similar to those provided for open fires and stoves, and these were dealt with in Chapter 2.1.

Gas boilers

Gas boilers may be either condensing or non-condensing. As with other fuels, the combustion system may be either open to the room in which the appliance is situated (open flued) or sealed from that room (room sealed). A wide range of subtle variants is possible as defined in BS DD 221[85].

Since condensing boilers are a more recent development than non-condensing, the majority of existing installations are expected to be of the non-condensing kind, and these are therefore dealt with first.

Non-condensing boilers still constituted the vast majority of the new and replacement boiler market in the UK at the end of the 1990s. Nevertheless, although modern non-condensing boilers (many of which have fan flues) may have bench efficiencies which are apparently within a few percentage points of those of condensing boilers, there is still, and always will be, a significant gap in performance. This does not prevent considerable debate over the merits of each type.

Open flued gas boilers
These can be floor or wall mounted with the flue rising to roof height or above (Figure 2.25). They burn air from the room in which they are situated which therefore needs a fixed ventilator to supply replacement air from the outside. They should now, arguably, only be installed in new work where no suitable outside wall is available for a room sealed boiler since burning room air can cause draughts and 'wastes' warm air. Unless they already exist and are in satisfactory condition, flues and chimneys can also be relatively expensive to build or to adapt with flue liners.

Many installations will have been constructed to rules which have now been replaced so far as new construction is concerned, although existing installations may not necessarily need to be updated to the new standards. Open flued appliances should have been provided with a route by which adequate air for combustion can reach them; and, like room sealed appliances, if in a compartment, they should be able to dissipate heat losses. Where a compartment was built to house a room sealed or open flued appliance, ventilation via permanent non-adjustable vents at high and low level should have been provided. If ventilation is to a space which is a room or hall, and the appliance is open flued, the space should have been provided with ventilation to the outdoors via an area of 450 mm² for every 1 kW in excess of 7 kW heat input to the appliance. Externally, any air inlet should have been at least 600 mm from a flue outlet[86]. The rules for new construction are contained, for example, in BS 5440-2[87] †.

A ducted supply of air for combustion can be provided from the outside, emerging close to the boiler. This reduces draughts and encourages outside air to be used directly for combustion.

---

† BS 5440-2 is under revision. The air change requirements given in the Standard are expected to change slightly.

Room sealed balanced flue and fan flued gas boilers

A number of types of boilers are in common use for new housing, including balanced flue and fan flued. The fan flued types are increasingly being installed as energy efficient replacements for time expired, open flued boilers.

The balanced flue type is a particular kind of room sealed boiler. Natural draught or buoyancy conditions suffice to draw air from outside the building and discharge combustion gases to a location close to the air inlet (Figure 2.26). This is most commonly achieved with the air inlet running concentrically around the flue ducting which passes through the external wall of the building. Having an air inlet and exhaust outlet close together generally ensures that any local pressure fluctuations caused by wind has an equal effect upon inlet and exhaust, and do not adversely affect the buoyancy within the boiler.

The fan flued boiler is another variant of the room sealed boiler. As the name implies, a fan drives the combustion air and exhaust flue gases instead of natural buoyancy. They have a much stronger draught than balanced flue appliances, and allow more control over the heat transfer process within the heat exchanger – a characteristic likely to be used with the more efficient boilers. Fan flued boilers tend to be

Figure 2.27 This condensing boiler has a single corrosion resistant stainless steel heat exchanger (Photograph by permission of Keston Boilers)

more efficient than other types because of lower flue heat losses, and usually have electronic ignition instead of a permanent pilot light.

Balanced flue and other room sealed boilers can be floor or wall mounted and draw combustion air from outside the building.

Because they do not burn room air, eliminating the need for continuous replenishment from inside the dwelling, room sealed boilers do not cause draughts as do the more traditional open and conventionally flued types. Greater flexibility in boiler siting is also available with the fan assisted units, together with the advantage of smaller terminals.

Building Regulations Approved Document J[88] and BS 5440-1[89] give a number of constraints for siting boilers. For example, terminals for balanced flue and fan flued gas appliances should not be sited close to windows etc where fumes could be drawn into the building.

Condensing boilers

A particularly efficient form of fan flued boiler is the condensing boiler. For both commercial and domestic applications, these boilers are claimed to represent a new generation of heating appliance. They can be used in most buildings, new and existing, large and small, and achieve worthwhile savings in energy. Most, if not all, will be room sealed units, drawing combustion air from outside the habitable space.

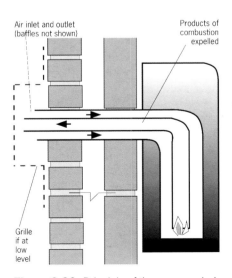

Figure 2.26 Principle of the room-sealed boiler

Condensing gas boilers

When first developed, condensing boilers normally incorporated two heat exchangers. The first heat exchanger removed the heat from the hot gases in the same way as a conventional boiler. The gases then passed over a second heat exchanger which removed further heat from some of the lower temperature gases, together with the latent heat released by the condensation process. As a result the exhaust gases were cooled to a point at which the water vapour in the gases condensed out. By extracting the latent heat from the water vapour present in flue gases, these boilers offered more energy efficient use of fuel. Flue gases from condensing boilers are of a much lower temperature than those from conventional boilers.

Some early designs of condensing boiler consisted of a conventional boiler with an add-on secondary heat exchanger. Each design would need to have been engineered to function efficiently as a package, with component parts correctly matched. For example, the controls that supervise combustion fan operation must be integrated with the burner control. Also, as with other types of boiler, the correct sizing of both primary and secondary heat exchangers is critical to performance. More recent developments have included condensing boilers with single heat exchangers of more durable materials than those used formerly.

Condensation occurs when the water vapour content of the flue gas cools on the surfaces of the secondary heat exchanger. The condensate formed in a gas fuelled appliance is slightly acidic, usually about 3.6 on the pH scale and rarely lower than 3. This is about the acidity of tomato juice. The secondary heat exchanger must be constructed of corrosion resistant materials and pipework be provided to carry the condensate to a drain or soakaway – this accounts for some of the additional cost of these boilers.

With natural gas, a particularly high proportion (about 10%) of the energy content of the fuel is able to generate latent heat. Even a fraction of this energy is worth recovering. Natural gas and LPG are the fuels best suited to the condensing process because of the high proportion of water vapour and, hence, latent heat energy released during combustion. Moreover the condensate formed is much less acidic with gas than with oil or coal, though condensing boilers operating on oil are now on the market.

Appliance efficiency is claimed to be improved by some 15% over that of a modern standard boiler of conventional design. Further efficiency improvements are achieved with a fan in the flue to make up for the lost buoyancy of the escaping gases.

Condensing boilers can achieve annual average efficiencies of 90%: much higher than other gas boilers. Both floor-standing and wall-hung condensing boilers are available, and all models have room sealed, fan assisted flues. Evidence from a large programme of demonstration projects funded by the the Energy Efficiency Office shows condensing boilers to be cost effective.

In cold weather, the exhaust gases from condensing boilers form visible plumes which are less buoyant than those from other boiler types due to the lower exhaust temperatures. While plumes also occur to some extent with conventional boilers, the plumes from condensing boilers may cause annoyance: flue terminals must be located so that views are not spoiled.

Payback of the additional capital cost over conventional boilers is achieved usually within a few years for large domestic installations, though for the smaller installations it will be longer. Price differentials are reducing, and Government grants may be available. There are some cases where condensing boilers are not economically attractive, but these tend to be the exception rather than the rule.

Gas fired condensing boilers started becoming available in the UK in the mid-1980s, but so far have not been quite as popular as might be thought. Although they are widely used in Europe in both dwellings and non-domestic buildings – for example, they have around 60% of the market in The Netherlands – by 1990 they still only accounted for 5% of the UK replacement market, and that market share is still growing only slowly.

The capital costs of condensing boilers can be more than those of conventional boilers[†]. For this reason having both condensing and non-condensing types in larger installations is often an attractive solution, provided the condensing boilers are sequenced to lead.

† Establishing exactly the additional capital costs is notoriously more difficult than establishing the energy savings. Costs can be considerably adjusted by quantity discounts and by grants for upgrading of installations.

Demonstrations by Building Research Energy Conservation Unit (BRECSU) in family housing, sheltered housing and non-domestic buildings have shown that condensing boilers can help considerably to reduce energy consumed, thereby saving on fuel bills and producing less carbon dioxide than conventional boilers. Although more expensive than their standard counterparts, the extra cost of condensing boilers can be recovered in a very short time: within a year in large commercial applications and in 2–6 years in housing. A boiler using less fuel will also reduce its carbon dioxide emissions by an equivalent amount; consequently, condensing boilers help to reduce global warming[(90)].

As well as being used in domestic situations, condensing boilers can also be installed in non-domestic buildings (Figure 2.28 and see the case study on the next page). Despite the low load factors on heating systems in office buildings, these boilers can reduce gas bills significantly and cost effectively. This is particularly true where condensing boilers are arranged to provide the base load in a multiple boiler installation. As with domestic buildings, condensing boilers in non-domestic buildings are more expensive than equivalent conventional or high efficiency gas boilers because of the extra cost of the heat exchangers. This extra cost is most easily recovered where there is a continuous demand for space or water heating and the boiler can be operated for long periods.

The maximum cost saving of a condensing boiler over a non-condensing one is around 35%, although around 13% will be normal. Replacing an old standard boiler with a modern condensing one may produce much greater savings, but not all of these are the result of the condensing operation; lighter heat exchangers, intelligent controls and fans all contribute to improvements in efficiency.

Back boilers

Back boilers, sometimes called side boilers, were used for many years in solid fuelled appliances to supply domestic hot water (see Chapter 4.2). They were commonly made of copper or bower-barffed iron.

Central heating has also been supplied by a boiler installed in the fireplace behind a gas fire, the boiler operating independently of the fire but using the same flue or chimney. As with ordinary open flued boilers, warm room air is used for combustion with consequent risks.

Electric boilers

An electric boiler can be used for domestic installations instead of a gas boiler to provide heat to radiators and a hot water tank. The main disadvantage is the high running costs, but, of course, no flue is needed.

For domestic use, two main types of electric heating are available: dry-core storage using high thermal capacity bricks, and wet storage. Both types are heated by off-peak supplies and require special wiring to carry the heavy loading.

Also available for domestic use, where there is no other alternative, is an on-peak non-storage type of boiler which takes up very little room – a fraction of the size of a

Figure 2.28 A condensing boiler installation in an office building

conventional boiler, yet still with capacity of up to 12 kW – and able to be connected in the usual way to a hot water radiator system.

For larger installations, and for industrial purposes, BS 1894[92] may be relevant.

Pipe and trunking runs

When the first systems were installed in Victorian times, circulation of hot water for space heating was by convection within either a large diameter single pipe system with a common flow and return, or a smaller diameter two-pipe system with separate flow and return. For small installations, gravity circulation is still feasible with careful design, albeit not necessarily desirable. It is less efficient and less responsive than pumped systems, and is not currently widely recommended. However, with the introduction of electrically driven pumps, small bore pipe systems became possible in the mid-1950s.

At risk of stating the obvious, size-for-size and flow rate-for-flow rate, high temperature hot water and steam pipes will carry a far greater heating load than will low temperature pipes; this should be borne in mind when considering replacement systems in existing buildings where space is at a premium, though safety

considerations may affect their practicability.

Medium and high pressure hot water circulation systems have tended to be used in larger buildings to take advantage of the higher temperatures possible, with pressure being provided by steam or compressed air. The system frequently depends on the provision of a flexible diaphragm within the installation to separate the circulated water from the pressure medium.

Boiler rooms

Boiler rooms, at least for large buildings, are normally sited at low level, either on ground floors, or in basements (if present). On the other hand, some boiler rooms may be found at the tops of buildings where fuel supply (eg gas) is easily arranged and where vibration can be isolated.

The space needed for a boiler room will depend upon many factors, though primarily on the floor area of the building for which the plant has to supply heat. For example, a building of around 10 000 m² floor area might have a boiler room of 50–60 m², whereas one of twice that size might have a boiler room of around 80–100 m² [31].

Flues, chimneys and terminals
Industrial chimneys

Some mention of industrial chimneys in brick was made in Chapter 0. Few of these massive Victorian chimneys now survive. The old rule of thumb for their design was that they were normally limited in height to ten times the external diameter of the base. The shaft of the flue was parallel, though the chimney walls were frequently battered in thickness from the base to the cap as the loads on the masonry decreased. Depending on the temperatures of exhaust gases, fire bricks were used to line either the whole or part of the height of the shaft.

Most industrial chimneys are now of twin walled metal, although a few may be found cased in reinforced concrete. Frequently the flues from more than one boiler are incorporated into a single shaft, which is then cased, or left supported from a framework (Figure 2.29)

Flues for boilers contained within the carcass of a building are normally sited at the highest part of the

building to minimise the amount of flue showing and the spillage of combustion products into open windows or air intakes. Where a flue is not directly over a boiler room, horizontal ducts incorporating fans may be seen.

Domestic boiler chimneys

Chimneys for open fires and for solid fuel burning appliances were dealt with in Chapter 2.1.

Open flued domestic boilers may be fitted to lined masonry chimneys or metal chimneys. Any chimney must be fit for the purpose and in good condition, and metal chimney components must be installed in accordance with the manufacturer's instructions. Sectional flue components should normally be linked with the socket facing upward; the components may be joined with a silicon rubber sealant where low temperature flue gases are involved. Flexible liners are not generally recommended for condensing boilers.

The general rule for open flued appliance terminals is that they are located either on the ridge or at least 600 mm above the roof level unless the slope is above 45° (Figure 2.30). More comprehensive advice is given in Approved Document J.

The ridge of a pitched roof is probably the best place, with possible alternative location points on the slope, or at the eaves or verge, or on the flat part of a roof. The height above the roof is crucial in preventing fumes migrating to low levels, either to that property or to neighbouring properties. Furthermore, the effects of bends in flues, and of terminals, which slow down the rates of flow must be taken into account and the appropriate equivalent heights calculated.

Flue pipes within roof spaces need to be kept clear of timber work and be adequately supported by metal strapping. In the site studies of quality in housing carried out by the former Defects Prevention Unit at BRE, a number of instances were observed where such support was totally inadequate (Figure 2.31).

Figure 2.29 Multiple flues from a large installation supported from a common tubular framework

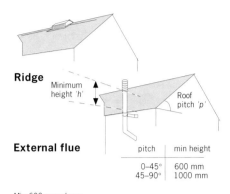

Ridge

Minimum height '*h*'

Roof pitch '*p*'

External flue

pitch	min height
0–45°	600 mm
45–90°	1000 mm

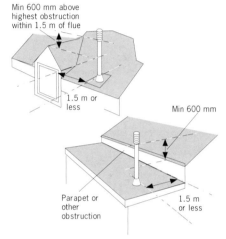

Min 600 mm above highest obstruction within 1.5 m of flue

1.5 m or less

Min 600 mm

Parapet or other obstruction

1.5 m or less

Adjacent construction within 1.5 m

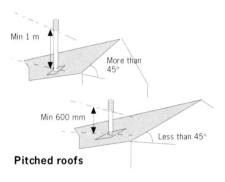

Min 1 m

More than 45°

Min 600 mm

Less than 45°

Pitched roofs

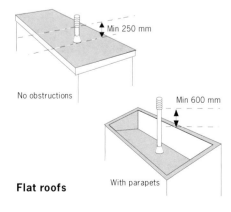

Min 250 mm

No obstructions

Min 600 mm

Flat roofs

With parapets

Figure 2.30 Locations for flues for conventional gas fired boilers and heaters

Chimneys for open flued gas appliances

The following quotation from the *Housing manual 1949*[20] gives a good description of the original provision for gas flues in many dwellings built during the 1950s and 1960s. Some of the advice is still relevant, though there are now more stringent requirements for siting terminals.

'As the flow of the products of combustion in a flue may be temporarily reversed under adverse pressure conditions, a draught diverter, if not already incorporated in the appliance, must be installed in the flue system and in the same room as the appliance, to prevent the possibility of downdraught disturbing the flames. This draught diverter will also function as a flue break and as a point of entry for additional air to the flue. In flued fires, however, no separate draught diverter is necessary as a flue break is provided in the appliance.

'The route of a flue pipe should be as short as possible, due regard being given to stability and the position of the terminal. The flue should rise progressively towards its terminal. Horizontal runs, right angle bends and tees should be avoided as they affect the efficient working of the appliance by restricting the flow of the flue gases. Long exposed routes should be avoided; the rapid cooling of the flue gases may cause sluggishness and condensation, particularly in flues from gas water-heaters.

'All gas flues should be provided with a terminal designed to offer minimum resistance to the products of combustion and to prevent the access of rain and blockage by birds or foreign matter: the unobstructed outlet area of the terminal should be at least twice the cross-sectional area of the flue. A normal 9 in by 9 in brick flue with the ordinary simple chimney pot is also satisfactory.

'To ensure a good updraught, the best position for the terminal is above the level of the ridge. The terminal may, however, be located at any suitable point on an external wall, well away from the eaves and from any projection such as soil and rainwater pipes. Flues should not terminate in any wells of buildings nor in re-entrants, as there may be adverse pressure conditions in such positions. The base of the terminals of flues which emerge from flat roofs should be built at least as high as the parapet coping; in no circumstances should they be adjacent to the inner face of the parapet wall and below the coping.

'As flues for gas appliances do not collect soot, they can be substantially smaller than flues for appliances burning solid fuel.

'Flues of brick, precast blocks, asbestos cement pipes, glazed and unglazed earthenware pipes and protected cast iron pipes may be formed in the fabric of a building during the course of construction. Precast concrete blocks are made in suitable dimensions for bonding into normal building structures. It is preferable to use blocks with spigot and socket joints as they minimise the possibility of internal mortar fangs being formed. Changes of direction in precast concrete flues should be made by means of blocks designed for this purpose'.

Since the 1950s, a raft of British and European Standards, and other practice documents, have refined and modernised these principles. Building regulations have also contributed to improvements in design: for example, in the mid-1960s with the requirement for lined chimneys.

Asbestos cement, of course, is not now used in flue linings, although some installations may have survived.

The practice of using precast concrete blocks built into the walls of dwellings for some gas appliances was a popular method of removing the obtrusive chimney breast from the ever-decreasing size of dwelling rooms. However, early designs of flue blocks had very narrow cross-sections, and required great care and diligence in construction. Later designs, and greater attention to installation methods, information and supervision, have improved the situation but some confidence has been lost in the technique. No examples were observed in the BRE site studies on quality in housing, and insulated flues now seem to be preferred. Many installations may, however, still exist (BS 1289-1[93]).

Condensate drains

Another special requirement of condensing boilers should be considered early in the selection process. Condensate has to be drained from the boilers and so a suitable drain location must be found. Normal plastics materials are needed for the drain connection (metal is not allowed); this should include a U trap and perhaps a tun-dish (to provide visible indication of condensate flow or drain blockage). If the drain pipe is exposed, it must be protected from freezing. A continuous fall to the drain connection of at least 3° is recommended. The cost of providing a drain is usually small.

The quantity of condensate produced will rarely exceed 50 ml/h for each kW of boiler input rating. (The theoretical maximum value is 150 ml/h per kW.) Actual volumes vary with operating temperatures.

Figure 2.31 An example of an unsupported heavy asbestos cement flue pipe discovered by BRE site investigators in the roof space of a newly built house in the mid-1980s

Pumps

Small pumps for powering central heating installations are mainly a development since the 1939-45 war. Previously, with gravity systems, no pump was required, provided the system was correctly designed: in particular whether pipe diameters were adequate to reduce transmission losses to an acceptable level. The development of suitable pumps was the single most crucial development in the use of small and micro-bore central heating systems, first coming into general use in the 1960s (Figure 2.32).

Temperature and humidity control

Room thermostats (or room-stats)

A room thermostat is a device for measuring the air temperature within a space and for switching space heating on and off (Figure 2.33). A single target temperature may be set by the user. A programmable room thermostat is a combined time switch and room thermostat which allows the user to set different periods with different target temperatures for space heating.

Figure 2.32 A small installation (in 1978) comprising a cylinder thermostat, a pump and a three-way motorised valve. The valve is controlled by both the cylinder thermostat and the room thermostat, and directs flow to whichever system is calling for heat. If both systems call, a shared position is selected, if neither call, the pump and boiler are switched off

Delayed start is a feature of a room thermostat which delays the chosen starting time for space heating according to the temperature measured inside or outside the building. **Optimum start** is a feature of a room thermostat or boiler energy manager to adjust the starting time for space heating according to the temperature measured inside or outside the building; it aims to heat the building to the required temperature by a chosen time.

Since a room thermostat is a device that operates a switch at a preset temperature, it could turn a boiler off when the temperature rises above a certain value and turn it on again when the temperature drops below that value. In practice some thermostats have a dead band such that the switch-on point is slightly below the switch-off point. This means that the control is less accurate, though it does prevent rapid cycling and wear of the boiler, and can help to improve overall system efficiency. There are two types of room thermostat:

- electromechanical
- electronic

The sensing part of the electromechanical type is either a bimetallic strip or a vapour capsule; of the electronic, usually a thermistor with the electronics setting the dead band and operating the switch.

Electronic thermostats can also be programmable, allowing different temperatures to be set at different times. Both types of thermostat can also have separate frost protection so that if the temperature drops below 5° C, say, the heating will come on.

If a house heating regime is to be controlled by a single thermostat, there is no ideal location for it. Usually it devolves into a choice between the hall or the main living room. In the hall it will maintain the whole house at some reasonable temperature which has to be determined by trial and error to suit the occupants. If the living area temperature is more important, the thermostat can be placed there, but other parts of the house may then be too cold.

Figure 2.33 A room thermostat under test at BRE in the mid-1980s

It is important that the thermostat receives no direct heat from, say, the sun or other sources, or be in a cold draught. In other words it needs to sense a representative temperature for the whole house or room.

Experiments to test the effectiveness of an electronic room-stat were carried out in the early 1980s in the BRE matched-house pair with simulated occupancy. A saving of 2–5% was found with the electronic room-stat over electromechanical alternatives. Results concerned with the electronic room-stat's effect on boiler cycling rates and room temperature variability indicated, respectively, no increase in the cycling rate and no improvement in variability. On balance the electronic room-stat performed satisfactorily[94].

Boiler thermostats

A boiler thermostat is fitted within the boiler casing to limit the temperature of water passing through the boiler. The target temperature may either be factory-fixed or set by the user.

New designs of central heating systems with small quantities of water in the systems are less tolerant of overheating than were older heating systems with large water capacity. Boiler thermostats for these newer systems have needed to be capable of greater accuracy than was formerly required.

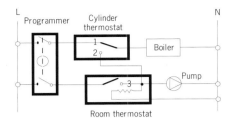

Figure 2.34 Partial integrated control for a system with pumped space heating and gravity fed domestic hot water, without using motorised valves

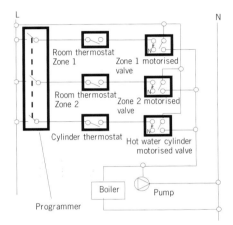

Figure 2.35 Fully pumped independent zone and hot water control

Interlocked (integrated) controls

If no heat is called for by the room or cylinder thermostat, the boiler can be switched off by interlocked controls. This does not always happen with normal controls and the boiler can be left cycling on its own thermostat, keeping only itself warm and so wasting fuel.

The simplest, albeit not necessarily the best, way to avoid this is shown in Figure 2.34 where, if no heat is required, the boiler is switched off by interlocked controls. This is not ideal in that the hot water temperature is not controlled when space heating is required, but it has the distinct advantage of not requiring any plumbing or additional devices. The next sections show the 'proper' way to do it.

Zone control

Separate upstairs and downstairs temperature controls are now required in new installations and a satisfactory way to do this is shown in Figure 2.35. Motorised valves do two things:
● they open up water valves
● they operate electrical switches to energise connections to other devices

Valves

Motorised valves turn water flow on and off; they operate electrically. A two-port motorised valve controls water flow to a single destination. A three-port motorised valve controls water flow to two destinations (usually for space heating and hot water), and may be either a diverter valve (only one outlet open at a time) or a mid-position valve (either one, or both, outlets open at a time). Valve movement will also open or close switches which are used to control the boiler and pump.

Automatic bypass valves control water flow. They are operated by the water pressure across them and are used to maintain minimum flow rates through boilers when alternative water paths are closed; for example, to maintain a water path when all thermostatic radiator valves (TRVs) are closed. Sensor units should be mounted horizontally rather than vertically to minimise this problem. For this reason too, it is normal practice to install at least one radiator without a TRV.

Thermostatic radiator valves use air temperature sensors to control the heat output from radiators by adjusting the flow of hot water. Valves driven by sensors within the body of a valve have a possible disadvantage in that heat from the radiator may be sensed rather than the temperature of air in the room.

Provided space heating is called for, TRVs operate on single radiators and hence can control the temperature in individual rooms. TRVs on their own cannot provide interlocked (integrated) control and some other means must be used to turn the boiler off when no heating is

required. One way of doing this is to include a centrally placed room thermostat which, by trial and error, can be set to turn the heating on and off at predetermined temperatures.

A comparison of room thermostats and TRVs controlling two separate heating systems was carried out by BRE in the early 1980s. They examined energy consumption and achieved comfort temperature levels in two near-identical field test houses with simulated occupancy heat gains. Under these carefully controlled conditions the room thermostats performed as well as the TRV systems[95].

Unless radiators are over-sized, TRVs can only reduce the temperature and hence are particularly useful where:
● lower temperatures are acceptable
● there are significant incidental gains locally from solar or other sources

Requirements for zone control of domestic heating systems were introduced into the building regulations in 1995. In some cases TRVs controlling individual radiators have been allowed as an alternative means of control.

Figure 2.36 A thermostatic radiator valve (TRV) used to control the heat output from a single radiator by controlling the flow of hot water. The photograph dates from the mid-1980s

Cylinder thermostats

A cylinder thermostat measures the temperature of the hot water cylinder and switches the water heating on and off. A single target temperature is set by the user.

A programmable cylinder thermostat is a combined time switch and cylinder thermostat which allows the user to set different periods with different target temperatures for stored hot water. The principles of operation are the same as with the room thermostat except that the cylinder thermostat is sensing the temperature of the hot water stored in a tank. It needs, then, to be placed up against the metal of the tank, about one-third the way up. This location is a compromise: if the thermostat was placed at the bottom of the tank where the cold water enters, the boiler would be heating up the tank continuously; if placed at the top, the boiler would switch off as soon as the first water to be heated reached the top, leaving the rest of the water in the tank still cold. A location about one-third up the height of the tank should prove the most efficient in operation.

Figure 2.37 Programmers can take many forms. This one dates from the early 1970s. More recent models tend to be digital rather than electromechanical

Weather compensation

Weather compensation is a feature of a boiler energy manager which adjusts the temperature of the water leaving the boiler according to the temperature measured outside the building. An outside weather compensator control is standard for most large conventional heating systems. An optimum start control may also be expedient and should be set as normal.

Frost thermostats (or frost-stats)

These are sensors that detect abnormally low air temperatures and switch on heating to avoid frost damage; they are arranged to override other controls. A frost-stat will always be required to protect a boiler installation sited in an unheated basement or outhouse.

Programmers

A programmer is a simple time switch operated by a clock to control space heating or hot water. The user chooses one or more 'on' periods, usually in a daily or weekly cycle.

Usually the programmer performs two functions, through two switches and a clock, to control space heating and hot water separately. The user chooses one or more 'on' periods, normally in a daily or weekly cycle. A mini-programmer allows space heating and hot water to be on together, or hot water alone, but not heating alone. A standard programmer uses the same time settings for space heating and hot water. A full programmer allows the time settings for space heating and hot water to be fully independent of each other.

In its simplest form a programmer is therefore a time clock set to switch on the electrical supply as required. Again there are electromechanical and electronic versions. Nearly all domestic systems have at least two settable periods of heating. The normal method of operation is to have the heating off at night and during the day but with heating periods to provide warmth when people are getting up and in the evenings. Some users have difficulty in setting these devices.

Optimum start control

In buildings which were intermittently occupied, manual firing of solid fuel boilers was the norm. After an initial heat-up period, heating systems would be run for as long as necessary, and control of the amounts of heat obtained was by the relatively simple process of greater or smaller amounts of fuel, and by hand regulation of the combustion air supply. Being a boiler fireman was a relatively skilled occupation.

The first attempts to economise on fuel were by firing up heating systems to achieve the ideal room temperature at a particular time and in relation to ambient external temperatures – hence the introduction of optimum start times; then to shutting down systems shortly before the building became unoccupied. Time controllers suitable for this type of regime are, of course, only feasible with automatic systems. The Property Services Agency achieved some success in the 1970s using optimum start control systems in government office buildings under their management. Considerable progress has occurred, though, since then; for example, with microprocessor-based devices.

Boiler management systems

A boiler energy manager is a device for improving boiler efficiency by weather compensation or load compensation, or both. Systems might also include optimum start control, night setback, frost protection, anti-cycling control and hot water override.

A **pipe thermostat** is a switch governed by a sensor which measures the temperature of water in a pipe. It is normally used in conjunction with other controls such as a boiler energy manager.

A **boiler anti-cycling control** is a device to introduce a time delay between successive boiler firing. Any energy saving has been attributed to a reduction in performance of the heating system, although there is some evidence that older commercial-scale installations may benefit from these devices†.

Boiler interlock is an arrangement of the system controls for ensuring that a boiler does not fire when there is no demand for heat. In a system with a combi-boiler (see Chapter 4.2), it can be achieved by fitting a room thermostat. In a system with a regular (ie non-combi) boiler, it can be achieved in wiring arrangements of the room thermostat, cylinder thermostat and motorised valves. It may also be achieved by a boiler energy manager.

Though not very common in the domestic market there are other more complex controllers and boiler energy management systems which can give self-learning optimum start, weather compensation, anti-cycling and multiple time–temperature periods for space and hot water heating. All of these are more relevant to larger buildings with higher heat losses.

Building management systems

Building management systems (BMSs)‡ have two main categories of application: either the control of building services or the monitoring of their operation, or both[96].

The control operation covers:
- time schedule control (programmable)
- optimum start/stop control (including frost and condensation protection)
- flow temperature compensation
- setback control
- maximum demand limiting and load shedding
- air conditioning and boiler sequencing
- plant selection (boilers and chillers) and chiller optimisation
- plant cycling
- remote manual override control of plant
- thermal store control
- standby plant operation

The monitoring operation covers:
- status
- plant alarms
- space conditions
- personnel access
- utility meters (gas, electricity and water) and fuel flow
- plant efficiency
- plant duty (eg hours run)
- data logging
- logging of trends over time
- maintenance report and work sheet generation
- energy report generation.

Many BMSs also incorporate functions other than boiler and heating control:
- lighting control (timed and photoelectric)
- lift control (via lift control systems)
- security alarms
- smoke and heat exhaust systems in case of fire (see Chapter 4.5)
- fire alarms
- smoke alarms

A BMS generally comprises one or more of the following:
- sensors feeding information into the system (eg temperatures and plant status)
- operator panels feeding instructions to the system
- output channels allowing display of information by the system
- actuators allowing alteration of services by the system (eg electric motors for dampers and valves)
- programmable computers to enable automation of some controls and to monitor functions

BMSs are normally monitored and controlled at a single central station. This location will usually contain both a printer and a visual display unit. It will also offer a means of entering instructions: a keypad with designated functions, a keyboard or a light pen. A reduced version of these facilities may be provided at the outstations. The minimum list of operator functions to be expected is:

- display (on VDU or print-out) of readings at a sensor point or group of sensors
- switching on or off plant selected from the central stations
- selection of control settings
- acknowledgement of alarm messages received

Manufacturers can supply variously extended versions of central station functions, including colour graphics display of plant schematics on a VDU, graphical plots of outputs changing with time, maintenance schedules and hard copy reports.

Where hard copy is to be produced on a printer, there is great advantage in being able to limit the output to essentials to avoid unmanageably large quantities of paper. The ability automatically to condense monitored data as it is collected at the central station (eg to form an arithmetic average) saves time and effort.

Wiring runs

Great care must be taken to ensure that none of the instrumentation wiring for a BMS system to and from outstations to the central station introduces electrical interference. In particular, the connections should never share conduit with mains cable or with other data digital transmission lines. The mains supply to outstations should not be liable to power surges or voltage spikes from large plant (eg lift motors). Outstations should be as far as local geometry permits from arcing contacts (eg heavy duty relays, or electric spark igniters). Manufacturers' requirements for earthing the equipment must be followed. Failure to follow these requirements can lead to persistent unreliability in data transmission or, exceptionally, in programmed operations.

† Further studies of these controls are needed before they can be properly evaluated.

‡ Where building management systems deal exclusively with energy matters, the alternative building energy management systems (BEMS) is sometimes used.

Main performance requirements and defects

Outputs required

Generally speaking boiler outputs should be quoted in kW, though British thermal units (BTUs) may still be found. Boilers having a rated output of under 45 kW are normally found in domestic construction, with larger outputs in non-domestic installations. Boiler sizes are usually specified to provide a small excess of capacity over and above the normal workload to provide for emergency demand. To run a boiler for long periods at small demand levels tends to be inefficient.

The ability to control a central heating system precisely is very important in its efficient use of energy, whatever the fuel used. Misuse or misunderstanding of controls has a detrimental effect on the energy efficiency of any system and may cause some householders to continue suffering from poor heating standards, and, possibly, condensation and mould. Advice to building occupants on energy and controls after a refurbishment may save around 10% on fuel bills[97].

In non-domestic situations, particularly with larger buildings, BMSs are now widely used. A survey for BRE of user experience with recently installed systems was aimed at identifying how well they performed, the benefits that they brought to building management and control, and the scope for improving systems. The survey covered 21 BMS installations in medium to large multi-storey buildings and building complexes. There was a mix of building types, including offices, hotels and hospitals, with services ranging from basic heating to sophisticated air conditioning. At each site the BMS manager was interviewed and the controls and the plant in the building were examined. Further interviews were then conducted with the suppliers of the BMS.

The survey found that the performance of BMSs had improved significantly since the late 1980s, and are far more readily accepted by users. The in-built monitoring and alarm facilities of BMSs make it easier now to detect problems[98].

Efficiency

The efficiency of domestic boilers can be compared by using a database entitled the 'Seasonal efficiency of domestic boilers in the UK' (SEDBUK). This gives the average in-use efficiency applicable to a particular boiler and should not be confused with other figures for efficiency sometimes quoted in product literature. The database was developed to assist in SAP calculations[†].

Much of the advantage achieved by replacing an inefficient boiler by an efficient unit remains unrealised if the building fabric is deficient in performance. Satisfactory overall economics of heating plant of course demand adequacy of thermal insulation, and control over fortuitous ventilation. The easiest element to insulate is the loft space in a pitched roof. Although over 90% of dwellings in England have some measure of thermal insulation installed in lofts, only in 1 in 6 of these dwellings is it up to the thickness of 150 mm now required by Building Regulations[7] [‡]. In Scotland, some 90% of dwellings with lofts have some thermal insulation, although 1 in 6 have less than 100 mm[9]. In Northern Ireland, of those dwellings with loft space, some 1 in 8 have no thermal insulation and about one-quarter have insulation thicknesses of less than 100 mm[10]. Thermal insulation of roofs was dealt with in *Roofs and roofing*, Chapter 1.4.

Case study

Replacement of an old gas fired boiler
In 1988, the annual gas consumption of the Ritz Hotel was 7.5 million kWh, which equated to 526 kWh/m². This is classified as poor according to criteria established by the Energy Efficiency Office (EEO). Following the replacement of the hotel's inefficient boiler plant and controls under a project funded through Contract Energy Management (CEM), the annual gas consumption dropped to 350 kWh/m² in 1990. In subsequent years, by close energy management, the figure has reduced still further to 322 kWh/m², a reduction of 40% compared with the 1988 value[99].

Coal fired boilers

Coal firing may still be preferred for some installations, and the modern coal fired boiler offers considerable improvement on its more traditional predecessor. The techniques of underfeed stoking and automatic de-ashing have been considerably advanced, and efficiencies of 80% are achievable compared with older designs at around 70%. This type of boiler, which has automatic stoking and de-ashing, is particularly suitable where 24-hour heating is required, as in sheltered housing. In one installation examined by BRE, and compared with standard coal fired boilers, operating costs had been considerably reduced over the old installation[100].

Choice of solution for upgrading controls and major items of plant

Any improvements to existing dwellings should, wherever possible, conform to building regulations applicable to new work and follow industry guidelines. An industry endorsed guide on upgrading controls in domestic wet central heating systems has been produced by the Government's Best Practice Programme[101].

The main improvements to existing domestic systems are likely to be:
- making the system fully pumped
- adding time and temperature controls for both space and hot water heating

For houses of more than one storey, each storey should have its own temperature controls. Separate time controls are also advisable but this depends on each storey having its own pipework system.

In domestic situations, boilers can now be fitted wherever it is most convenient (Figure 2.38); in former times the location of boilers was constrained in a number of ways.

Condensing boilers are particularly suited to the needs of large establishments, both for the provision of heating and for domestic hot water. When operating at low loads, the condensing boiler will be far more efficient than its standard counterpart. However, when only domestic hot water is required, water from the boiler is needed at a higher temperature and the benefits are reduced.

Condensing gas boilers used for heating sports and recreation buildings provide running cost savings of 15–20% with paybacks of under 3 years. They provide high efficiencies using conventional system design, and are easy to install and maintain while being environmentally friendly[102].

Management of the heating regime

Using BMSs to control heating (and also air conditioning, ventilation, lighting and other services) in large buildings is growing rapidly. Advice is available on how to specify and select a BMS to help building occupiers make cost effective decisions appropriate to their particular requirements[103].

Unwanted side effects

Noise and vibration

The pump is perhaps the main source of noise and vibration in small domestic central heating systems. While some noise and vibration at the source has to be accepted (eg with some kinds of boilers), proper mounting of the pump is essential in reducing noise. A partial but probably less effective solution is to insert a short length of flexible piping into the heating pipework to isolate the pump from other parts of the system.

When a central heating system fires up intermittently, the expansion and contraction of the pipes and radiators as they heat up and cool down is often accompanied by loud clicking and knocking noises. Invariably a noise is due to friction at a point of restraint, and the sudden release of a restraint is accompanied

Case study

Inadequate temperature control in intermittently heated office building
Operating experience in the BRE low energy office (LEO) highlighted significant failures in reaching target temperatures in intermittently heated buildings. In the BRE LEO this failure was attributed to a combination of plant sizing, design and commissioning faults in installed controls, and the action of thermostatic radiator valves. In this building, integration of the heating and ventilation controls increased useful heating system output, at design temperatures, by 30%. This, combined with the use of more appropriate controls including the BRE optimum start algorithm BRESTART, greatly improved comfort levels.

by a loud report. It is not always easy to locate precisely these sources of noise, or to find effective remedies. It helps, therefore, if the pipes, where they pass through walls, joists etc, are kept out of direct contact with the structure by means of sleeving or packing material; also, pipes should not be gripped too tightly by their supporting clips. Radiator brackets in particular may need attention, perhaps by the insertion of short pieces of PTFE tape or another lubricating medium between the radiators and suspension brackets.

Electrical interference

Building management systems can be susceptible to electrical interference which, increasingly, is being produced by sources such as IT equipment, industrial and scientific equipment, and radio frequency transmitters. BMSs must be designed with built-in immunity to quite high levels of electrical disturbances and be properly installed so that the disturbances to which they are exposed do not exceed their immunity thresholds. Once a system has been installed, eliminating electrical interference can prove difficult and costly. The problem is covered by the Electromagnetic Compatibility Directive which came into force in 1992. Equipment that suffers from or causes electrical interference must be withdrawn from service[104].

Figure 2.38 A wall hung condensing boiler (far right) integrated into a kitchen layout. In domestic situations the boiler can be sited in the most convenient position (Photograph by permission of Keston Boilers)

Carbon dioxide production

Although carbon dioxide production varies with the efficiency of the boiler, it also varies significantly with the type of fuel used. A boiler using LPG produces, on average, around 25% more carbon dioxide than a mains natural gas boiler, and one burning oil around 45% more.

Commissioning and performance testing

For replacement boiler installations generally, reference should be made to the manufacturer's instructions. Information will also be found in the relevant British Standard (BS 6798[105]).

Flues and terminals for gas appliances

BRE surveys during the years 1980–90 found a number of instances of flue pipes for gas appliances touching sarking felt, combustible insulation in roofs, timber wedges and chipboard floors; pipes were seen both totally unsupported in roof spaces, or supported by timber wedges between pipe and combustible construction materials. Building regulations requirements for separation between flue pipes and combustible construction were commonly not met.

If flue pipes are not properly supported, not correctly spaced from combustible materials and not sleeved where required, there is a risk that joints will open and leak flue gases (eg into the roof space) and that combustible materials may be subjected to heat (BS 5440-2[87]).

Terminals must be sited so that flue gases can disperse freely, and be guarded or shielded where they might otherwise be damaged or cause damage or injury. In practice this means they should be sited no nearer than 300 mm to a window, door or other ventilator, and 1.5 m above or below another terminal in the same wall. Plastics rainwater downpipes are particularly vulnerable and should be protected.

Furthermore, terminals, whether for natural or fanned draught appliances, should be located no closer than 850 mm below a plastics gutter, eaves or verge soffit, and not be closer than 450 mm below a painted soffit; otherwise an aluminium alloy sheet shield should be specified, not less than 750 mm long, fitted to the underside of the gutter or soffit above the terminal[106].

BRE investigators have also found air vents for open flued appliances with too little clear ventilation area. If the vents are sited where they cause draughts, occupants will often block them, and, in doing so, create conditions for producing potentially-lethal carbon monoxide.

Ventilation may be direct (ie via a vent in the external wall of the room containing an appliance) or indirect (ie via an adjacent room, hall or suspended floor void). If air reaches the appliance via 3 or 4 vents in series, the sizes of the vents, or the clear ventilation area of each vent, should be increased. Site studies have shown that this has not always been done.

Flues for non-domestic boilers normally discharge at high level, well out of reach of pedestrians, and clear of openable windows in the same or adjacent buildings (Figure 2.39).

Various kinds of terminals and cowls are available which afford some added protection in storm conditions, particularly when the appliance is switched off.

Other checks

During commissioning, all boiler and heating circuit flow rates should be checked. For a non-domestic installation, it may also be necessary to measure the boiler's excess air level and to adjust, where possible, to the manufacturer's design value. Clearly the efficiency test measurement point must be upstream of any dilution air; for acceptance tests, efficiency by the flue loss measurement can be obtained by combining this measurement with flue gas temperature. This method is only approximate because it assumes that all the flue gas leaves the appliance in a saturated condition. Ideally, the condensate production rate should also be measured over a period of steady operation. This can give a valuable check on boiler operation.

Low pump speeds tend to increase boiler cycling and slow down heat recovery times.

Where a replacement boiler is fitted up to an existing heating circuit, the existing heat emitters (ie radiators) are often generously sized relative to the building's new heat requirements resulting from the greater efficiency of the new boiler and any improvements in insulation since original construction of the building. These now-oversized radiators can meet the same load at a lower temperature, and so lead to a slight further improvement in boiler efficiency.

Figure 2.39 In this building on the BRE site, the existing radiator system was reused when new gas boilers replaced the mains distribution system. The stainless steel flues terminate in such a position that emissions are carried clear of opening windows

Health and safety

The temperatures of heating surfaces, and the possibility of contact by and sensitivity of a building's occupants to high temperature surfaces, is clearly an important consideration in selecting radiators. Where temperatures exceed 80 °C, some form of shielding or casing may be required. Low surface temperature radiators are available.

The risks of building occupants (especially small children in dwellings) suffering burns from heating equipment should be eliminated, or otherwise minimised as far as possible. Controls should not get too hot, and there should be no feeling of discomfort from contact with or close proximity to radiating surfaces.

The following may help in preparing criteria for a performance specification.

● No part of any heating appliance or distribution system which can be contacted with an external probe 110 mm long by 25 mm diameter (eg an adult finger) or 70 mm long by 12 mm diameter (eg a child's finger) shall at any time be at a surface temperature greater than 80 °C†

● No control of any heating appliance shall at any time be at a temperature ≥ 60 °C if it is of a material which is a good diffuser of heat, or ≥ 65 °C if it is a poor one

● Every non-sealed domestic heating appliance should (ideally) incorporate a control which shuts down the system at a water temperature of 95 °C. For a sealed system this value is set at 110 °C for operational reasons

† These suggested criteria for accidentally touching surfaces are tentative, based on child and adult finger sizes. The temperature quoted is debatable and different values are given by different authorities. It is probably around the hottest reasonable radiator temperature, but could of course burn on skin contact of greater than about one-tenth of a second.

The Gas Safety (Installation and Use) Regulations require there to be a *'safety control designed to shut down the appliance before there is a build-up of a dangerous quantity of the products of combustion in the room concerned'.*

Spillage of combustion products from an open flued combustion appliance can be potentially hazardous, since some of the products of combustion are toxic and combustion appliances are frequently installed in habitable rooms. However, the major cause of death is by asphyxiation when carbon dioxide replaces the oxygen in the atmosphere. Spillage is difficult to predict because of the variety of different types of combustion appliance and because of variations in room size, airtightness and extract ventilation rates[107].

Experiments were carried out on open flued gas and oil boilers in a test house at BRE Garston to find out what parameters affect combustion gas spillage. The experiments consisted of testing the pressure differences and extract fan flow rates at which spillage occurred. The results showed that, in a large number of cases, spillage will result if an open flued gas appliance and air extract fan are running concurrently. Spillage was not encountered with the oil boiler which had a pressure jet burner. Any fan installed in the same room as an open flued gas appliance or an oil fired appliance with a vaporising plate burner should have a maximum capacity of 20 l/s, and the fan and the appliance should together pass a relevant spillage test[108].

Boiler explosions

Another risk to building occupants is that from boiler explosions, with steam being generated following blockages in the system or inadequate levels of feed water. Boiler explosions were at one time not uncommon: for example it was estimated in 1876 that between 100 and 200 persons were killed every year in boiler explosions, though most of these occurred in industrial plant rather than in boilers used for heating purposes. Matters only

improved when the insurance companies took on a more proactive role[18], and boiler explosions are far less disastrous now, though arguably no less frequent.

In 1992, for example, there were 79 incidents, of which 45% involved a back boiler to an open fire. Blockages may occur when frost has caused ice to form in the system, or may simply be a long term risk from hard water deposits on pipe walls. During 1992 water freezing in boilers caused around 30% of explosions. All installations should have a device to allow pressure to escape before explosions occur. Valves weighted by doughnut shaped rings of cast iron were in widespread use in the interwar years, but there are now more sophisticated devices such as fusible plugs melting at around 96 °C.

Figure 2.40 This boiler unit, having been catastrophically damaged, has been removed for diagnosis and replacement

Should the boiler be replaced?

A dwelling seen by a BRE investigator in 1999 had a 29 years old gas fired, fan flued, cast iron central heating boiler (Figure 2.41) which had survived the change from town gas to natural gas, and was still functioning satisfactorily even though the manufacturer had stopped producing boilers several years earlier. The electromechanical control unit had been replaced and the bearings in the forced draught fan unit were sometimes a little noisy at change from idling to operating speed, as might be expected. The prognosis was that the installation was still serviceable – and would remain so until a major part fails. The seasonal efficiency, of course, at about 65–70%, would not compare favourably with more modern designs such as condensing types at around 85%, but it was installed in a well insulated house, and the owner observed that an awful lot of gas could be burned for the cost of a new condensing boiler, even if a small local authority installation grant was available to offset some of the cost! This characterises a potential dilemma for the industry – national and individual priorities do not always coincide.

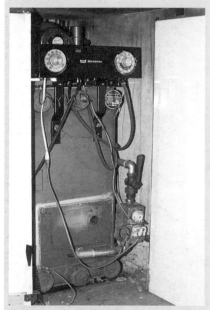

Figure 2.41 A cast iron, gas fired, fan flued 29 year old boiler still in use and still acceptably efficient as far as the owner was concerned

Durability

Unpublished BRE investigations dating from the early 1980s put the expected lives of cast iron boilers at around 25 years (Figure 2.41), with around 10–20 years for welded mild steel boilers. The life of oil storage tanks will depend on material and the degree of protection, if of steel, but around 20 years seems probable (Figure 2.42). Automatic stoking apparatus for solid fuelled boilers lasted around 12 years. For the most part, however, objective evidence on longevity seems to be in short supply, but recent anecdotal evidence seems to support these figures.

In general, though, controls and boilers would appear to have an economic life of about 10 years. Recommended design temperatures for heating systems have been increased in recent years and old installations may not be capable of meeting present needs, and need replacing in any case for that reason.

'Airlocking' has been found to occur either by entrainment of air through faulty design of the system or by the generation of gas by corrosion within the system. Air can be drawn into the system through the vent pipe if this is not positioned sufficiently high above the topmost section of the circulation system to balance the suction developed by the circulation pump; it may also be taken into the system by circulation of hot water through the vent pipe and expansion tank due to using a pump of excessive pressure head. 'Airlocks' caused by corrosion within the system are due primarily to hydrogen generation, although certain forms of bacteria may produce nitrogen within the system.

With condensing boilers, both stainless steel and aluminium have proved suitable materials for the condensing heat exchanger, although manufacturers' assurances on lifetime should be sought. Cast iron does not resist corrosion from condensate. Some appliances may be less forgiving of bad system design or installation. In all cases, manufacturers' recommendations for commissioning and maintenance should be observed.

In heating installations, pressed steel has largely superseded cast iron for radiators, and thin gauge copper tube has replaced black iron or galvanized steel for pipework. The combination of copper tubing, steel radiators and cast iron boilers in the same system has caused concern, and defects developing after comparatively short periods of use have in the past produced wide speculation on the advisability of using mixed metals in this way.

In a closed heating system, the initial dissolved oxygen content of the water is reduced fairly rapidly to a value of 0.1 ppm or lower. When iron reacts with de-aerated hot water, ferrous hydroxide is formed. This is dissolved until an equilibrium is attained and the reaction ceases unless the ferrous hydroxide is removed. One process by which this occurs is a reaction in which the ferrous hydroxide is converted to the more insoluble magnetite leading to the release of hydrogen gas. This probably occurs in an all steel system although at a very low rate, but the rate of the reaction is increased by the presence of copper. Hydrogen generation may, however, be significantly suppressed by inhibiting

Figure 2.42 An oil storage tank with its surrounding brick wall, both to provide mechanical protection and to contain possible leakage after being holed by corrosion

the formation of ferrous hydroxide or preventing the copper going into solution.

Inhibitors can be specified to give added protection, though different compositions are needed for different systems and there can be special additives in commercially available inhibitors. Sodium benzoate, sometimes specified, is a potential food for micro-organisms. Biocides and wetting agents may be also added.

Perforation of radiators by corrosion is relatively rare. Where it occurs in steel radiators, it can be due to pitting corrosion of the internal face after entrained air is taken into the system. Sulfate-reducing bacteria associated with the production of hydrogen sulfide have been identified in failed systems. These bacteria are widely distributed in clay subsoils and gain easy access to heating systems; the risk of attack is small, though, and can be eliminated by treatment with a biocide. The system should be flushed with the biocide before filling, emptied and refilled with mains water; this is preferable to dosing the system which might encourage the establishment of a resistant strain of the bacterium.

Aluminium tubing with external dimensions and wall thicknesses identical to light gauge copper may also be found, as indeed may also aluminium radiators. The tubing is a composite extrusion of aluminium/manganese alloy on the outside and a zinc/aluminium alloy, 0.05 mm thick, on the inner surface. Accelerated tests with this tube in a system which incorporated a copper boiler and copper tubing, have shown that the zinc-alloyed inner layer had a sacrificial effect, producing saucer shaped pits which did not extend through the external wall of the tube. Durability depends as much on preventing external corrosion (by painting or lacquering) as on ensuring that the oxygen content of the circulated hot water is low.

Thin walled, coated carbon steel tubing to BS 4182 (now withdrawn) may just possibly still be seen in small bore, closed circuit central heating systems. Although the initial fill of water has tended to produce some surface attack in the bore, the corrosion risk ceased when the available oxygen in the water was reduced. On the other hand, corrosion of the external surface of the pipe could be severe if the protective plastics wrapping ceased to function.

Cast iron radiators and screwed iron pipework may be serviceable despite their ageing appearance, since these are very durable. The replacement of sound cast iron radiators and large diameter pipework can normally be justified only if the replacements are of equal or better quality.

Although many protective cages fitted over boiler flue outlets are of stainless steel, many are not, and they are subject to corrosion, particularly from damp exhaust gases (Figure 2.43).

Maintenance

Regular maintenance of most designs of boiler will be needed to ensure continued safe and satisfactory operation, especially with regard to the possible leakage of combustion products. Regular maintenance is certainly necessary for oil fired boilers where flame adjustment is critical. Legislation to ensure that an inspection is carried out at least annually is in force for certain forms of housing tenure, but it is good practice for all installations.

Assuming that recommended maintenance procedures are followed, the cost of maintaining a condensing boiler is not likely to be significantly higher than for conventional boilers. For all types of boiler, it is possible that some ancillary components, such as the combustion fan and safety controls, may need to be replaced during the boiler life although high levels of reliability should be expected.

Spares for boilers, pumps and controls more than 10 years of age may not be obtainable, although parts may be available for older solid fuel and natural draught boilers.

If a condensing boiler is fitted as part of a general upgrade to improve the energy efficiency of a central heating installation, older, larger radiators can be retained. This is despite the fact that insulation added to the building fabric might tend to make them redundant or oversized since condensers work well at low circulating water temperatures.

Figure 2.43 Protective cages to exhausts need to be inspected for corrosion. Moisture laden air from the exhaust gives rise to onerous conditions. Rust from these chromed units can badly stain walls beneath

Work on site

Storage and handling of materials

It is important that items vulnerable to bad weather are properly packaged and stored in appropriate conditions. Materials and fittings for boilers and water services are particularly attractive to thieves.

Workmanship

Fixing copper tubes may be achieved by using copper alloy holderbats for building into, or screwing to, the structure; also by using strap clips of copper, copper alloy or plastics, or purpose made straps or hangers. BRE site studies have shown that the fixing of clips has often been inadequate.

The following points are made where flue pipes pass through roof voids.

● Supports for flues should be robust (eg steel strapping of an appropriate size) and well fixed

● No combustible packings should be used between flue pipes and fixings

● Sockets should not be installed inverted

● Flues should be supported under every socket, at every change of direction, and throughout their length at intervals not exceeding 1.5 m

● The installation should be designed and completed so that the flue pipe discharges to the external air

The bayonet-type spigot and socket joints of proprietary insulated flue pipes need to be properly aligned to effect a gas-tight seal. Obviously all pipes should run as close as possible to the vertical.

Inspection

One of the considerations which has given BRE investigators most concern on site inspections is the number of instances where metal flues for gas appliances passing through roof spaces have not complied with requirements. It is necessary to carry out three-dimensional checks (eg of the proximity of flues in roof spaces to truss bracing or other combustible parts such as combustible insulation, sarking, vapour control layers etc).

The problems to look for in relation to flues are:
◊ flue pipes inadequately supported in roof spaces
◊ flues too close to roof timbers
◊ fire stops inadequate where flues pass through structures
◊ joints in flue pipes inadequately made
◊ terminals too close to windows, doors, soffits and rainwater pipes

The problems to look for in relation to pipework and vents are:
◊ inadequate diameter or capacity of pipes
◊ pipework liable to airlock
◊ pipes inadequately clipped
◊ stop valves leaking or ineffective
◊ pipes in unsuitable positions to isolate, drain down or flush out
◊ obtrusive pipework
◊ central heating systems not properly ventilated
◊ water pipes vulnerable to damage or in unknown locations
◊ vents for combustion air too small
◊ unacceptable noises in the boiler installations and pipework
◊ boilers or controls obsolescent or inefficient (eg not set to take advantage of low tariff periods)
◊ hot pipes touching other pipes which may cause vibration
◊ uninsulated hot and cold water pipes touching or too close, which leads to heat transfer (Figure 2.44)
◊ missing equipotential earth bonding in steel duct frames
◊ vents for heat dissipation too small
◊ vents badly sited and possibly blocked off by occupants
◊ appliances not provided with isolating valves for ease of maintenance or replacement

Figure 2.44 In this new installation in a refurbishment project, the uninsulated hot and cold water pipes are touching. Not only will the hot pipe be cooled but the cold pipe will be heated, leading to inefficiencies

Chapter 2.3

Floor and ceiling heating, and electric storage radiators

Although heating for floors and ceilings tends to operate on different principles – floor heating always implies a degree of heat storage while ceiling heating does not necessarily do so – it is convenient to include them both in this chapter. Electric storage radiators are also included because they operate on a similar principle to electric floor heating, with similar advantages and disadvantages.

As with all forms of electrical heating, efficiencies usually quoted do not take primary energy use into account.

Characteristic details

Floor heating

Floor heating was dealt with to some degree in *Floors and flooring*, though from the perspective of the flooring rather than the heating. However, some appropriate text from that book is repeated in this chapter for the convenience of readers.

Heated floors are not a recent innovation. The Romans developed ingenious systems for passing the products of combustion from a furnace through ducts under mosaic covered floors built of brick or stone, called hypocausts, though the thermal efficiencies they achieved must inevitably have been low. Many centuries later, in Victorian times, warmed air was frequently introduced to rooms through cast iron grilles set into the floor finish, although this perhaps can hardly be described as floor heating – rather as heating through the floor (see Chapter 2.4). Later still, systems for

heating the fabric of the floor were reintroduced, and, in consequence, the temperatures reached need careful control if deterioration of the flooring finishes is to be minimised.

Two basic kinds of underfloor heating laid in screed or slab have been used in the 1980s and 1990s:

- low pressure hot water in steel or copper pipes, or, more commonly, polypropylene
- electric elements laid either directly or in conduit

Ceiling heating

For many years it was believed that radiant panel ceiling heating was more efficient than other forms of heating in achieving comfort levels, though there is little evidence to support this view[18]. Of course, where heating pipes were laid in suspended upper floors without suspended ceilings, some of the heat would inevitably find its way downwards by radiation.

Hot water or steam heated pipes, or electric heating elements, were used in suspended ceilings. In later years, though, the hot water or steam

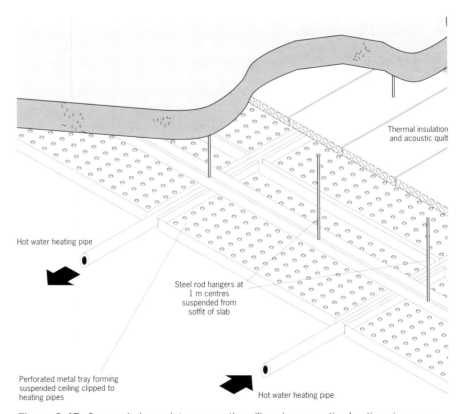

Thermal insulation and acoustic quilt

Hot water heating pipe

Steel rod hangers at 1 m centres suspended from soffit of slab

Perforated metal tray forming suspended ceiling clipped to heating pipes

Hot water heating pipe

Figure 2.45 Suspended proprietary acoustic ceilings incorporating heating pipes were popular during the 1960s and 1970s. The metal trays enabled heat to be distributed over a large area while the acoustic blanket also acted to reduce heat losses into the ceiling void

was principally distributed by tubes sandwiched between metal panels. Thermal insulation quilts were incorporated into, or sometimes laid over, the panels (Figure 2.45). More recently, electrically conducting membranes sandwiched between thin electrically-insulating sheets have been developed. Another form of ceiling heating often found in large industrial type buildings is the unit radiant heater, powered mostly by gas or occasionally by electricity.

Electric heating panels and tubular ceiling heaters are fairly easy to install and can have individual time and temperature controls. The big disadvantage is the high cost of on-peak electricity. At one time, wall mounted, tubular, oil filled panel convectors were common, but they are not so frequently seen now.

Electric storage radiators
To overcome high running costs, night storage heaters use off-peak electricity, releasing the heat generated gradually during the following day. Controls include a temperature sensitive flap or fan which is needed to give a heating boost towards the end of the day. The main disadvantage is that a change in the weather cannot always be predicted during an operational cycle and so the air temperature may be too high or too low; when too low, extra heat output from other sources might be required, perhaps using on-peak electricity. Also, some heat may be emitted from panels at night, leading to higher than wanted night-time room temperatures.

Space requirements for major items of plant (dimensions etc)
Hot water floor and ceiling systems are supplied by the kinds of boilers which were described in Chapter 2.2. Electric floor and ceiling systems are supplied off the mains, and space requirements can be accommodated in the normal electrical distribution system. Electric storage radiators can be obtained either with or without fans to assist in distribution of stored heat. Units used during the 1970s were very much larger than those used in later years.

Main performance requirements and defects
Strength and stability
The main problem with electric storage radiators is their weight, particularly where they are considered for installation in old buildings with suspended floors with inadequate loadbearing capabilities. BRE site investigators have seen them installed in old dwellings where the timber joisted floors were decidedly 'bouncy'. When installed against walls normal to joist span, that is to say near the bearings, they are self-evidently less of a problem than when installed against walls parallel to joist span, particularly when they are placed at mid-span.

Outputs required
Modern underfloor heating systems have a number of significant advantages over radiator systems, and for many applications underfloor heating represents the ideal solution. The low temperatures are well suited to condensing boiler operation. On the other hand, the least favourable application for a condensing boiler would be one with high temperature radiant panel heat emitters and short, intermittent, heat requirement. Although a reasonable efficiency may be achieved, payback periods will be longer.

As well as simplicity of design and installation (resulting in fewer operational problems), electric heating potentially offers several benefits including reduced capital costs, low maintenance overheads and increased flexibility. These benefits need to be balanced against other factors such as the lower cost of fuels associated with more conventional systems.

Suitability of substrates and finishes
Underfloor heating
Lightweight cementitious screeds are not appropriate for use with underfloor heating because of their insulation value. Ordinary cementitious screeds are appropriate for use with underfloor heating, although they can show excessive

drying shrinkage caused by the heat drying out the material too quickly. Anhydrite screeds are very appropriate for use with underfloor heating, but operating temperatures should not normally be allowed to exceed 50 °C. Although the screed can take higher temperatures, the finish may not.

So far as finishes are concerned, synthetic resin material is rarely used over underfloor heating but can be used provided the base is sufficiently strong, crack free and dry. Surface temperature should not exceed 28 °C. Most paints and seals are suitable, except of course bitumens.

Mastic asphalt and pitchmastic floorings are not in general suitable for underfloor heating, though there are examples of their being used in the past. Even the hardest grades soften at ambient temperatures above around 30 °C. The effects of underfloor heating are exacerbated by the presence of textile floor coverings which act as insulants. Mastic asphalt can, however, be used as a damp proof membrane below floors which are heated, provided the asphalt is protected with a layer of thermal insulation.

Textile coverings will invariably have an insulating effect on the flooring substrate, and the effects of heating should always be ascertained. Cork is not very suitable for use over underfloor heating because of its thermal insulation properties, although the thinner tiles may prove to be satisfactory if the surface temperature does not exceed 27 °C. Unbacked PVC can also be used where surface temperatures do not exceed 27 °C. Backed PVC is not recommended because of the insulating effect of the backing which will reduce the efficiency of the heating. Special grades of adhesive should be specified for these conditions because of the risk of enhancing plasticiser migration.

Underfloor heating has been used successfully under composition block flooring up to 48 °C. Normally, though, systems should not be designed to run at temperatures exceeding 28 °C. Early examples date from the 1930s. Indeed, there is some indication that the material was at first developed with underfloor heating in mind, as the thermal movements of the material were comparatively small. When laying on screeds containing heating elements, it is essential that sufficient drying time is allowed for the screed to shrink before fixing blocks.

Timber floorings are not the best materials for using over underfloor heating owing to their thermal insulating properties, their propensity to dry out and shrink when the heating is on, and their ability to pick up moisture and expand when the heating is off.

Ceiling heating

Ceiling heating is not appropriate for all circumstances, and there have been complaints, particularly from sedentary occupiers, of being too uncomfortable.

Health and safety

In the case of hot water based systems, the temperatures reached in the pipes may be as high as 54 °C, or even more. It is generally recommended, for a variety of reasons such as degradation of flooring and adhesives, and indentation and shrinkage, that the floor surface temperature should not exceed 27 °C.

Ceiling heating systems are generally out of the physical reach of occupants, and can usually, indeed may need to, operate safely at higher temperatures.

Durability

Although heating pipes buried in screeds are now not permitted under Water Regulations and British Standards, nevertheless many examples have been discovered by BRE investigators where this has been done.

Steel or copper pipes embedded in concrete were at one time thought to possess a high risk of failure from corrosion, but in practice this has not proved always to be the case. In addition, the introduction of annealed copper pipe and, later, plastics, enabling the virtual elimination of buried joints, further enhanced confidence in the method. (Figure 2.46). Nevertheless, problems have occurred with corrosion of metal pipes necessitating removal of floor finishes.

It will be found that in situ floor finishes, such as terrazzo and granolithic, are more likely to curl and crack when heating elements are incorporated into a newly laid floor because of the increased drying shrinkage and the thermal gradient imposed across the section of the floor. Clay, concrete and terrazzo tiles usually present no problem, provided that the base has been dried out by using the underfloor heating before the tiles are laid. However, it must be switched off and cooled before they are laid.

Sheet and tile finishes of linoleum, cork, rubber and PVC, and the adhesives normally used with them, can behave acceptably provided that the base is dry. Floorings such as asphalt, pitchmastic, thermoplastic tile and vinyl tile are more likely to suffer permanent indentation on a heated floor than on an unheated one, more especially where the flooring is covered with a textile material.

It has been reported that some plaster ceilings have failed when heating pipes, embedded in the suspended floors above (ie ceilings and floors were part of the same elements), were recorded at temperatures up to 85 °F.

Figure 2.46 Defective low pressure plastics hot water heating pipes exposed by removing the screed

Maintenance

It is not possible to maintain buried pipes and cables without digging up the floor. This is the main reason why they should not be installed without means of access.

Work on site

Inspection
The problems to look for are:
◊ water leaks from floor systems
◊ softening of floor finishes
◊ deterioration of ceiling finishes
◊ textile or other unsuitable floorings covering heated floors
◊ non-withdrawable cables and pipes

Chapter 2.4

Warm air systems and air conditioning

Although a necessary part of air conditioning in most situations, for the purposes of this book the requirements for mechanical ventilation are dealt with separately in Chapter 3.

Warm air space heating systems developed in late Victorian and Edwardian times usually involved large diameter iron pipes heated by steam, contained in a perforated enclosure, either in the walls or the floors, and protected by decorative cast iron grilles, so that distribution of the warmth was by natural convection. Although used to some extent in later periods, the system is now less commonly seen.

Figure 2.47 A heating installation suspended in the roof of a factory building

Later developments have included the the fan operated unit heater, often found slung from the roofs of industrial buildings (Figure 2.47). Smaller installations using similar principles were developed for domestic use from the mid-1960s[18].

Furthermore, since the 1950s the use of forced warm air curtains at the entrances to buildings has increased, enabling large numbers of people to enter or leave in cold weather unhindered by revolving, sliding, pivoting or swinging door leaves. These curtains encourage high heat wastage.

Air conditioning is undoubtedly a product of the twentieth century. The term implies the pretreatment of ventilation air to control and modify its temperature and, sometimes, its moisture content. The air may also be filtered and cleaned of contaminants. In countries with warm, humid, climates, its use has been principally to counter the effect of ambient environmental conditions. Since air conditioning usually involves at least one or more of these other factors, in addition to heating and cooling, it tends to be relatively expensive in terms of fuel consumption when compared with naturally ventilated buildings, using about 50% more primary energy.

Although the term 'air conditioning' was not used until the early years of the twentieth century, the techniques had already been employed on a significant scale in industrial applications in the USA. Further applications were still confined largely to the USA during the 1920s and 1930s, all essential large scale applications being in large

buildings such as department stores. In the UK, there were only isolated examples until the 1960s, and Billington and Roberts estimated in 1982 that less than 20% of new office buildings in the UK had full air conditioning[18]. The number of domestic applications was then minute. Although that situation still remains, it seems set to change. As people get used to having air conditioning in their cars they will surely come to expect it at home too, and the industry is already marketing domestic air conditioning systems to satisfy that demand.

Some of the terms used to describe modern heating, ventilating and air conditioning (HVAC) systems are given in the feature panel opposite.

Mechanically serviced buildings can have a significant effect on environmental issues such as global warming (through emissions of carbon dioxide resulting from energy use) and ozone depletion (through the use of, say, chlorofluorocarbons).

Air conditioning was introduced in response to the perceived need to cool modern buildings, which have tended to suffer from high solar heat gains from oversized windows, or, conversely, poor natural daylighting encouraging the use of many energy intensive lighting appliances in compensation. Increased concern over the adverse environmental impact of energy use has stimulated the design and construction of energy efficient buildings, many of them suited to natural ventilation.

HVAC system descriptions and definitions[110]

Window ventilation and radiator perimeter heating
Air enters and leaves the building via a combination of opening windows and trickle ventilators to provide ventilation throughout the year, and to dispel excess heat when required. The radiator system provides heating during the winter. The indoor temperature is always higher than the outside temperature. In winter, ventilation with cold outdoor air can cause discomforting draughts. At any time of the year, air entering through windows is unfiltered and carries with it any fumes and contaminants present in the outdoor air. Similarly, open windows provide a direct route for outdoor noise intrusion to the office space. The effectiveness of natural ventilation is limited in deep plan spaces.

Window façade ventilation and radiator perimeter heating
This is similar to window ventilation and radiator perimeter heating, but has an extra façade located 600–1000 mm outside the window wall. The void between the inner and outer façades is sectionalised with one devoted to bringing air into each room at low level, and the other to collecting and dispelling used air from high level in each room. The arrangement improves the effectiveness of the ventilation and reduces the adverse effects of wind and outdoor noise. As with window ventilation, the system affords no means of filtering the air supply.

Mechanical extract ventilation, window supply and radiator heating
This is similar to the two previous systems but some ventilation is assured as the used room air is positively removed by means of fans which induce some outside air to enter via the windows regardless of the prevailing weather.

Mechanical supply and extract ventilation with radiator heating
This is a development of the previous system. Outdoor air is now also positively brought into the room from a central air handling unit which is normally equipped with supply and extract fans, air filters and heater to temper the supply air in cold weather. The air supply is not, however, cooled.

Mechanical displacement ventilation with radiator heating
Displacement ventilation involves introducing of a low velocity air stream at low level within the space, its temperature being only slightly below the desired room temperature. The supply air forms a pool of cool air in the lower part of the occupied space. Internal heat sources such as people or equipment warm the air surrounding them to create convective plumes which entrain air from this cool pool and cause an upward movement. This warm air rises and forms a layer at the ceiling where it is removed by a high level mechanical extract system. The effect of the convective plumes is not only to cool but also to remove contaminants from the breathing zone as the air is drawn over the occupants. Dehumidification of the supply air may be required in summer to maintain a suitable humidity within the space.

Mechanical displacement ventilation with static heating and cooling
This is a development of the previous system with additional static cooling provided actively by chilled ceiling panels or beams to meet local cooling requirements. The static cooling panels are fed with lightly cooled water provided by a central chiller. Warming is also by static means, usually in the form of perimeter radiators. The system is frequently used with displacement ventilation.

Ventilating chill/heat beams
This system involves providing a supply of lightly cooled outdoor air from a central air handling unit to long beam-like metal boxes hung from the ceiling. These can be exposed or concealed. The air supply out of these ventilating beams is arranged in order to induce air from the room through finned beams concealed within the boxes. Hot or cold water is circulated through the finned beams to heat or cool the air as required.

Four-pipe fan coil units with central ventilation
This is a conventional form of air conditioning. Fan coils are boxes containing a fan, air heating and cooling coils, and an air filter. The coils are served with hot and cold water from central boilers and chiller plant.

Room air is continually circulated through them, being heated or cooled as required. The units are usually concealed in ceiling voids but can be floor mounted. Fresh air is normally delivered separately to the room from a central air handling unit where it is filtered and tempered, although it is possible for it to be supplied directly through units standing adjacent to an external wall.

ATM zonal air conditioning
Air treatment modules (ATMs) are large floor-standing boxes (usually housed in purpose built cupboards) which contain a number of fan coils. They can also be smaller individual units which are housed in distributed plant rooms. In both cases air is conveyed between the ATM and the rooms it serves through flexible ducting concealed in the ceiling void. Changes in room partitioning can be accommodated by changing this flexible duct configuration. The ATM can be accessed for maintenance without entering the space. Outdoor air is supplied through the ATM to rooms usually from a central air handling unit, as with conventional fan coils.

Terminal heat pump with central ventilation
Terminal heat pump units are arranged in a manner very similar to the fan coil units described above. In this case however heating and cooling of the air supply is achieved by a small reversible heat pump built into each unit. The heat pump is actually a refrigerant circuit that presents either its evaporator to the airstream to cool it or its condenser to heat it. Surplus heating or cooling produced by the heat pump is dispelled to a tepid water ring main circulating through all the units. This can balance energy use between rooms that are on opposing heating and cooling cycles, as can occur for some periods of the year. When the circulating water system becomes too cold, it is warmed by a central boiler system. When it is too hot, a central chiller plant is brought into operation.

VAV air conditioning with radiator perimeter heating
This is a conventional central air conditioning system having a mixture of outdoor and recirculating air which is filtered and cooled at a central air handling unit, from whence it is ducted to the rooms. The flow of cold air delivered into each room is varied by a local variable air volume (VAV) terminal box – essentially a modulating air damper – to match the cooling needs of the room. The main fans are controlled to adjust the overall airflow to the building based on the consensus of all the terminal boxes' demands for air. The VAV boxes and their ductwork are concealed in ceiling voids.

Variable air volume (VAV) air conditioning with terminal re-heat
This is similar to the previous system but with a local heater in each VAV terminal providing the warming, instead of a radiator system.

Fan assisted terminal VAV
In this case a fan is added into each VAV terminal to mix circulating room air with the central ducted variable flow air supply. The fan assists in maintaining a constant airflow volume and avoids some of the problems that can arise with basic VAV systems when a minimal cooling requirement can lead to problems in maintaining a good air distribution pattern.

Low temperature air fan-assisted terminal VAV
This is a development of the previous system where air from the central plant VAV system is delivered at 8–10 °C or less as opposed to 12–14 °C for traditional VAV systems. As less air is needed to convey a given quantity of cooling, this can permit smaller ducts and hence requires less service space. The low temperature air is often cooled by an ice store in conjunction with a chiller which is operated night and day either to make ice for the following day, or to supplement the cooling available from the ice. In this way the size of the refrigeration plant can be made far smaller than for alternative forms of air conditioning system.

Variable refrigerant flow (VRF) rate
This technique is becoming more popular in domestic situations. It involves passing liquid refrigerants through a distribution system. The technique is also widely known in the trade as variable refrigerant volume (VRV).

Figure 2.48 Air handling ducting installed in a very highly serviced building. The suspended ceiling has yet to be fixed

Natural ventilation offers an alternative to mechanical ventilation, with the potential bonus of providing greater occupant control; it also avoids the perceived health risks associated with some air conditioned buildings. As a result of these concerns, there are signs that, increasingly, clients and developers are seeking naturally ventilated solutions to building design. It can provide year round comfort, at minimum capital cost and with negligible maintenance[111]. (See also Chapter 3.1 for a summary of the main issues with respect to natural ventilation.)

Characteristic details

Space requirements for major items of plant (dimensions etc)

Plant room sizes typically can be around 1 m² per 25 m² of floor area served for some of the smaller installations, subject of course to a minimum smallest area for items of basic equipment. There is, though, a considerable pro rata reduction for the very largest installations.

Additional clearance may be required round electrical apparatus for safety reasons.

Small unit air conditioners, designed to be fitted to individual rooms, perhaps in window apertures, may also be encountered, although not yet on a significant scale in domestic situations.

Pipe and trunking runs

Ducts for the transmission of air at a range of velocities, in common use between the wars, were made mainly by either of galvanised sheet steel or, for the smaller diameters, in asbestos cement. It was only after the 1939–45 war that other materials, such as aluminium and, later, plastics, came into common use. At first, bends for metal ducting were fabricated from flat sheet; but flexible ducting of annular spiral-wound sheet with interlocking joints and various linings to improve thermal insulation or noise suppression, is now in common use.

Generally speaking, circular ducting will be used for the higher velocities, and rectangular for the lower but this rule is not universal (Figure 2.48).

It is not only the space taken up by the duct itself that is important but also the space around the duct necessary for access. BS 8313 gives the requirements for clear space for access[112]. Additional space may also be needed where input and extract ducts cross.

Supply air inlets and outlets

So far as terminals are concerned, a wide range of experimental conditions has been studied by BRE investigators, including different types of air supply terminals (grilles and diffusers) at various locations in rooms (Figure 2.49). The range of air supply rates and source temperatures, within which comfortable conditions can be maintained, has been determined. Broadly the results indicate that where the source temperature is 26–28 °C the diffuser should be located high up in the room but for temperatures greater than 28 °C a low level location is better. In general the type of terminal is not critical, and the effects of room size, ancillary heating and furniture do not alter the basic usefulness of warm air distribution systems used in these circumstances[113].

The relevant sections in Chapter 2.2 describe flues for the dispersal of combustion products.

Energy storage

Loads on heating and cooling plant in buildings can often be reduced, and at the same time occupant comfort can be improved, by storing excess heat or unwanted coolness within the building fabric. This excess or unwanted energy may result from diurnal variations in external air temperature or be produced by sources of incidental energy such as sunlight, people, lighting and IT equipment. A potentially attractive way of storing the excess energy is to exploit the latent heat of fusion of materials with a melting point around room temperature. These materials – referred to as phase change materials or PCMs – allow the thermal storage of a building to be increased substantially without an undue increase in building mass or volume. They were referred to in Chapter 1.2.

Figure 2.49 An air supply terminal: a form of register, set into a wall, comprising a cone which moves in and out of a nozzle to allow greater or less airflow

Main performance requirements and defects

Choice of solution for major items

For warm air heating in domestic situations, the boiler, which can be fired by a variety of fuels, though in practice mainly gas, provides the primary heat to warm the air which is then ducted to vents in the rooms. Electric systems may also be found, with a storage medium often being used to take advantage of off-peak electricity prices.

Early inexpensive warm air systems relied on in-house air circulation to heat outer rooms such as bedrooms. This never worked well and it proved far better to have ducts to every room. Sound insulation problems were certainly no worse with ducted systems. The advantages of warm air systems include rapid warm up of the spaces and no freezing. When used in conjunction with a fairly airtight house and mechanical ventilation they may allow heat recovery and, possibly, air conditioning. Domestic hot water is normally supplied by a separate system.

One of the most difficult of building types to heat properly, when it is necessary to do so, is the high rack storage warehouse. These buildings can be as much as 20 m high. Of course, the requirements for the system depend to a large degree on the requirements for temperature and humidity of the contents to be stored. The main problems with more conventional heating systems relate to possibly unacceptable temperature gradients within the high space, and a solution commonly encountered is to jet the heated air from the duct nozzles or registers in such a way that ducted air mixes with existing air in the building. One system using this method showed a variability of temperature of $\pm\,0.5\,°C$ over the whole height of the 18 m high space. Suitably sited fans could be expected to have a similar result. The power consumption required to achieve the mixing, either by high speed jets or subsidiary mixing fans, may need to be included in considerations of overall energy use.

The following points should be considered when preparing specifications for new air handling systems or altering existing systems.

- Ducts should be constructed from durable materials: any sheet steel galvanised to BS 729 minimum weight 335 g/m^2 or given two coats of zinc paint to BS 4652

- All joints in ducts and plenum connections should be mechanically locked, made airtight and the whole duct thermally insulated externally

- Registers and grilles should be manufactured in durable materials, with supply registers having opposed multi-blade dampers for balancing. Closeable dampers with shut-off and balancing actions should be provided in kitchens and bedrooms

- Registers in floor finishes are not normally acceptable

Plant can take many different forms and the choice available to specifiers is considerable. For example, absorption refrigeration equipment monitored at three different sites by BRE during 1992 each contained a different type of absorption chiller used for comfort cooling:
- ammonia/water
- water/lithium bromide, indirect fired, single effect
- water/lithium bromide, direct fired, double effect

Potential conflict between HVAC and SHEV systems

There may be conflicts between a HVAC system, as described in this chapter, and a smoke and heat exhaust ventilation (SHEV) system (described in Chapter 4.5). In fact, the relationship between the two is very complex. There is a fundamental difference between a HVAC system, which is designed to put heat into a building and a SHEV system which is designed to remove it. It is unlikely, therefore, that equipment designed solely for one system can operate for the other system also, although theoretically feasible with sophisticated building management or control systems. In non-integrated systems, therefore,

Case study

BRE tests on absorption cycle chillers
Investigations were carried out over a six month period to determine the realised performance of six relatively new commercial models of absorption cycle chillers, three single effect and three double effect. The single effect chillers consisted of a small (17 kW) gas fired ammonia/water unit and two identical medium sized (707 kW downrated to 422 kW) steam powered water/lithium bromide units. The double effect chillers were small (88 kW) gas fired water/lithium bromide units. The energy efficiency of chillers is normally rated by the coefficient of performance (COP), which is the ratio of the cooling capacity of the chiller to the input power required to operate it, at a particular set of design temperature conditions. A more informative, but seldom used measure, is a seasonal COP which indicates the average performance of the chiller over a typical cooling season. BRE measurements suggest that the seasonal COP of the ammonia chiller was 0.2–0.3, with a maximum short term (15 minutes) COP of 0.44–0.48. In contrast, the manufacturer's published performance data suggest the chiller is capable of a maximum COP of 0.5 and a seasonal COP of 0.4. The seasonal COPs of the steam driven lithium bromide chillers were 0.34–0.39 with maximum short duration COPs of 0.41–0.45. The maximum short term COP for two of the double effect chillers was 0.66–0.74. The study suggests that the way in which the efficiency of chillers is presented in order to aid selection and realise their full potential needs to be reviewed[114].

the HVAC system must shut down when the SHEV system is triggered. In addition, motorised smoke dampers should be installed in ducts to prevent the migration of smoke to undesired parts of the building. BRE experience is that this provision is often overlooked[115].

In North America, integrated HVAC and smoke control systems are commonly used, and are increasingly likely to be used in the UK for tall multi-storey buildings. These allow smoke detectors to over-ride the standard HVAC controls in order to prevent smoke movement from the compartment, or more commonly the floor, of origin to other neighbouring spaces or escape routes. This may be done in a number of ways, involving positive use of the HVAC system rather than simply closing dampers. (See, for example, *Design of smoke management systems*[116].)

Noise

Domestic warm air systems were usually considered to be noisy, partly due to the ducting throughout the installation which transmitted noise as well as the warmed air.

The following points should be considered when preparing specifications for new, or alterations to existing, air handling systems.

● Heating ducts should be designed so that they do not create noise on expanding or contracting

● Ducts should be lined internally to provide sound attenuation

Air conditioning plant, particularly small units, tends to be noisy. While some amelioration of the noise may be possible by soundproofing the plant room, individual elements cannot be treated in this way.

Other unwanted side effects

Using refrigeration for air conditioning in UK non-domestic buildings leads to the release of more than 4 million tonnes of carbon dioxide per annum. So far as domestic buildings are concerned, the market is only just beginning to grow. However, there may be other problems with domestic warm air systems. These commonly suffered from condensation on cold surfaces, caused by intermittent operation together with rapid increase in air temperature. Temperature gradients also tended to be higher than with radiator systems.

Chlorofluorocarbons (CFCs) formerly widely used as refrigerants in air conditioning systems are being phased out following the Montreal protocol. The problem is that they are ozone depletants, and they also contribute to global warming.

It was estimated in 1989 that 4,370 tonnes of CFCs were used in the UK, of which 1,360 tonnes were used in air conditioning systems. Although the majority of systems used R22, a hydrochlorofluorocarbon (HCFC) some systems at that time were still using R12 or R500; most UK centrifugal systems used R11. Many smaller commercial chillers used R12, with some use made of R500 or R502. Most larger commercial chillers and some supermarket systems used R22. The majority of domestic refrigeration equipment used R12.

Regardless of the environmental issues, changes to the Montreal Protocol and EC regulations made it prudent not to specify CFCs and HCFCs because of likely problems with their future availability, and so systems using these refrigerants are no longer being specified. This avoids the need for premature replacement or conversion of refrigeration plant. Although it was thought at one time that some CFC refrigerant would continue to be available from recovery and recycling of the CFC stockpile, this now no longer seems to be the case, and existing systems need to be converted to use alternative refrigerants.

One possible alternative, R22 (HCFC22, ODP = 0.05) was at one time considered as the first choice transitional refrigerant for chillers in building services, indeed it is currently the most widely used refrigerant in the UK, but it now seems as though this substance has already been banned in parts of Europe. Other environmentally 'safe' alternatives, such as absorption chillers, should also be considered. R134a and other HFCs[117]. were once considered to be environmentally safe alternatives, but concern about their high global warming potentials mean that they are unlikely to provide a long term solution.

It is now expected that hydrochlorofluorocarbons (HCFCs) will also be phased out soon. Ammonia will be considered for some new air conditioning installations as it has the advantage of having a zero ozone depletion potential, as well as being an efficient refrigerant. However, because special safety measures are required, being toxic and inflammable, ammonia is not always a realistic option. One of the hydrocarbons such as propane may be a possibility, but these too are inflammable[117].

The substitution of CFC and HCFC refrigerants has been examined in BRE Information Paper IP 6/98[118]. This paper also describes some case studies.

Commissioning and performance testing

Since many air conditioning systems are commissioned by means of performance specifications, it is necessary to proof test them after installation for conformity with the specification. Much of this will of necessity be carried out in situ, but type testing of individual parts or subsystems can of course be done in the laboratory (Figure 2.50).

Air conditioned buildings tend to be deficient in short term response times in alleviating of discomfort conditions, which may help to explain why occupants often seem less satisfied with them than might be expected. With suitable management, today's controls can already deliver high levels of comfort and energy efficiency. However, they and the systems they control are often too complex for the average user, and need to be designed for better and easier manageability[119].

Externally, intake and outlet positions must be sited to avoid pollution and be covered in mesh against entry of birds etc. Inside the building, grilles must be sited to avoid short circuiting and draughts, and to heat the whole space. Extract grilles should incorporate a dust filter and, when fitted in a kitchen, a grease filter.

Health and safety

Occupants of a building should not suffer discomfort from airflow from heating registers. Since the temperature of the registers will be directly affected by the temperature of the air passing through them, the temperature of discharged air arguably ought at no time to exceed a suitable maximum, say 60 °C. Air should still be adequately mixed to achieve required room temperatures.

Legionnaires' disease

Legionnaires' disease gets its name from the American Legion Convention of 1976, when a number of legion members died from what were then unknown causes. After intensive research the cause was identified as the bacterium now called *Legionella pneumophila*, a strain of bacterium found in all waters, though usually at low concentrations[120]. The risk of infection depends on personal susceptibility, though the incidence might seem to be comparatively low with the number of cases reported annually in the UK in the low hundreds. On the other hand, these figures might need to be increased tenfold to account for under-reporting. The bacterium is carried in aerosol emissions, largely from defective or unmaintained air conditioning plant and domestic water systems. Infection is by inhalation of those aerosols.

A Public Health Laboratory Service survey found the bacterium was present in 50–75% of the hot and cold water services in the 180 buildings of various types that were examined – mostly in hot water systems, particularly in calorifiers, and large calorifiers at that! The bacteria grow quickest in temperatures of 20–46 °C[121].

Regular chlorination at 1–2 ppm is *'the most consistent and effective treatment for legionella'*[122,123].

Figure 2.50 The air conditioning systems test facility under construction at BRE

Figure 2.51 Monitoring biological particles, including airborne bacteria, fungi and house dust mites in a domestic environment

Sick building syndrome

Sick building syndrome (SBS) is a phenomenon experienced by people in certain buildings. The key symptoms are irritation of the eyes, nose, throat and skin, together with headache, lethargy, irritability and lack of concentration. Although present generally in the population, these symptoms are more prevalent in some buildings than in others, and diminish over time when the afflicted person leaves the building concerned. The syndrome can be discriminated from other building-related problems such as physical discomfort, infections and diseases resulting from long term cumulative hazards such as asbestos and radon. SBS is most often, but not exclusively, described in office buildings, particularly those that are air conditioned.

It is easy to regard SBS symptoms as minor since, apparently, no lasting damage is done. The symptoms are not trivial to the people who experience them on a regular basis at their place of work. Apart from being distressing to people experiencing the symptoms, SBS reduces productivity, increases sickness absence and is generally debilitating. Other likely effects are on unofficial time off, reduced overtime and increased staff turnover. If a building

gets a reputation for being 'sick' it can be difficult to rehabilitate its reputation, even if the building itself is improved. In extreme cases buildings may close for a period or even be abandoned.

The evidence is that the numbers of buildings and people affected are not small. A study carried out in the late 1980s[124] which questioned over 4,000 workers in 46 office buildings built during the 1970s found that more than 80% reported symptoms of illness – lethargy, irritation of nose throat and eyes, and headaches – which they associated with their working conditions. It was reported that naturally and mechanically ventilated buildings had fewer complaints than air conditioned buildings. Currently, air conditioned buildings are generally associated with higher prevalences of SBS: in the UK approximately 55% of staff in such buildings are affected – many only mildly, but to the extent that they perceive a negative effect of the office environment on their productivity.

The variation in symptoms during the course of the working day and week, and the slow increase in symptoms when a person starts work in a building, effectively rule out infection as the mechanism; it is more likely that a combination of

allergens, plus irritant or toxic effects, account for the symptoms. A contribution from the physical environment (eg temperature and humidity) must also be considered. In fact a very wide range of possible causes has been suggested. Current evidence indicates that no single factor can account for SBS. In all probability there is a different combination of causes in different buildings.

It can be helpful to view the causes of SBS at four levels.

● The building: many aspects of the design, construction and location of the building, and its services and furnishings, may contribute to SBS in a wide variety of ways – from the site-dependent microclimate, through shell design (ie depth of space and floor-to-ceiling heights) to the services and fitting out

● The indoor environment: the effects of the building and site will generally be mediated by the indoor environment

● The organisation: an organisation which occupies and operates a building may contribute to SBS; for example, via the quality of building maintenance and work force management

● The individual: reported experience of SBS varies from one person to another within a building for a number of reasons which would include personal control over the environment, constitutional factors, behaviour, and current mental and physical health

Ventilation systems

Air conditioning (where there is mechanical cooling of the indoor environment) is not the cause of SBS since it is possible to have healthy air conditioned buildings. However, it is almost certainly a contributory factor in many buildings, and the basic fact is that air conditioned buildings do have a higher overall prevalence of SBS symptoms.

There are many reasons why air conditioning could, in theory, result in SBS: systems which are malfunctioning, badly maintained, badly designed or ill controlled can reduce the amount of outside air supplied to a level below a specified minimum, and the distribution of fresh air within the occupied space may be inadequate. Ill maintained systems might cause hygiene problems. The fact that air conditioning often places control of ventilation and temperature out of the hands of the occupants is also a common source of complaint. Recirculation of return air need not cause problems and it need not be present for SBS to occur. It does however present an additional risk of exposure to pollutants and of system malfunction.

Ventilation

Air conditioned buildings generally have higher air change rates than naturally ventilated buildings. There is little evidence that symptoms are reduced by increasing the ventilation rate if it already meets current guidelines (approximately 10 l/s per person, in the absence of tobacco smoking). Poor distribution of air within the occupied space may be more significant than the ventilation rate itself.

Temperature and humidity

SBS gets progressively worse as temperatures increase, although the relationship has not been well quantified. Keeping to about 21 °C should minimise this risk but it is important to remember that air conditioned buildings tend to have higher symptom prevalences. Risks also increase at low humidity but again the quantification is

incomplete. There is little point in raising relative humidity (RH) to a level higher than 40% to prevent SBS, and 20% RH is probably too low; the effects over the intervening range are unclear. Humidification systems can themselves present a risk if they are not maintained in a hygienic condition. Temperature, humidity and air movement can also alter the rate of emission and deposition of pollutants from materials in the building, and from people.

Air pollutants

Many non-biological pollutants can be present in buildings, although generally at low concentrations except in some industrial workplaces. There is concern that mixtures of many pollutants, each at low concentration, may have adverse effects that are not fully understood or predictable, but hard evidence is lacking. Symptoms are more evident where there is tobacco smoking, but evidence on other pollutants is that only some cases can be attributed to individual air pollutants. However, routine checks for organic gases and vapours, particulates, combustion products and ozone, can pick up these problems. There is some evidence to associate SBS with viable airborne particulates (eg pollens, fungal spores or fragments, bacteria and viruses) but these cannot always be implicated in individual cases.

Indoor surface pollution (ISP)

ISP refers to particulates and adsorbed vapours present on or in surface materials (eg desk tops, chair covers and flooring materials). The level of particulates to which office workers are exposed can be much higher than ambient airborne levels, since people create their own 'dust cloud' by stirring up settled dust in the course of their work. Particulates should not be regarded as simply air pollutants, effective only when inhaled, since they can be transferred directly to the skin or ingested with food or drink; these aspects have hardly been explored at all. ISP is still not widely known as a cause of SBS but the evidence for this potential

cause is at least as good as the evidence for any other cause. For example, controlled experiments show that intensive office cleaning can reduce the prevalence of SBS symptoms, and that this is in some way related to allergens (including dust mite allergens) present in ISP. Apart from the implications for office cleaning, it may be that the relatively constant environment in air conditioned offices promotes viable conditions for mites and micro-organisms generally.

Light and noise

Although natural light is usually preferred, and people seated near windows tend to have fewer symptoms, the effect of light on SBS has not been established. Similarly, although the capacity of noise to cause annoyance, distraction, tiredness and headaches is accepted, there is little evidence that 'sick' buildings are more noisy than 'healthy' ones. Infra-sound is unlikely to be a serious problem at the levels encountered in offices, but could contribute to symptoms in some buildings. Working at a VDU has been found to be correlated with experience of SBS symptoms but, again, the reason for this correlation has not been established.

Individual control of environments

One clear factor in the incidence of SBS is perception of lack of personal control over the physical environment or the work environment. Currently available evidence does not clearly establish whether this is because controllability is an important variable in its own right or if control gives individuals the ability to create environments which suit them. The notion of control extends to privacy and the number of people sharing an office.

Management and organisational factors
To claim that bad management causes SBS can be seen as either claiming an obvious truth (ie problems in the workplace are always due to bad management) or propagating an unfair nonsense (ie it is not the management, it is the environment in the building). The correct balance between these views can only be struck by establishing in specific terms what management could have done which would have avoided the problems. Broadly speaking, management can be seen as contributing to SBS if it does not act effectively to create a good indoor environment or if it does not establish a good organisational environment for avoiding stress and for dealing with complaints.

Characteristics of individuals
SBS depends on a number of individual characteristics, being more likely to be reported by women, by staff employed in more routine, low-paid jobs, and by those with a history of allergies. These characteristics should not be seen as causes but as factors which make an individual either more sensitive to the environmental challenges which cause SBS or more likely to be exposed to those challenges. Psychological factors are generally more likely to affect the reporting of symptoms rather than the symptoms themselves.

Pest infestation
Rodents
There is a risk that rodents can enter buildings and, if suitable precautions are not taken, gain access to various parts of the buildings through holes in walls for trunkings, and even inside the trunkings themselves.

Ducts and trunkings should be tightly built in wherever they pass through walls, floors, ceilings and foundations. Seals should not be so tight that they can cause noise from thermal movements and vibration. If possible, ducts and trunking should also be compartmented with wire mesh where they pass through partitions. Rodents are unlikely to gain access through or into partitions if the diameter of the holes is no more than 5 mm greater than that of the ducts passing through.

Cockroaches
Cockroaches are identified as one of the pest species of most concern to the health of building occupants, particularly the accumulation of persistent cockroach allergens in dwellings. Higher internal temperatures in buildings can be expected to lead to certain species of cockroaches breeding at faster rates. This could therefore lead to more severe infestations in future.

There is further evidence that cockroaches pick up and carry micro-organisms on their body surfaces and inside their guts, and are capable of spreading these organisms throughout a building. They may also carry bacteria. Allergic reaction to cockroaches is one of the two most common allergies in people with asthma, the other being to house dust mites. Cockroach allergens have high potency in producing histamines: some reactions have been reported which can be life threatening.

Cockroach infestations are most prevalent in metropolitan areas and multiple occupancy dwellings, including tower blocks. Building design features such as ducted services and voids appear to provide harbourages which encourage the establishment and spread of cockroaches, and may hinder disinfestation.

Good design and management of buildings are important in preventing or restricting cockroach infestations. Avoiding repeated applications of insecticides may inhibit them in building up resistance to treatments. Cockroaches are very dependent on moisture for survival, so improved ventilation and heating regimes, sealing access routes and voids, maintaining plumbing systems, and taking adequate measures to prevent condensation within harbourages are important means of reducing the risk of infestations.

Mites
The growth in populations of house dust mites depends on temperature and humidity in their micro-environments, and on the age, use and cleaning of soft furnishings, especially mattresses. In the UK, house dust mites are most numerous in the summer but the allergens they produce can persist year-round.

Allergens from mite faecal material can trigger allergic reactions. The most important of these is asthma, but mites have also been implicated in perennial rhinitis, eczema and sick building syndrome. In the case of asthma, there is clear evidence that mite allergens induce sensitisation in susceptible individuals and also cause acute (sometimes severe) reactions in sensitised people. Mites are less visible than moulds, but probably constitute a more important public health issue.

Most approaches to controlling mite populations include cleaning: either intensive household cleaning or special techniques such as steam cleaning or liquid nitrogen treatment. Alternatives include physical barriers (eg mattress covers) or removing unnecessary soft furnishings (eg carpets). Chemical acaricides, sometimes combined with a particle aggregating agent, can improve the success of cleaning. There has not been total success with any approach, and eliminating mites does not immediately reduce exposure to the allergenic faecal particles. These measures are clearly outside the scope of building standards.

Mites do not survive low humidity; it has been suggested that, where feasible, indoor absolute humidity should not exceed about 45% RH at room temperatures to prevent proliferation of mites. The critical level can be achieved by using mechanical dehumidifiers, but there can be problems in practice with these devices; for example, the background noise and their reduced effectiveness at lower indoor air temperatures.

Mechanical ventilation (MV) systems are better at reducing winter humidity in homes, compared with windows, by increasing air change rates and removing moisture. The reduction in humidity has been associated with a reduction in dust mite numbers in experiments in Denmark. Widespread application in the UK would require evidence from UK homes that full MV systems are more effective in controlling humidity than the current requirement in new homes for extract ventilation only, and that the systems are used as intended and properly maintained over the long term.

Durability

Some unpublished BRE studies dating from 1980 put the lives of air conditioning equipment in large installations at around 15 years, with 10 years for smaller units; there were differences, though, for various parts of the installation. It is not thought that longevity has changed much since the 1980s.

Maintenance
Legionella

For the prevention of growth of legionella, it will be necessary to maintain vigilance, with possible use of biocides, particularly in cooling towers, and maintenance of the seals of water filled traps to prevent aerosol suspensions being emitted to the atmosphere. Specialised cleaning services are available. Cooling towers should be kept in regular use, and will need disinfecting and cleaning if out of use for more than one week.

Sick building syndrome

Finding a cure for SBS depends on appreciating the range of possible causes and the information given earlier (pages 86–88). Simplified practical guidance for employers has been provided by HSE[125]. Ideally, SBS should be dealt with by prevention rather than cure and thus by addressing the process of building design, construction, installation, commissioning, operation and maintenance.

In existing buildings, the first step should be diagnosis. SBS is a complaint of people, not buildings, and can only be diagnosed by assessing the building occupants, not by examining the building itself. By using a standardised questionnaire and staff survey method, the results of a survey can be compared with a wider database. The ideal is to use proactive monitoring of buildings to head off any problems before they become unmanageable. Complaints can be characterised and put into perspective.

Until a more complete picture of the causes of SBS emerges, action should be taken on the basis of existing knowledge. While it is difficult at present to be definitive about the causes of SBS, indoor climate 'risk factors' as already described can be identified and avoided. The evidence that these risk factors are direct causal agents is limited. It is also important to remember what may be behind a poor indoor climate, including those factors listed in the 'Inspections' checklist at the end of this chapter.

Problems with these factors will sometimes be obvious, but the existence of a problem does not guarantee that removing the problem will reduce the prevalence or severity of SBS symptoms. In fact there are no easy answers. There are many possible causes of SBS and they are interrelated and interactive. SBS is a multifactorial problem which demands a multidisciplinary approach: a comprehensive view and systematic checking of possible causes, and solutions that are designed to be applied to specific problems and conditions.

Plant

Full environmental servicing systems are almost bound to call for more maintenance than other partial systems, and are also likely to call for complete renewal more than once during the life of the building carcass. Though it is clear that the need for servicing should be minimised, the chief requirement is ease of access with as little disturbance as possible to the normal use of the building.

Wherever there is a likelihood of future extension or upgrading, the

Figure 2.52 Air handling equipment retrofitted to laboratory accommodation. Although this equipment is easy to maintain, it cannot of course be placed in locations accessible by the public

design of the building should recognise that possibility. This avoids disproportionately costly changes to the structural or spatial enclosure system at a later date. Whether this was done in older buildings is doubtful, for who could have foreseen the phenomenal growth in demand?

Ducts are commonly of thinner construction than carcassing, and are hardly ever properly sealed to walls through which they pass. For junctions in timber suspended floors, small access traps can be cut with an annular plane; large ones would need to be framed.

The reduction of carbon dioxide emissions through the optimisation of building services calls for an increase in the efficiency of both the plant and the controls. Major contributions to high energy consumption are wear and tear of plant, poor maintenance, and failure of parts of the air distribution system which do not significantly impinge on plant output. Sticking dampers and faulty sensors might not be considered serious problems by building services engineers; however, they may affect the efficiency of the system.

As part of a programme which looked at alternatives to the vapour compression refrigeration cycle in plant in the early 1990s, BRE reviewed the factors that operators and maintainers of absorption chillers considered to be the main obstacles to the future use of this technology. The review was based on experiences with older generation chillers. The main obstacles to the uptake of absorption chillers were found to be:

● a shortage of building services engineers who fully understood absorption chiller technology
● a perception that there would be long waiting times between fault identification and rectification
● a widely held view that the inherent energy efficiency of the absorption cycle is poor[126]

Up to 75% of UK refrigerant sales is believed to be used for plant servicing, which implies a high rate of loss. Refrigerant losses can occur through operational leakage, plant failure and during maintenance. Stringent precautions must be made to minimise these losses. Relevant codes of good practice should be followed. In particular, precautions should be taken in the areas of plant design, maintenance and decommissioning[127].

Motorised smoke dampers in ducts need to be tested regularly to check that they will close in fire conditions, and reopen successfully after the test.

Work on site

Inspection

The problems to look for in relation to plant include:
◊ inadequacy of regular maintenance
◊ missing equipotential earth bonding in steel duct frames
◊ vermin in the ducts
◊ noise from vibrating ducts and plant
◊ air output from grilles is too hot
◊ draughts and poor mixing of incoming air
◊ missing motorised smoke dampers in ducts
◊ defective condensation drains on heat exchangers
◊ air intakes for HVAC systems too close to boiler flue terminals

The particular factors to look for in relation to sick building syndrome include:
◊ deep building plans
◊ open plan offices of more than about 10 work stations
◊ changes of use of building and office partitioning after commissioning
◊ sealed windows
◊ large areas of soft furnishing, open shelving and filing
◊ new furniture, carpets and painted surfaces
◊ lightweight thermal properties
◊ poor insulation
◊ poor provision for daylighting
◊ uncontrolled solar gain
◊ no separately ventilated spaces for smoking, photocopying etc
◊ services not designed for easy maintenance
◊ air inlets close to exhaust or outdoor pollution sources
◊ inadequate air filtration
◊ luminaire types and positions giving high glare and flicker
◊ inadequate commissioning and, when necessary, recommissioning
◊ poor building services maintenance affecting hygiene and operational standards
◊ insufficient office cleaning
◊ poor general management (eg of staff complaints)
◊ low general satisfaction of staff with their jobs and their organisations

Chapter 2.5

Geothermal and heat pumps

These two topics are taken together because it is usual to find heat pumps to be associated with geothermal low grade heat sources.

A heat pump is a *'machine operating on a reversed heat engine cycle to produce a heating effect. Energy from a low temperature source, eg earth, lake or river, is absorbed by the working fluid, which is mechanically compressed, resulting in a temperature increase. The high temperature energy is transferred in a heat exchanger'*[128]. There is another similar but not identical definition in BS 6901[129]. The operation of a heat pump has been likened to that of a refrigerator, though saving and using the heat instead of throwing it away.

Although the principle of the heat pump was identified during the nineteenth century, heat pumps were not developed on a practical scale until the 1930s. Experiments continued following the 1939–45 war and a few buildings were later equipped with heat pumps, though it is understood that none reached performance expectations[18].

There are comparatively few installations in the UK, but commercial heat pumps suitable for use in dwellings are currently more widely available in France and Germany.

Since the 1950s, experimental work has continued, and the principle has always been considered to be a viable long term proposition. The Royal Festival Hall was originally equipped with a heat pump installation taking low grade heat from the River Thames, though it proved uneconomic and was subsequently abandoned.

Heat pumps suitable for domestic heating have been investigated by BRE, but have not found widespread application for domestic heating. They were used experimentally in the late 1970s for reclaiming heat from waste water, and others were used for both space and water heating and ventilation air heat recovery. The most successful application was an electrically driven heat pump using air as its heat source[130].

However, heat pumps have been successfully used in particular applications such as office blocks. In one office building, both heating and cooling was provided by two heat pumps and thermal storage, and local control systems provided lighting, time and temperature control. In another case, energy was extracted overnight by heat pumps operating on off-peak electricity, and stored in a thermal storage vessel. The heat was then circulated during the day through a secondary pumped water circuit[131].

Heat pumps are likely to be most cost effective and have the most promising markets in commercial premises and dwellings with high hot water use[132].

Figure 2.53 Heat pump installation in the Integer House on the BRE site, drawing energy from 50 m underground

Heat pump at a swimming pool

Monitoring over 12 months showed that the performance of a retrofitted gas engine driven, heat pump dehumidification system for a 508 m² swimming pool had more than exceeded design expectations by achieving a 70% saving in heating energy compared with the system formerly in use. Further assessment revealed that future gas engine heat pump systems using fresh air ventilation could offer even greater savings.

Analysis indicated that only a gas engine heat reclaim heat pump, or an electrical equivalent, could have saved more energy; the gas engine heat pump system gave good performance and financial return on investment. Moreover a relatively short pay-back period of 4.8 years on the capital invested was predicted.

Heat pump in a school

An air-to-air heat pump was installed for space heating of a classroom. Its operation was monitored by BRE investigators over two heating seasons and the seasonal coefficient of performance (COP) was found to be 2.06. The major cause of this low COP was cycling. Despite this, it was a cost effective alternative to storage radiators, the internal rate of return being 11.4% assuming no future increase in energy costs in real terms. Though some sound insulation was fitted, the machine was too noisy and cold air was delivered to the classroom during defrosting. It was considered that these problems could be overcome and the COP increased by using purpose built machines.

Characteristic details

Fuel

Work done in the mid-1980s on domestic heat pumps showed that the market had been dominated up to then by the electrically driven air-to-water types. However, most of the dwellings suitable for heat pump installations have been heated by gas, oil or solid fuel. BRE measurements showed that the electric heat pump could not compete with gas heating and, although just competitive with oil, solid fuel and electricity, the performance advantage offered to offset the extra capital expenditure was likely to be marginal in most domestic installations[133].

Main performance requirements and defects

Outputs required

Most applications have achieved a coefficient of performance (COP) of around 3; that is to say, the output is something like three times as great as the input.

In the mid-1980s BRE studies found that savings in the running costs of electric heat pumps, when compared with most conventional central heating systems, did not justify their extra capital cost. Once a 40% reduction in capital cost, or a 43% increase in performance, or smaller improvements in both are achieved, the heat pump will be cost effective as a replacement for oil fired heating. The fact that the market has not yet responded seems to indicate that these conditions have not yet been achieved[134].

Noise and other unwanted side effects

Noise, as a deterrent factor, is probably a main consideration in the operation of heat pumps and needs to be checked with manufacturers before selection.

Commissioning and performance testing

The capacity and performance tests for air-to-air and air-to-water heat pumps are given in BS 6901.

Durability

Durability depends on the original type and quality of pump, and use which the installation receives. Undoubtedly there have been considerable disappointments with a number of installations and, to date, most of these have been treated as being experimental. Available figures indicate that lives of around 10–15 years may be achieved.

Maintenance

Since heat pump installations are relatively uncommon, it is essential that routine maintenance is put in the hands of competent and experienced firms.

Work on site

The problems to look for are:
◊ missing manufacturers' handbooks
◊ missing equipotential earth bonding in steel duct frames
◊ noise
◊ low coefficients of performance

Heat pump at the Royal Botanic Gardens, Kew

The Sir Joseph Banks Building at The Royal Botanic Gardens, Kew, uses a heat pump system to drive its air conditioning system. The system operates during the summer months to extract heat from the building and during the winter as a heat pump to extract heat from outside the building for use within the building interior. The decision to use a heat pump was strongly influenced by the availability of an excellent low grade heat source in groundwater flowing in the gravel beds around 3–4 m below the site. During the winter, heat is extracted from one well and discharged at a lower temperature into another well. During the summer, this process is reversed, with water being extracted from the latter well and discharged at a higher temperature to the former.

The heating system within the building uses low temperature fan driven convectors since the temperature of the water is very much lower than with boiler driven systems. The performance of the system was monitored by BRE in conjunction with the former Property Services Agency. The coefficient of performance (COP) of the system is greatly influenced by the costs of running the pumps to circulate the considerable volumes of groundwater involved.

Chapter 2.6

Solar energy

Solar energy has both beneficial and adverse effects, but the beneficial effects normally predominate in the UK. It is described in this chapter mainly in relation to water and space heating. Photovoltaic systems which generate electricity are also included in this chapter although they have only a superficial similarity to collection panels for heating water.

Solar energy can provide free natural ventilation, daylight and heat. Its effective use, in refurbishment and in new buildings, creates good working environments and reduces fuel bills, and so minimises global environmental impacts[135]. BRE has long been involved in the development of methods of harnessing this energy, and several experimental buildings were constructed on its site at Garston from the late 1970s onwards.

The intensity of solar energy at the earth's surface is about 1 kW/m². Solar radiation is conventionally divided into the short wave range, 0.29–4 μm (micrometre –1 μm = 1/1000 mm), and long wave, 4–100 μm. The short waveband contains about 99% of the solar energy reaching the surface of the earth. About half of this is in the visible range with wavelengths of 0.4–0.7 μm (Figure 2.54, and *Roofs and roofing*, Chapter 1.5).

Not all of this energy is, of course, usable. Average usable daily totals for the UK vary throughout the year, ranging from 4–5 kWh/m² in summer to 0.2–0.6 kWh/m² in winter.

Adequacy of sunlight provision

Sunlight has long been considered important in the design of UK housing. The criteria and design tools used in the end of the twentieth century were most likely based on *Sunlight and daylight: planning criteria for the design of buildings*[136]. This contained some guidance on the penetration of sunlight into the spaces around buildings, as well as on the insolation of rooms. The methods were based on potential sunshine amounts (as though the sky was always clear) rather than the more realistic probable sunshine, although it could be argued that the planning criteria took this into account.

Obstruction of solar access should be considered in broad terms, taking account solar altitudes and azimuths, land forms and slopes, trees and overshadowing from objects outside the site boundary. Within these constraints, forms and arrangements of buildings giving high levels of useful solar gains can be explored. For housing, this can have a great impact on the pattern of subdivision into plots and their relationship to roads and services. For complex arrangements of large buildings, the juxtaposition of built forms on the site may be the dominant factor.

A helpful starting point for design, but not necessarily the only criterion for solar access, is the north-to-south spacing of buildings in relation to their heights and any ground slope. Other things being equal, north-to-south spacing needs to be greater at higher latitudes where solar altitudes are lower. UK mainland latitudes range from 50–59°N; at these extremes, maximum solar altitudes at the winter solstice (December 21) are 16° and 8°. For parallel, east-to-west terraces or blocks, it will usually be impracticable to design for significant solar access on this date[137].

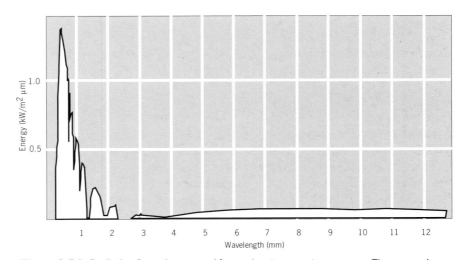

Figure 2.54 Radiation from the sun and from a low temperature source. The curve shows the distribution of energy in the solar spectrum at sea level. The dips in intensities are due to selective absorption by substances (eg particulates) in the atmosphere

Figure 2.55 A conservatory can provide useful passive solar gains, heating the internal structure of the building for re-emission during the hours of darkness

Collection of solar energy

The energy emitted by the sun can be collected and used both passively and actively in several ways.

● Using the so-called greenhouse effect. The sun is allowed to shine through a glazed wall or roof, striking a floor or wall within the space enclosed, which is then heated. Some of this absorbed heat is re-radiated from the surface, but, because surface temperatures are quite low, re-radiation takes place in the long waveband. There is a buildup of surface temperature as solar energy is absorbed in excess of long wave emission from that surface. The air within the space is then heated by contact with the warmed wall and floor surfaces. Of course, these surfaces cool down during the night, but, if they are of sufficient mass, will still retain some warmth

● Using water filled solar collectors

● Using photovoltaic (PV) cells

PV systems have no moving parts, and are non-polluting. The main drawback to their wider application is cost.

The other side of the coin is that too much sunshine might have adverse effects, and these also are explored in the following sections.

Characteristic details

Choice of systems for heating
Passive

Heat gains from sunshine are not the result of other energy use and ideally, therefore, should be made as large as possible whenever space heating is required. The benefit that can be derived from solar heat gains is determined by the design of the building as well as the climate of its location. This approach is often referred to as passive solar design. Passive solar design sets out to make the best use of available sunshine through controlling the form and fabric of the building.

Solar radiation can be a significant source of heat gains, even in an existing house in the UK climate. In a new building, the designer can

deliberately set out to minimise artificial heating requirements by controlling the area and orientation of windows and by the use of special architectural features such as conservatories and other judiciously placed glazing. At the simplest level, it means ensuring that glazing is predominantly on south facing aspects and limiting overshading from adjacent buildings. It can often be done without incurring significant extra cost. It is beneficial when applied to the layout of a new estate development where there may be opportunities for influencing orientation and positioning of dwellings, but there may also be opportunities available in rehabilitation work.

Passive solar designs which have very large areas of glazing need to be carefully analysed to ensure that they do not cause overheating in summer. Overheating may be avoided by limiting the amount of glazing and by the appropriate use of shading, ventilation and thermal mass for storage purposes.

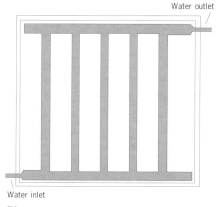

Water outlet

Water inlet

Plan

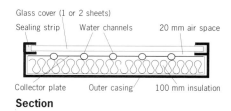

Glass cover (1 or 2 sheets)
Sealing strip   Water channels   20 mm air space

Collector plate   Outer casing   100 mm insulation

Section

Figure 2.56 Typical solar water heating panel

Active

Active solar heating equipment takes two main forms:

● water filled solar panels feeding heat directly into the building's space heating or domestic hot water heating systems
● photovoltaic cells feeding electrical current into the building's electrical system

A typical collector module is shown in Figure 2.56. Short wavelength solar radiation falling on the collector panel is transmitted through the glass and absorbed by the black surface of the collector plate. This raises the temperature of the plate and any fluid circulating through the water channels. Most of the long wavelength radiation emitted by the plate when warm is retained within the assembly because glass is a poor transmitter of this radiation. Some heat is, however, lost in this way and some is lost also by conduction through the insulation behind the plate. The operating efficiency of a solar panel decreases, then, as the plate temperature rises.

A simple system using natural gravity circulation is shown in outline in Figure 2.57. As the temperature of the water in the panel rises, its density decreases, establishing a circulation head and progressively heating the water in the cylinder. Circulation is either by gravity or by pump; the latter gives more freedom in positioning the storage tank since, for the former method, the panel has to be sited below the cylinder.

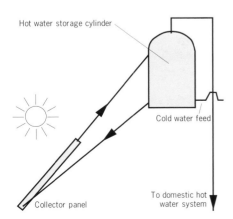

Figure 2.57 A simple solar collection system

The collector plate is usually made in steel, copper or aluminium with the water channels formed by pressing from integral sheet, with or without attached welded tubes.

Photovoltaic installations produce low voltage direct current (DC) from panels which can be supported independently on the building or as an integral part of its surface. To power appliances the electricity needs to be converted to alternating current (AC) using an inverter. Some installations now generate sufficient power to make it worthwhile to feed the excess into the National Grid, though the regulations and other criteria which need to be complied with are considerable.

PV systems have been installed in many different kinds of buildings in the UK, and experience in their operation is growing. There are several different kinds of materials forming the active layer of collectors, mainly based on compounds of silicon.

Most installations are fixed rather than being tracked or moveable to follow the sun. They will also have been specially designed, though smaller installations may be available as ready designed and manufactured packages.

Storage devices

Hot water is usually stored in a cylinder. A conventional indirect cylinder of about 200 litres capacity will be suitable for a domestic installation with about 4 m² of panel; the cylinder and all connecting piping should be well insulated. Cylinders are available with two indirect coils, the lower for solar heated water and the upper for connection to a conventional boiler. As in a domestic central heating system, provision must be made for expansion of the water. This may involve a header tank at the highest point in the system or a diaphragm-contained expansion cylinder. The latter is normally the easier to install when panels are roof mounted and, for a domestic system, requires a capacity of 2–5 litres, depending on the system water capacity.

Figure 2.58 A conservatory offers one useful means of exploiting the heat of the sun

Some passive solar installations require massive thermal storage to even out the gains during the day and losses during the night. The most quoted example is the Trombé wall, named after the Frenchman who first exploited the principle in buildings on a substantial scale. The principle, which takes the form of massive masonry soaking up heat during the day and gently emitting it at night, was adopted by Victorian gardeners whose greenhouses were normally built against substantial south facing perimeter walls. Modern conservatories use these same principles (Figure 2.58). More heat is gained to the interior through the glass than is subsequently re-radiated back through it, unless it is removed by ventilation of the heated air.

Solar shading devices

Brief mention of solar shading devices was made in Chapter 1.7 of *Walls, windows and doors*. Although solar gains may be beneficial, there are also some potentially adverse effects from too much sun:

● overheating
● glare
● fading of furnishings

Shading may be fixed or adjustable, and will normally consist of one or more of the following types[138]:
- external shading by canopies, overhangs, shutters and screens
- modified glazing by tinting, reflecting or absorbing
- blinds or louvres fitted into the glazing (between sheets)
- curtains, blinds or louvres fitted internally against the glazing

Some mention of shutters was made in *Walls, windows and doors*, Chapters 5.3 and 10.3, although in neither of these cases was the primary use for solar control purposes, as used in summers in continental Europe.

Tinted glazing may be:
- absorbing (body tinted) or
- reflective (coated)

A wide variety of colours is available, some of which may be imperceptible unless and until compared with a view through an open window.

Comparatively recent innovations have been in the development of the so-called 'smart' glazings:
- electrochromic and liquid crystal glazing, which darkens when electrical current is applied
- photochromic glazing, which darkens under sunlight
- thermochromic glazing, which turns milky when it warms up

Figure 2.59 Photovoltaic installation on the BRE Environmental Building. A comparatively large surface area is needed to produce a significant amount of power, especially during winter conditions

All three types can control solar gain, but admit daylight and useful winter solar gains, but they are not effective in controlling glare.

Main performance requirements and defects

Outputs required
Heating
Hot water heating panels should ideally face the noon sun, but their exact orientation is not critical, any convenient direction in the SE–SW quadrant being satisfactory; shading by adjacent buildings or trees should be avoided if possible. The panels should slope at 20–60° to the horizontal, but, again, the exact angle of inclination is not a critical factor. A normal domestic installation will require 4–6 m^2 of panel area or about 1 m^2 for every 50 litres of daily hot water demand. The total panel area is the most important factor in determining the heat collected annually. A correctly designed solar system will even provide water on a few days in summer at the required tap temperature, say 55 °C, and no further heating will be required.

The British Board of Agrément publishes two Methods of Assessment and Test relevant to solar collectors[139,140].

Under maximum radiation, domestic PV systems produce peak outputs in the range 2–5 kW for each system, while larger installations for non-domestic applications in the range 10–100 kW. The area required will vary according to design, with an average of around 6–10 m^2 per kW, depending on orientation, inclination, efficiency of the system etc[141]. Power output is variable, with peak outputs during the summer in cloudless conditions, but low outputs, as is to be expected, during winter. As one example, the array on the south facing wall of a 1997 BRE office building produces around 0.5 kW even on cloudy days in midwinter, although this is only a fraction of the total requirements of the building (Figure 2.59).

Shading
To reduce solar overheating, a shading device should have a low total solar transmittance (TST) value which can be found quoted in commercial literature for various forms of glazing. Since the need for protection will vary with the seasons, some form of adjustable shading may be the best option, although potential durability and maintenance problems must be anticipated and taken into account. Fixed shading, for example an overhang, can offer protection against high angle summer sun, but lets through low angle winter sun which may be more beneficial.

Manual control of shading devices is normally specified in dwellings and offices, but automatic control is more usual in larger buildings.

Unwanted side effects
Reverse flow in a solar heating system
Circulation in a hot water collector panel occurs only when solar input is sufficient to heat the water in the panel to a temperature greater than the water in the cylinder. Provided the bottom of the cylinder is at least 600 mm above the top of the panel, there should be no reverse flow when there is no solar input.

Overheating
Overheating of internal spaces is most likely to be a problem where:
- windows are large and face the southern half of the sky
- rooflights are horizontal
- the building has high internal heat gains
- the building needs to be cooler than normal

External shading is nearly always better than internal in preventing unwanted heat gain. Some electrical controls operating through a BMS may introduce unacceptable delay in response to need, and this may prompt users into interfering with the system.

Glare

If direct sunlight causes problems of glare, tinting the glazing may not help much because the sun is so bright. Thin light coloured blinds may not help much either, since they themselves may become uncomfortably bright under sunlight. Low angle sun in mornings and evenings onto east and west facing façades offers the most intractable problem with shading since canopies are virtually useless, and blinds cut views through windows.

Commissioning and performance testing
Heating systems

In the United Kingdom about 3.5 GJ of solar radiation fall on each square metre of south facing roof in an average year, obviously with far more during summer than winter. A domestic solar water heating system with 4 m² of panel area may be expected to operate at an annual efficiency of about 35%, supplying about 5 GJ (1400 kWh) of heat over the year; this is roughly 40% of the domestic hot water requirements of the average household. Agrément certificates may be available for components.

Components for PV systems are normally tested to the requirements of the International Electrotechnical Commission (IEC), which publishes type approval criteria for photovoltaic devices.

In view of the rather limited experience of many installers, in addition to tests on individual components it will be important to test the complete installation. If the system is to be connected to the National Grid, tests specified by the electricity supply authority will need to be carried out.

Shading

Louvres which retract as well as rotate give better all round performance. Although opaque louvres are most common, heavily diffusing louvres can give reasonable glare control while letting in some extra light.

Tinted glazing has little effect on glare from the sun, and some types cut a considerable amount of incoming daylight. Conventional low-emissivity glazings have little impact on solar heat gain, though newer forms may offer improvements.

Window films tend to have similar performance characteristics as tinted glazing, but offer a relatively low cost retrofit option. Special UV films are available for use in museums and shops to reduce the fading of fabrics, paintings and drawings[138].

Health and safety
Heating

A pressure valve is a necessary safety precaution in a sealed domestic hot water system and may be conveniently combined with a pressure indicator to check that there are no leaks in the system. The outlet of the valve should be connected via an atmospheric break to a suitable drain point or container. In an open system, where a header tank is used, no pressure release valve need be fitted.

Shading

Shading against overheating and glare have already been referred to under 'Unwanted side effects'. Glare is particularly crucial when operating visual display units, and suitable conditions are now defined in legislation[138].

Some people are adversely affected by the striped pattern caused by venetian blinds, with closely spaced slats being worst.

Durability
Heating

Using dissimilar metals in a plumbing system can cause corrosion. In a solar system, the main risk occurs where aluminium panels are used with copper tube; the panels can suffer pitting corrosion even in the absence of physical contact of the two metals. Precautions which can be taken include the use of a corrosion inhibitor in the water circulating through the panels but it may be preferable to use a sealed circuit filled with distilled water and suitable anti-freeze. The long term effectiveness of these precautions, however, is not yet reliably established. Systems with aluminium panels should be regularly examined for leaks. Steel panels can be used with more confidence in indirect systems because of the experience gained in domestic central heating systems. Copper panels are, of course, completely compatible with copper tube.

Since there are no moving parts in the collection cells of PV systems, they are expected to last for a considerable time. However, the different formulations do have different expected lives, with some being guaranteed to maintain minimum output for longer periods (say 10 years) than others before renewal. Auxiliary parts of the system, particularly an inverter, may need to be easily accessible and renewable.

Shading

Many adjustable solar shading devices, whether manually controlled or motorised, installed in new-build schemes seen by BRE investigators between 1970 and 1990, have broken down within relatively short periods of time: typically not more than five years, and usually two or three, warranty or no warranty. This applies both to external and internal equipment. These devices must be robust, not vulnerable to wind damage, free from corrosion, and able to withstand the occasional rough handling by users not fully conversant with their modes of operation.

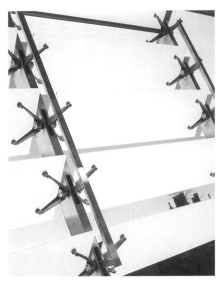

Figure 2.60 Solar shading louvres on the BRE Environmental Building

Maintenance
Heating

Natural circulation in a hot water panel system has the advantage of simplicity in design and fewer maintenance problems than a pumped system.

In an indirect system, anti-freeze (of a type approved by the local water authority) can be added to the water which circulates through the panels. Both direct and indirect systems can be designed in which the panels drain whenever the pump is switched off. However, this is not normal practice because the fresh water continually being circulated through the panels can form scale and hence impair the efficiency of the panels. Special anti-freeze solutions may be available for use with aluminium but their service life is limited to between two and five years, after which time the solution should be renewed. A heat transfer oil may provide an answer to problems of both freezing and boiling.

When maintenance is undertaken on or near PV systems, the solar collectors must be covered so that power is cut off. **It is not possible to switch them off**. Stationary systems will require maintenance of their weatherproofing if they form part of the fabric of the building, and the operation of tracked systems will depend on the maintenance of their moving parts.

Shading

Moveable or powered solar shading devices need regular attention if they are to perform satisfactorily. BRE experience has been that these systems have a very mixed record, with some breaking down after two or three years. Later designs may be more robust (Figure 2.60).

Work on site

Workmanship

The prime requirements when hot water panels are mounted on an existing pitched roof are clearly that they should remain fixed under all conditions and that the roof should remain weathertight. Because of the many differences in construction of both panels and roofs, it is impossible to recommend detailed fixing techniques. but some important considerations can be mentioned. If the panel is to be mounted on top of the roof covering, the roof must remain structurally safe with regard to the fairly considerable extra weight. In this situation, panels with aluminium casings might corrode when in contact with the cement in concrete roof tiles; all panels are at risk in high winds unless very securely fixed. The weight of the panels will be less of a problem if tiling is removed and the panel is fixed either on top of, or between, the rafters. This will also reduce the risk of condensation between panels and tiling which causes corrosion attack, but it creates substantial problems for maintaining weathertightness. Careful attention must be paid to weatherproofing, the panels being treated as if they were skylights and flashed into the roof. (For further information see *Roofs and roofing*, Chapter 2.1.)

Where PV systems are incorporated into the fabric of the building, either on wall or roof, similar attention must be paid to weatherproofing details.

Supervision of critical features

As already mentioned, PV systems must be covered before maintenance work is to be undertaken on them to reduce the risk if electric shock.

Inspection
The problems to look for are: **Solar heating** ◊ corrosion of metal parts ◊ loose fixings ◊ weathertightness details leaking ◊ risk of freezing **Solar shading** ◊ summer overheating ◊ glare from the sun ◊ lack of daylight ◊ lack of maintenance of moving parts

Chapter 3   Ventilation and ducted services

This third chapter deals with the provision of ventilation in buildings, whether by natural or by mechanical means. Although there is obviously a lot in common between ventilation in domestic and in non-domestic situations, there are sufficient differences to justify distinguishing between domestic and non-domestic situations. They are also treated separately in the Building Regulations Approved Document F[143].

Adequate ventilation is essential for the well being and health of building occupants. Traditionally it was achieved by natural means – by providing fresh air (Figure 3.1). Fresh air is needed to:
- provide sufficient oxygen
- remove excess water vapour
- dilute body odours
- dilute, to acceptable levels, the concentration of carbon dioxide produced by occupants and combustion processes
- remove or dilute other indoor pollutants

Air is required in order to provide oxygen for breathing, but is less than that needed for dilution of body odours, and much less than the ventilation needed to dilute tobacco smoke and other pollutants to acceptable levels of odour perception.

The ideal, of course, would be to remove or eliminate all sources of indoor pollutants from the building, but in practice this is not possible because of the very wide range of building and furnishing materials, cleaning products, processes and activities found in, and essential to, buildings. It is to be hoped that some

reduction in indoor pollutants produced by materials and substances may be achieved in the near future; for example, by manufacturers declaring polluting ingredients and characteristics of their products, and by producing low emission products. However, ventilation will remain the primary means of controlling indoor pollutants in buildings for the foreseeable future.

Poor ventilation, combined with inadequate heating and insulation, may produce condensation problems which lead to a buildup of

mould and fungi. BRE has, in recent years, spent much time and resources on this topic, and produced extensive guidance for use by building owners and occupiers[45,144]. The ventilation requirements for homes are covered in current building regulations, and various means of providing adequate ventilation are described there. It is important that heating, ventilation and insulation are considered as a composite design criterion; this will help to overcome problems of condensation and mould growth that might otherwise occur later.

Figure 3.1
Natural ventilation by means of opening lights in windows – the traditional and the most widely used method. Ideally windows should provide adjustable degrees of opening for fine control of ventilation rates, some protection from driving rain when left open, a reasonable degree of security against unauthorised entry, and a reasonably airtight seal when closed. Some designs do none of these things particularly well. This example was investigated by BRE to determine the cause of air and water leakage

Opening windows for ventilation may present both security and safety risks, and let in outside noise; inflexible control of windows can waste energy too. Therefore 'background' ventilators provide a desirable or alternative additional means of ventilation, and are effectively required in kitchens and bathrooms under building regulations if there is no other acceptable means of ventilation.

Strictly speaking, the regulations in England and Wales merely require a building to be adequately ventilated from the point of view of its occupants. Approved Document F, Section F1, provides guidance on how to meet that requirement. Scottish and Northern Ireland regulations are different in that they give deemed-to-satisfy guidance[145,146].

Although extract fans are obviously mechanical devices, they are referred to in Chapter 3.1 rather than in Chapter 3.3: they do not operate on a continuous basis, and so are treated for the purposes of this book as an auxiliary to natural ventilation systems. Air conditioning was dealt with in Chapter 2.4 and is not covered in Chapter 3.3 although, clearly, there are ventilation as well as heating and cooling aspects to consider.

Some aspects of smoke control in buildings take the form of natural and forced ventilation, but for the purposes of this book, smoke control is dealt with in Chapter 4.5.

Chapter 3.1

Natural ventilation, through windows, trickle vents and airbricks

Natural ventilation is defined as ventilation driven by the natural forces of wind and temperature. Its use should be intentional and controlled. It should not be confused with infiltration which is the unintentional and uncontrolled entry of outdoor air, and the exit of indoor air, through cracks and gaps in the external fabric of the building.

Depending on the circumstances, good design can provide natural ventilation either for an entire building or for particular areas of a building (Figure 3.2). Sometimes, natural ventilation can be augmented by the intermittent use of

Figure 3.2
A large proprietary ventilator set into sloping patent glazing and providing positive natural extraction for a part of an industrial building. At one time, vast numbers of these were fabricated in asbestos cement, though later models, such as this, were in sheet metal

local extract fans for wet or polluted zones such as photocopying rooms and kitchens, bathrooms and sanitary accommodation.

Buildings on green field sites pose fewest problems for natural ventilation. In urban sites, the entry of noise and external air pollution, and their effects, can be reduced by appropriate window design and location of internal spaces such as offices.

Buildings in areas of severe exposure (eg on hills and coastal sites) generally need smaller or fewer ventilation openings than those that are sheltered. Shelter can be provided by adjoining buildings as well as by natural or other features such as wind breaks. Shelter belts (plantings of trees and shrubs to act as wind breaks) are usually placed to face the direction of the prevailing wind. The resistance depends on the density of the foliage and is greatest immediately behind the shelter belt. Artificial wind breaks can also provide shelter. Ventilation design for low-rise buildings near tall buildings should take into account the effect of higher wind speeds at ground level as well as changes in wind induced pressures on building façades.

The mechanisms of natural ventilation

Natural ventilation in a building is created by pressure differences between inside and outside induced by wind and temperature (Figures 3.3 and 3.4).

Wind pressure on a building depends on wind direction, the shape of the building and wind

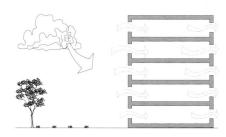

Figure 3.3
Wind pressure induced natural ventilation

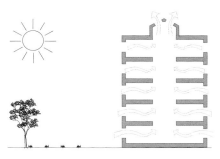

Figure 3.4
Temperature induced natural ventilation

speed. Temperature difference between indoors and outdoors causes density differences in the air which, in turn, cause pressure differences. Airflow rates through openings are not linearly related to pressure differences so, for any opening, wind induced and stack induced flows cannot readily be added together (see Chapter 3.2). The pressure generated (wind or temperature, or both) is used to balance the resistance to airflow of all the openings on the air route through the building.

The total wind pressure acting across a building is roughly equal to the wind velocity pressure. The mean values for buildings with different heights for a meteorological wind speed of, say, 4 m/s at a height of 10 m are shown in Figure 3.5. If the building is isolated, the full meteorological wind speed is used. In suburban areas and city centres, where actual wind speed is lower, local winds at a height of 10 m will be reduced to 2.4 and 1.3 m/s respectively, with correspondingly reduced pressures.

The following basic design guidelines help to improve natural ventilation:
● shallow plan forms are better than deep ones
● shading (preferably external) minimises summer overheating
● an airtight building envelope minimises unwanted air infiltration
● trickle ventilators provide controllable background ventilation
● openable windows provide controllable and draught-free ventilation but, for night cooling, they should be lockable in a secure position
● precise control of local ventilation should be allowed to the building occupants
● using low-energy lighting and IT equipment avoids unnecessary heat gains
● office machinery should be situated near local extract ventilation

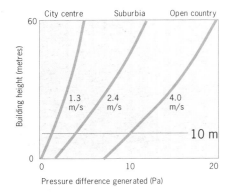

Figure 3.5
Wind induced pressure generated across a building for 4 m/s wind speed

Whether or not these guidelines have been followed in existing buildings may largely be a function of their age, with only those buildings built in the last quarter of the twentieth century having any real prospect of conformity. For those built in earlier years, it may be a question of either inadequately considered ventilation through opening windows, or of mechanical ventilation, a practice which has grown largely since the 1950s (see Chapter 3.3). Refurbishment of older buildings involving replacement windows may provide an opportunity to install trickle vents.

Good design is based on the principle that adequate ventilation is essential for the health, safety and comfort of building occupants, but that excessive ventilation leads to energy waste and sometimes to discomfort. In a naturally ventilated building, air enters either by design (eg openable windows) or adventitiously from uncontrolled leakage through infiltration.

'Build tight – ventilate right'

The aim of good design is therefore to minimise uncontrolled (and usually unwanted) infiltration by making the building envelope airtight while providing the required ventilation with fresh air in a controlled manner (Figure 3.6). It should be emphasised that a building cannot be too tight – but it can be under-ventilated. For an overall successful natural ventilation strategy, therefore, the three issues of building tightness, good ventilation for occupants, and natural ventilation design have to be brought together in an integrated manner[147,148].

To apply the principle of good design for natural ventilation (ie 'build tight - ventilate right'), sufficient information is available on ventilation requirements to satisfy safety and health criteria. However, criteria relating to comfort, especially those associated with odour, metabolic carbon dioxide and summer overheating are not so amenable to solution, and this is an issue where specialised advice may need to be sought[149].

New units of air quality, the 'olf' and the 'decipol' have lately been debated. These units are based on the degree of acceptability of odour and sensory irritation[†].

In the non-domestic field, and in those instances where more effective control over ventilation is needed than can be obtained by natural means alone, there is a trend towards using what has become known as 'mixed mode' systems. Mixed mode systems take full advantage of natural ventilation but which is augmented with artificial means when the need arises. There was further description of some of these techniques in Chapter 2.4.

Characteristic details

Windows

Windows should:
- ventilate effectively but not cause draughts, at least during the heating season
- provide sufficient glare-free daylight
- keep out excessive solar gain
- provide good insulation and avoid condensation on frames (ie not create a thermal bridge)
- allow occupants to adjust finely the openable area
- be simple to operate
- be secure when open

A window is the most obvious controllable opening for natural ventilation, especially in summer. The 1995 edition of Building Regulations Approved Document Part F recommends an openable area of at least one-twentieth of the floor area for rapid ventilation of habitable rooms in domestic buildings. (No minimum area is specified for other rooms.)

† The validity of using 'olfs' and 'decipols' to determine and define ventilation rates and compare pollution emission rates is at present under discussion, in relation to their derivation, their theoretical basis and the suggested methods of application. A procedure has been developed for testing buildings by using a panel of trained people to rate air quality directly in decipol. A pilot study of the procedure has recently been carried out in several European countries but there are practical limitations to its use[150].

Figure 3.6
Natural ventilation in a multi-storey office building

Figure 3.7
A variety of window shapes and sizes, mostly double hung sashes, in this much altered but jettied, originally half-timbered, building in the Middle Temple, London

The traditional way of meeting at least some of these criteria for many years (the period from Georgian times to the 1914–18 war) was the double hung, sliding sash window. It could be opened fairly easily from just a crack top and bottom, to 50% of the window area. If it were set back into a reveal or protected by a hood mould or cornice, it was reasonably rain tight. Single glazing was not a good insulator, however, and poor fit of the meeting rails made sure that the broad equivalent of a modern trickle ventilator operated most of the time. It was just possible to clean the outside from inside the building, albeit with an element of risk. Taking all things into consideration, it served its purpose well enough, and it is still worthy of being kept in repair (Figure 3.7). Insulation can be improved by fitting an inner (double) opening window or by fitting suitable draughtstripping.

Windows can contribute to the balance of heating and ventilation (and therefore the control of condensation) only if the building

occupants find the controls easy to reach, convenient to use, and precisely controllable so that open areas can be adjusted to best advantage in windy conditions. The arrangement of fixed and opening windows should aim to minimise draughts by providing ventilation to the upper levels of a room, preferably at least 1.75 m above finished floor level. The small top hung opening light had the advantage of small degrees of opening which it is difficult to match with, for example, large pivots. Fine control of ventilation is, however, also possible by incorporating unobtrusive trickle ventilators in the heads of frames (see the later section of this chapter on trickle ventilators).

There are two main reasons for this 1.75 m criterion. In the heating season, having a vent entirely at a minimum 1.75 m above the floor reduces the risk of draughts. For thermal comfort in summer, though, some part of the opening window should be at least 1.75 m above the floor level and some part at a lower level. Cool air then enters at low level and warm air leaves at high level.

Extract fans

As already noted, extract fans can be used as an auxiliary ventilation device to natural ventilation through windows.

Mechanical removal of stale or polluted air is achieved either directly by extract fans or a single fan and duct system (described in Chapter 3.3). At the same time fresh air will be automatically drawn into the room from other rooms or from the outside via leaky windows, trickle ventilators, cracks and other openings. Building regulations now provide for all new kitchens and bathrooms (even if adjacent to an external wall) to have extract fans fitted or to be connected to a mechanical extraction system (see Chapter 3.3) or passive stack ventilation system (see Chapter 3.2). Internal toilets also require extract systems, and toilets with external walls may be provided with extract fans as substitutes for opening windows.

Where an open flued heating appliance is to be installed in a kitchen already fitted with an extract fan, adequate fixed ventilation to the outside must be provided, sized to accommodate both the maximum volume of the fan and the requirements of the heating appliance.

Location of extract fans
An extract fan can be fitted:
● on an external wall, expelling stale air directly through the wall. The cavity must be closed tightly where the unit passes through
● on an internal wall, with a duct to the outside through which stale air is expelled
● in a window, single or double glazed (Figure 3.8)
● in a ceiling, with a duct through the roofspace to the outside of the house

In a kitchen the extract fan can be in the form of a cooker hood, the outlet of which should always be connected to the outside either directly or via a duct (Figure 3.9). A recirculating system will not perform adequately and does not comply with building regulations. Cooker hoods can be wall mounted, fixed under cupboards or built in as part of a canopy or special kitchen cabinet.

All systems can work adequately so long as the location is correct in terms of air paths; choice will depend on such matters as room layouts and wall space available.

Single sided ventilation of a deep office building

Tracer gas measurements of the local mean age of air were carried out by BRE investigators at different locations within an office room 10 m deep ventilated by single sided ventilation. The results showed that fresh air was generally distributed over the whole depth, suggesting that, if ventilation effectiveness is important, natural ventilation may be incorporated into deeper open plan rooms and wider buildings than the present rules of thumb imply. The adequacy of the total supply of single sided ventilation to avoid overheating in summer was also assessed, indicating that comfortable temperatures could be maintained[151].

The exact location of extract fans should be:
● as high as possible in the room
● as close as possible to the source of pollution (ie the cooker or hob in the kitchen, or the shower in the bathroom)
● so that the fresh air entering the room does not 'short circuit' to the extract fan
● not directly above a cooker hob or where the temperature could rise above 40 °C (except in the case of cooker hoods)
● so that the control cord or other switch (if fixed) can be easily operated

Single sided ventilation
Single sided ventilation (SSV) occurs when large, natural ventilation openings (such as windows and doors) are situated on only one external wall (Figure 3.10). Exchange of air takes place by wind turbulence, by outward openings interacting with the local external airstreams and by local stack effects.

Rules of thumb for SSV indicate that windows with openable areas of at least one-twentieth of the floor area can ventilate spaces to a depth of about two and a half times the height of the room (ie a depth of 7.5 m for a 3.0 m high room).

In non-domestic situations, to provide adequate ventilation for cooling in summer, traditional guidance recommends using cross-ventilation for naturally ventilated offices deeper than 7.5 m, but this is often not practicable in many modern designs for office buildings. The case study alongside shows that greater depths may be feasible in individual circumstances.

Figure 3.9
A cooker hood incorporated into a kitchen installation

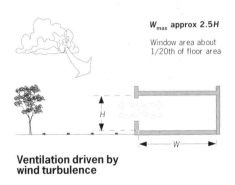

W_{max} approx 2.5H
Window area about 1/20th of floor area

Ventilation driven by wind turbulence

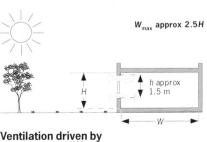

W_{max} approx 2.5H

h approx 1.5 m

Ventilation driven by temperature difference

Figure 3.8
A fan fitted into domestic window glazing

Figure 3.10
Single sided ventilation

Cross-ventilation

Cross-ventilation occurs when inflow and outflow openings in opposing or adjacent external walls have a clear internal flow path between them. Flow characteristics are determined by the combined effect of wind and temperature difference. Cross-ventilation also depends on windows (or other openings) on opposite sides of the building being opened sufficiently; this needs the cooperation of occupants. The effectiveness of ventilation in deep spaces can be affected by internal partitions and obstacles. This generally affects only air movement which can then be increased locally, by ceiling fans for example.

A general assumption may be made that spaces can be cross-ventilated to a depth of about five times their height (ie 15 m depth for a 3 m height). *CIBSE Guide A1* [152] and BS 5925 [153] give simple equations to estimate single sided and cross-ventilation flows. If heat gains are lower, BRE research indicates that deeper spaces can be ventilated naturally [154].

Trickle ventilators

Trickle ventilators were originally developed for the domestic market, but are now available in a variety of designs suitable for use in other building types.

Domestic

Trickle ventilators can provide controllable background ventilation. Minimum open areas needed for these ventilator are 8000 mm^2 in habitable rooms, and 4000 mm^2 in kitchens, bathrooms and WCs.

Non-domestic

In winter, ventilators with an open area of 400 mm^2 per m^2 of floor area will usually provide adequate background ventilation in multi-celled buildings. Controllable trickle ventilators can be a very effective way of doing this (Figure 3.11). The simplest trickle ventilators are just holes or slots cut into the frames of windows, and protected from the weather by a shield, and a hit-or-miss sliding shutter to give the occupants a degree of control.

Laboratory tests were carried out at BRE in two identical adjacent deep-plan offices. Trickle ventilators were installed in one room and the other was used as a control. Figure 3.12 shows typical results obtained.

Figure 3.11
A trickle ventilator set into a timber window head

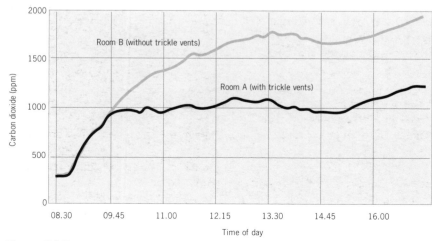

Figure 3.12
Typical experimental results showing carbon dioxide concentrations during a working day in two adjacent rooms occupied by three people under typical external conditions

Trickle ventilators for two office-type buildings

Manually operated trickle ventilators were installed in two typical refurbished office buildings used mainly for clerical work. Both offices were in general use Mondays to Fridays, and unoccupied over weekends. The trickle ventilators were slot ventilators located either in the window frame or incorporated into the pane of the window. In the latter cases, the ventilators had flanges of the same thickness as the glass and lay in the same plane, forming a structural part of the glazing. Both types were provided with a damper mechanism that allowed user control of the ventilation opening – a significant and positive feature that makes natural ventilation attractive to the user.

The BRE studies showed that ventilators with an openable area of 400 mm^2 per m^2 of floor area were capable of providing the fresh air required to maintain carbon dioxide levels at or below 1000 parts per million (ppm) – which equates to a ventilation criterion of 8 l/s – for average external weather conditions, and should be able to provide the necessary background ventilation to satisfy average office occupant densities of 10 m^2 per person [155].

Airtightness of existing buildings

Air infiltration cannot be designed for, nor indeed can it be precisely built into, a building; it may therefore be considered as an unwanted addition to the running costs of the building. Moreover, infiltration is neither a reliable nor an energy efficient substitute for properly designed ventilation. Good design should therefore separate the mechanisms for providing fresh air to occupants from the adverse and unpredictable effects of air infiltration. This demands good ventilation design coupled to a clear and workable specification for an effective and maintainable airtightness layer.

A building with an airtight envelope provides identifiable benefits to those that own, maintain and occupy it.

● Energy savings, since energy costs for space heating may be up to 20% less than for an equivalent but leaky building. Also, sophisticated energy saving heating control systems and heat recovery systems can be economically viable options in airtight buildings

● Enhanced comfort, since draughts and localised cold spots are minimised in an airtight building. Providing controlled ventilation (eg at high levels) ensures adequate fresh air for occupants with a minimum of draughts

● Reduced risk of deterioration in a properly ventilated but tightly constructed building, otherwise air leaking out of the building will tend to pull warm and moist internal air through the fabric of the walls and roof[156]

For windows, airtightness is very dependent on building use and degree of environmental control indoors. It is generally agreed that the very highest levels of airtightness, given in BS 6375-1[157], are only necessary with windows where there is full control of temperature and humidity by mechanical ventilation. This is usually found only in the non-domestic field.

At present there are no recommendations or standards for the airtightness of the outer envelopes of buildings in the UK, although there are some standards for individual components, notably windows. Some countries do have standards for the entire envelope. At the time of writing, an airtightness standard is under consideration for the Building Regulations for England and Wales.

Adventitious ventilation rates in dwellings

Natural infiltration rates vary widely in the UK, even in apparently identical properties. Dwellings with mean natural infiltration rates as low as 0.2 ach are the exception rather than the rule; 0.7 ach is probably typical. Natural infiltration rates cannot be assessed purely by visual inspection; some form of measurement is needed (Figure 3.13).

The construction features identified as increasing adventitious ventilation rates include gaps between walls and window or door frames, poorly fitting windows and doors, draughtproofing which is absent or inadequate, gaps round service pipes through walls, suspended ground floors, and penetration of upper floor ceilings (eg hatches, pipes and cables). Less often mentioned are gaps in walls; for example, at first floor joist level, particularly where joists are built into the inner leaf of a cavity wall. The contribution of air permeation through wall materials themselves when painted or wet plastered, but not necessarily when bare or dry-lined, is generally small compared with the role of cracks and gaps.

Natural infiltration rates are difficult and expensive to measure directly. Air leakage rates, under artificially applied pressure differences, can be measured quickly and easily using fan pressurisation. The air leakage rate may be regarded as a measure of the sum total area of all the cracks and openings in a dwelling's envelope through which air is exchanged.

Figure 3.13
A commercially made fan pressurisation system being used to measure air leakage in a dwelling. The fan is set up in an airtight panel within the open front doorway. The measuring instruments are hung from the door leaf

Although air leakage rate cannot be interpreted directly in terms of infiltration rate, it can be used as a good indicator. Generally speaking, a dwelling with a mean natural infiltration rate of 0.2 ach will have an air leakage rate, at an applied pressure difference of 50 Pa, of about 4 ach. Although the infiltration and air leakage rate are expressed in the same units of air changes per hour, they have completely separate meanings because of the different pressure conditions under which they are measured.

National and regional differences in construction techniques may lead to considerable differences in airtightness. A detailed description of air leakage characteristics in over 400 UK dwellings, including some recently built, is contained in *Airtightness in UK dwellings: BRE's test results and their significance*[149].

Adventitious ventilation rates in non-domestic buildings

It is sometimes helpful to compare the envelope leakage characteristics of a building with those of other buildings of different shape and size to obtain an idea of what rates can be achieved in practice. Fan pressurisation techniques are now available in which the leakage rates of whole buildings can be measured at a specified pressure differential

(Figure 3.14). It has been shown that this leakage rate, divided by the permeable external surface area of the building, gives a measure or index of the construction quality of the building fabric with respect to air leakage.

Figure 3.15 shows results of fan pressurisation measurements on five large commercial buildings in the UK, and compares them with buildings tested in Canada and the USA, and Sweden. The Low Energy Office (LEO) building is shown as being twice as airtight as the other UK office built in a more conventional manner, and as tight as the North American offices.

Broadly speaking, a building with an envelope leakage index of 5 (ie 5 m^3/h per m^2 of envelope area) at 50 Pa applied pressure difference can be classified as of a good, low leakage standard, while an index of 10 is average. A building with an index of 20 is poor, and likely to give rise to complaints about uncomfortable draughts and difficulties in keeping warm.

Results from two old industrial hangar-type buildings showed them to be over twice as leaky as one built within the last decade under current UK Building Regulations. But comparison with tight Swedish industrial buildings shows that a further five-fold reduction is possible.

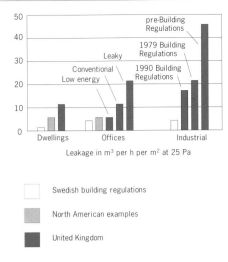

Figure 3.15
Envelope leakage indices of buildings

Main performance requirements and defects

Outputs required

The first performance standard for the ventilation of buildings, and for many years the only one, was set in 1902 when 600 ft^3/h (5 l/s) per person was required in the first Cotton Cloth Factories Act. Prescriptive requirements also date from around the same time, when 2.5 in^2 (400 mm^2) of ventilation area per pupil was required to be provided in school buildings[18]. Current guidance from CIBSE recommends a minimum of 5 l/s, and preferably 8 l/s, per person for dealing with carbon dioxide and body odours, although much greater rates are needed to deal with diluting odours from tobacco smoking.

It is comparatively rare to find a building where natural ventilation was modelled at the design stage, and subsequently tested to verify the model and the design assumptions, although there have been some[158]. Accepting that the external envelope of a building allows both intentional and unintentional penetration of air, the unintentional component is often very difficult to predict; it can vary substantially between one building and another even though they may be of nominally similar design. In some circumstances this adventitious leakage through the building fabric is a source of energy waste and even

Figure 3.14
Apparatus for measuring air leakage rates for an entire building (BREFAN)

discomfort. There is a move towards producing more controlled ventilation of living spaces in buildings. This can be achieved by more airtight construction, by better draughtproofing, and by introducing background ventilators and mechanical ventilation at source. These go some way towards satisfying the opposing requirements of reducing heat loss and providing sufficient ventilation to prevent condensation and to reduce the concentration of indoor pollutants.

Ventilation rates

Where possible, occupants should be given control over their local environments; this can contribute significantly to their acceptance of the environments. The benefits of producing a brief and simple occupants' guide to using the building should be considered. Ventilation controls should be ergonomic and respond rapidly, and their use should be explained to occupants.

The design of a naturally ventilated building should reflect different requirements for winter and summer occupancy. In winter, excess ventilation needs to be minimised, but background ventilation can be provided in the form of trickle ventilators to meet occupants' needs for fresh air. In summer, ventilation may need to exceed what is required solely to satisfy occupants' needs to avoid overheating. As part of this process, the distribution of fresh air is important within the habitable spaces, and can enhance comfort conditions and freshness.

Dwellings

Ventilation of new dwellings is covered by building regulations. The Approved Documents for the Building Regulations for England and Wales now include recommendations for background ventilation for habitable rooms, and mechanical extract ventilation for bathrooms and kitchens. Regulations for Scotland and Northern Ireland recommend similar provisions. Although building regulations are not retrospective, consideration

should perhaps be given to bringing all ventilation provision in dwellings up the minimum standards recommended by the Approved Document or the regulations in Scotland and Northern Ireland.

In situations where natural ventilation is not possible (eg in internal and basement rooms) and in cases where natural ventilation would be too difficult to control or would result in excessive noise penetration, the basic requirements will often not be continuous. It will therefore be important to establish the patterns of occupation and to consider the ventilation requirements in relation to the patterns, wherever such situations are likely to arise as a result of external conditions or of design choices (Figure 3.16).

Where warm or cool air is chosen as the means of heating or cooling, the basic ventilation requirement will

be unaffected, but will have to be met in association with the thermal comfort requirement.

The main provision for ventilation of habitable rooms has remained more or less unaltered since the Model Byelaws introduced the requirement for a minimum opening area of one-twentieth of the floor area, with the further provision that some part of the openable area should be higher than 5 feet 9 inches above the floor. Since the Model Byelaws, providing ventilation has been progressively revised; there are now provisions for background ventilation and for mechanical extract or passive stack ventilation (PSV) systems. (PSV is explained in Chapter 3.2.) Background ventilation rates for both domestic and non-domestic situations were given earlier in this chapter.

Air change rates of between 0.5 and 1 per hour for a whole house are

Figure 3.16
Local air extract fans fitted through the brick soldier courses of the external wall of this block of flats. Many sound attenuated installations of this type were fitted to reduce noise levels from adjacent roads when ordinary windows would otherwise need to be open for ventilation

generally recommended; they can usually be obtained by trickle vents, extract fans or whole house ventilation systems. This 'controllable ventilation' will maintain good air quality without wasting fuel. The objective should be to remove the main pollutants, particularly excess water vapour, at source.

Approved Document F of the Building Regulations for England and Wales gives guidance on minimum ventilation standards for all rooms. The Regulations for Scotland and Northern Ireland are slightly different. (The relevant Building Regulations should be consulted in all cases.) Table 13 of *Thermal insulation and ventilation*[159] provides further information, though some of it has been superseded by amendments made to Approved Document F in 1995.

Approved Document F lists opening areas for rapid ventilation according to whether there is an opening window in rooms such as bathrooms or according to floor areas for habitable rooms. Background ventilation, however, is listed according to specific open areas which may be provided as controllable trickle ventilators in windows or as airbricks in walls (see also Chapter 4.1 of *Walls, windows and doors*[22]). The *CIBSE Guide A* gives design values for various types of building. A summary of background ventilation practice for dwellings is given in *Background ventilation of dwellings: a review*[160].

Habitable rooms are recommended to have an openable window (minimum area one-twentieth of the floor area) and a trickle vent (minimum 8000 mm^2). There is more discussion of this topic in *Walls, windows and doors*.

Kitchens should have a mechanical extract fan (extracting at a minimum 60 l/s, or 30 l/s if in a cooker hood) operated on demand and a trickle vent (minimum 8000 mm^2), or a mechanical extract system capable of operating continuously at 1 ach. The cooker hood should vent externally and not recirculate air. PSV systems may be used as an alternative to extract fans (see Approved Document F[143] and BRE Information Paper IP 13/95[161]).

Bathrooms and shower rooms should have a mechanical extract system (extracting at minimum 15 l/s) operated on demand. Separate toilets should have an openable window with a minimum area of one-twentieth of the floor area; or a mechanical ventilation system extracting at a minimum 3 ach or 6 l/s, operated on demand (with 15 minutes minimum overrun).

Until 1995 there were no regulations covering ventilation in utility rooms, but these are now covered in Approved Document F. Due to the likely generation of water vapour in these rooms, an openable window, trickle vent or humidistat controlled extract fan are recommended to combat condensation and mould growth. Alternatively, whole-house mechanical ventilation systems are allowed under the regulations.

Ventilation requirements for heating appliances that burn room air are covered by the Building Regulations for England and Wales. The Regulations for Scotland and Northern Ireland vary slightly. There is a risk of combustion product spillage from appliances in rooms where extract fans are fitted. This risk has already been referred to in Chapter 2.2

Extract systems must be adequately sized for their function, as recommended in Approved Document F, 1995; the values are given in Table 3.1.

The relationship between fan size and room air change rates can be easily calculated by multiplying the volume of the room by the air changes per hour to give the required capacity of the fan (eg in m^3 per hour).

Extract fans in windowless bathrooms and toilets should operate automatically on the light switch and have a minimum 15 minute overrun after the light is switched off. It is not unknown for fans to be installed the wrong way round, and checking by placing a piece of paper over the aperture will soon confirm the correct operation: suction will obviously hold the paper in place.

An extract fan may also be operated by a humidistat which causes the fan to keep running until the humidity level in the room reaches a predetermined level. Solid state sensors are preferred to the cheaper plastics strip-type as they suffer less from contamination by grease etc. Three systems can be used:

- a built-in humidistat which turns the fan on and off in relation to a variable, preset level of relative humidity
- a built-in humidistat which regulates the speed of the fan in relation to the difference between internal and external relative humidities. This allows good control of internal humidity with minimal energy use and noise level, particularly on start-up when humidity levels are low
- a remote humidity sensor and controller, operating as above. It is important, though, to locate a remote sensor close to the main source of water vapour.

Other types of control can be appropriate; for example, passive infrared (PIR) movement detectors in WC compartments in a non-domestic building, so that operation of the fan is in proportion to the frequency of use of the compartment.

Table 3.1 Summary of minimum extract rates for room forced ventilation	
Room/system	**Minimum extract rate**
Kitchen extract fan	60 l/s or PSV
Kitchen cooker hood	30 l/s
Utility room	30 l/s or PSV
Bathroom (including shower room)	15 l/s or PSV
Toilet	6 l/s

A manual override switch to turn the extract fan on when required should always be included. Electrical wiring of the control must ensure that the humidistat controller turns on the fan when required and that accidental manual control does not make the humidistat inoperative.

In existing dwellings, airtightness varies over a very wide range[149]. Some, probably most, existing dwellings are sufficiently leaky not to need any background ventilation from trickle vents, but a significant proportion of the existing housing stock is sufficiently airtight to need some additional ventilation. The problem is that it is almost impossible to tell which are airtight and which are leaky without doing a fan pressurisation test. The solution recommended by BRE, and which is also incorporated into the building regulations, is that all dwellings are fitted with background trickle vents which are closeable; the occupants of airtight houses then have the means of obtaining more ventilation or of allowing closure in windy conditions to reduce draughts.

The dangers of inadequate ventilation rates are perhaps most apparent where flueless appliances are being used. Although these are not necessarily part of the permanent service equipment of buildings, nevertheless they need mentioning here. Nitrogen dioxide and carbon monoxide are increasingly recognised as dangerous indoor pollutants. They are generated by a wide range of unvented domestic combustion appliances. The largest contributors of nitrogen dioxide are gas stoves, unvented gas convectors and paraffin space heaters. For carbon monoxide, no particular appliance is especially worse than any other but poor tuning, poor ventilation and very slow combustion rates can lead to an order of magnitude increase in carbon monoxide emissions and indoor concentrations likely to prove fatal.

Non-domestic buildings

Ventilation rates for non-domestic buildings were introduced in the 1995 edition of Approved Document F, whereas previously reliance had been placed solely on other authoritative recommendations. The concept of an occupiable room was introduced and defined as *'room in a non-domestic building occupied by people, such as an office, workroom, classroom, hotel bedroom, but not a bathroom, sanitary accommodation, utility room or rooms or spaces used solely or principally for circulation, building services plant and storage purposes'*[143].

The requirement for rapid ventilation was established for windows with an opening area not less than one-twentieth of the floor area, as with domestic ventilation, together with background ventilation levels which were quoted earlier in this section. Alternative provisions for ventilation were specified in BS 5925[153] or in *CIBSE Guides A and B*[162,163].

Requirements for a sedentary person in an office are shown in Figure 3.17. Natural background ventilation is not necessarily the preferred solution for control of other air contaminants since local extract ventilation may be more practicable.

Renewed interest is being shown in designing for natural ventilation in non-domestic buildings as a means of avoiding the need for air conditioning. To provide adequate ventilation for cooling in summer, traditional guidance recommends cross-ventilation for naturally ventilated offices deeper than 6 m, although this is often not practicable in many designs for office buildings.

Exclusion of insects

Following the provisions of earlier legislation, such as the 1991 and 1995 editions of Building Regulations Approved Document Part F, trickle ventilators would have been installed in window frames or airbricks with hit-or-miss controls provided. If these vents were slotted, the

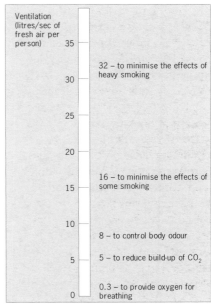

Figure 3.17

Fresh air requirements for a sedentary person in an office

Figure 3.18

A large wasps nest built into the eaves of a 1960s house. Access was made by an overlarge hole for the warning pipe to the central heating header tank

minimum open dimension should have been 5 mm for each slot, or, if they were holes, 8 mm for each hole. These dimensions were not intended to exclude insects.

The majority of the more troublesome larger varieties of insects can be excluded by fitting a 3–4 mm mesh over ventilation slots, but they can always take advantage of gaps between fascia and soffit boards, or where warning pipes are a loose fit in drilled holes (Figure 3.18). Care should be taken to ensure that any smaller mesh size introduced as a remedial measure does not reduce the ventilation opening size below that given by building regulations.

Proximity of obstructions

In the mid-1980s, it was suggested that deflecting wind from tall buildings could increase air infiltration rates in a neighbouring low-rise block, thereby increasing its associated heat loss. BRE investigators examined the ventilation and space heating loss of a three-storey low-rise office block located near a taller nine-storey slab building (Figure 3.19).

Wind tunnel measurements showed that a tall building near to (and in line with) a low building markedly changed the distributions of wind pressures on the latter. Analysis confirmed that this near-proximity did have an impact on ventilation and space heating compared with an isolated building.

● Ventilation rates were reduced by as much as 35% for winds blowing normal to the front (broad) face of the low-rise building

● The average ventilation rates expected during the heating season in the low-rise building were reduced by about 15% if the buildings were on an open site and by 19% if sheltered

● Overall space heating requirements in the low-rise building were marginally increased by 3% as a result of higher convective losses offsetting the reduced ventilation losses[164]

Seasonal differences

When naturally ventilating a building, two distinct strategies have to be developed: one for the winter and the other for the summer.

The key issue for winter ventilation is controlling indoor air quality. Natural ventilation as a strategy for achieving acceptable air quality is essentially based on the supply of air to a space to reduce (by dilution) the pollution concentration in the space. If building-related pollution sources are avoided, the only requirement is for a minimum of 5 l/s (and preferably 8 l/s) per person. Trickle ventilators will meet this requirement.

Summer ventilation is usually linked to the control of internal temperature and thermal comfort. In summer conditions, for example, one may actually wish to create a draught. Control of the airflow rate is not in itself so important. To avoid overheating, the airflow rates may need to exceed what is required solely to satisfy occupants' needs. As a result, sizes of openings are totally different from those appropriate to winter ventilation. As part of this process, it is important to ensure adequate distribution of fresh air within the space to enhance comfort conditions and freshness.

An air movement of less than 0.1 m/s is regarded as still air. The maximum acceptable level of air movement is about 0.8 m/s when cold draughts are detectable and the point at which papers on desks start to be disturbed[111].

Current best practice in non-domestic buildings

It is inevitable that design practice will change in the future as pressures increase on the environment and efforts are made to build for a more sustainable future. Although forming only a small proportion of buildings in the year 2000, the numbers of environmentally friendly buildings will grow. And while opportunities for upgrading existing buildings to new standards will be few, it is worth considering some of the new trends such as those epitomised in the first case study on the next page.

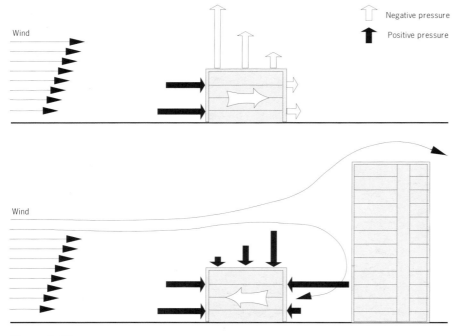

Figure 3.19
A tall building can deflect high speed winds down to ground level and affect the pressure distribution on a nearby low-rise building. This could change, or even reverse, the ventilation flow through the low-rise building. The actual pressures depend on wind direction and height and spacing of buildings. It is pressure differences, rather than absolute pressure, that drives flows

Cooling by night ventilation

Night ventilation is a low energy strategy for cooling a building, providing a more comfortable environment for the occupants during hot daytime periods. It works by using natural (or mechanical) ventilation to cool the surfaces of the building fabric at night so that they can absorb heat during the day[165].

Although the method can be used to some extent in domestic situations, it is particularly effective for office buildings due mainly to the way they are used, enabling relatively high airflows to flush out the interior of the building at night when it is empty of people. In the UK, temperature diurnal range is medium to large and maximum day external temperatures are relatively low during the cooling season from May to September. This combination of external conditions allows cool night air to be used to remove the heat absorbed by the thermal mass of the building during the day.

Night ventilation can affect internal conditions during the day in four ways:
- reducing peak air temperatures
- reducing air temperatures throughout the day, and in particular during the morning hours
- reducing slab temperatures
- creating a time lag between external and internal temperatures

The system works by providing large vents in the external walls which are controlled by the occupants during the day, and by the building's energy management system or manually during the night. Night ventilation can be applied successfully to older buildings due for refurbishment.

Energy efficient office of the future

The Environmental Building is a new office block at BRE, Watford. Completed in 1997, the building has three floors and covers a gross area of 2000 m² with a carbon dioxide emissions target of 20–30 kg per m² of floor area per year. It is intended to represent best practice at the year 2000 and to meet the energy and environment targets anticipated for the early part of the twenty-first century.

The building is naturally ventilated using a combination of single sided, cross and passive stack ventilation. It also includes exposed structural concrete ceilings with a regular waveform (ie sine-wave) design and incorporating voids which act as heat exchangers (Figure 3.20). These concrete ceilings have two significant purposes in relation to ventilation:
- to increase the surface area and, therefore, the thermal mass of the building
- to provide air pathways for cooling and ventilation

During the heating season, background ventilation is provided through trickle ventilators. In addition, the occupants can open low level or high level windows. For the summer, large areas of openable glazing are incorporated on both sides of the building to provide through draughts. Although these are controlled by the building energy management system (BEMS), the occupants are able manually to open windows in their local area. When there is not enough wind to drive ventilation in the building, pressure difference is created by the ventilation stacks attached to the south side of the building. If the solar generated stack effect is insufficient, fans inside the stacks are activated. The ventilation paths in the sine-wave roof are used during the night to provide pre-cooling.

Figure 3.20
Concrete ceilings in the BRE Environmental Building. These ceilings increase both the surface area and the thermal mass of the building, and provide air pathways for cooling and ventilation

Measurements carried out during the summer of 1997, one of the hottest years on record, showed internal temperatures stable and lower than the outside air temperatures that reached, during August, in excess of 33 °C over successive days. The measured indoor temperatures fully satisfied the design brief, not exceeding 28 °C for 1% of the year and 25° C for 5% of the year. Preliminary reports have indicated good indoor air quality and good occupant satisfaction.

Night ventilation in a 1950s office building

A typical 1950s four-storey office building situated near the coast had its curtain walling replaced. The building owner's project team had wanted it to remain a naturally ventilated building with the occupants given the option of night cooling. Ventilation was provided by 850 mm × 600 mm bottom-hung ventilators installed around the perimeter of the building. A mesh screen and perforated external louvres were placed in front of the ventilators to provide 24-hour security and weather protection. The ventilators could be opened manually by the occupants at the end of the day to provide natural night ventilation.

The building was examined by BRE investigators and the system tested. When the ventilators were kept open overnight (Monday to Friday) on the third floor and closed on the first, night cooling was clearly evident on the third floor with its lower overnight temperatures; more importantly, though, the daytime temperatures on the third floor remained below those on the first floor for most of the following working day. The ventilators were closed on the weekend days. It was also found that internal temperatures had reached similar values on both floors by Sunday. The internal temperature at the start of the working day, 08.00, indicated the effectiveness of night cooling the previous night. It also produced satisfactory comfort levels for the occupants at the start of each working day[166].

The following must be considered when designing for night cooling.

● Building mass: the effectiveness of building mass in storing 'coolth'. For night cooling, an ideal building mass is about 600–800 kg/m^2 of floor area

● Variation of effectiveness. It is known that variation of outside air temperature over a 24-hour period may not necessarily follow the same pattern as the previous or the following 24 hours. Likewise, the thermal behaviour of a building may not be the same as for the previous or the following 24-hour period: residual heat or coolth may be stored from the previous 24-hour period. Consequently the effectiveness of night cooling can vary from day-to-day and should be considered during design

● Controllability. The effectiveness of night ventilation is dependent on the difference between the ambient temperature and the internal temperature. But if the night-time ambient temperature is too far below the internal temperature, the building may be too cool the following morning. This is important in the UK where uncontrolled night cooling can cause uncomfortably cool conditions in the morning. Furthermore, temperatures can vary significantly between various locations which means that effectiveness of night cooling can be site-dependent

Although not of direct assistance to those concerned with existing buildings, a simplified night cooling design tool is available[167] for use when the basic form and organisation of the building is evolving. User input is limited to a few key variables such as glazing ratios, orientation, internal gains, ventilation rates and thermal mass. This technique allows the designer to explore very rapidly the effects of a range of design variables, and it can also be used to assess the effects of potential alterations to existing buildings.

Noise and other unwanted side effects

Traffic noise entering through windows can be an important factor for those buildings situated in busy streets or near airports. In severe conditions it may be necessary to install acoustically attenuating ventilators to alleviate the problem, but thermal comfort in summer may still be problematical.

Some guidance on attenuation of noise on levels expected in urban environments is available as well as on the sound reduction performance of various window configurations. Acoustic baffles can improve the sound performance of open windows and, at least for new designs, siting the opening windows on the quiet side of a building (eg in an internal courtyard or on a building façade away from the main road) will help[168].

There can be conflict between good security and good ventilation. Large, open windows may present a security risk, especially on ground floors. However, by adopting particular window opening designs, or by separating the ventilation element from the window, natural ventilation can be provided without compromising security. This is particularly important for night ventilation.

Health and safety

Ventilation needed for satisfying health criteria is mainly set by the requirements for:
● human respiration
● dilution and removal of contaminants generated within occupied spaces

Safety criteria relate mainly to eliminating or minimising the risk of explosion resulting from airborne contaminants, both from gases and from dust suspensions. The risk levels are set by the higher and lower explosive limits for the relevant gases and vapours[147].

Many pollutants are found in indoor environments. These include formaldehyde, wood preservatives and other volatile organic compounds (VOCs); living organisms, sometimes referred to as viable particulates (eg bacteria, moulds, dust mites); non-viable particulates and fibres (eg man-made mineral fibres and asbestos); radon; combustion products (eg nitrogen dioxide and carbon monoxide); and lead along with other non-viable particulates from vehicle exhausts. While there are now satisfactory procedures for dealing with many of these, others are increasingly causing concern.

The issues tend to differ between homes and other buildings. For example, in some office buildings the occupants report a range of minor building-related symptoms (SBS). This is not commonly reported in homes, but it is not clear whether this is due to differences in design (eg the presence or absence of air conditioning) or differences in response (eg individual householders improving their own environment compared with groups of office workers taking complaints to a building manager). Similarly, Legionnaires' disease and Pontiac fever, associated with wet cooling towers and domestic hot water systems in complex buildings, are not usually attributed to water services in homes.

In homes, many issues can be dealt with by providing satisfactory ventilation, by careful choice of materials and by good construction practice.

Pollutants

Several substances can be considered as indoor pollutants, though they may easily be commonly occurring substances, and it is only their concentration that renders them polluting to the indoor environment. Other substances may be dangerous to health even in moderate concentrations.

Common pollutants generated within buildings can include naturally occurring gases (eg carbon dioxide, ozone, water vapour and methane), products of combustion (eg carbon dioxide, carbon monoxide, oxides of nitrogen and sulfur), VOCs (eg formaldehyde), particulates and fibres, and environmental tobacco smoke (ETS).

Carbon dioxide is produced by people and animals, and by burning fossil fuels (eg cooking with gas). It needs removing from the house and replacing with oxygen. Carbon dioxide may also arise from landfill and from the ground in chalk areas.

Water vapour is always present in the air in varying proportions. Many activities within the house produce additional water vapour: cooking, bathing and showering, clothes drying etc, and cause problems of condensation on windows and other cold surfaces, mould growth and deterioration of materials, furnishings and decorations.

Smoke is produced by tobacco smoking and, possibly, by cooking. The former is strongly implicated in a number of health problems.

Smells (other than from smoke) are produced by people, and by food preparation and cooking; although they are not necessarily a hazard to health, they are unpleasant and need to be diluted or removed.

Formaldehyde is released from many building materials including chipboard, medium density fibreboard (MDF), adhesives and urea formaldehyde foam cavity insulation. Formaldehyde could be inhaled or ingested either as a gas or as a hydrolysable component of dust. BRE investigators measured the amounts of formaldehyde extracted from dusts using mild acid hydrolysis, and found that dusts from materials containing formaldehyde based resins, notably urea-formaldehyde (UF) cavity fill, released most formaldehyde. The extractable formaldehyde contents of dusts from 60 houses were measured and found to vary over a twentyfold range. Dusts in two houses were found to be high in formaldehyde from aged UF cavity fill; these cases were not associated with high gaseous formaldehyde levels[169].

Formaldehyde vapour is an irritant and may produce discomfort in the eyes, noses and throats. Exposure to high levels of formaldehyde (eg in the workplace) can lead to sensitisation. (Sensitisation means that a reaction to a particular substance could occur; a reaction

could be triggered, for instance, by a lower exposure than that which led to the sensitisation in the first place.) However, this is very unlikely to occur from the low levels found in most buildings. With ventilation rates above 0.5 ach concentrations of formaldehyde are not enough to cause problems.

In urban areas, the contamination of buildings' internal climates from external pollution, in particular from vehicle emissions, is considered to be a major barrier to using natural ventilation[170]. Added to this is external contamination from other nearby sources including building ventilation exhausts and boiler flues.

A study was carried out by BRE investigators of the factors which influence indoor levels of nitrogen dioxide and personal exposure. Seventy-two homes were selected on the basis of type of area (inner city, suburban or rural) and cooking fuel (gas or electricity). Passive sampling diffusion tubes were used to measure nitrogen dioxide concentrations in the bedroom, living room and kitchen, and immediately outside the home. Also, personal exposure was measured by a diffusion tube worn by each occupant. Data on each dwelling and its occupants (particularly those undergoing personal monitoring) were obtained using questionnaires and diaries.

The main factors which appear to influence nitrogen dioxide levels in the home, and personal exposure, are the use of natural gas for cooking and the number of people in the household. A winter survey might show the effect of heating fuel. Inner-city and suburban areas have higher outdoor concentrations than rural areas, but neither area nor outdoor levels significantly affect indoor or personal concentrations. Kitchen and bedroom windows are more likely to be open during the day in rural areas. Personal exposure in habitable rooms was greater in living rooms than in kitchens or bathrooms. The results probably reflect the amount of time spent in each room (more being spent in living rooms than in kitchens) and higher

concentrations in living rooms than in bedrooms. The number of hours spent in the bedroom, hours spent outdoors, and kitchen nitrogen dioxide concentration add to the variances. Approximately 70–75% of personal exposure to nitrogen dioxide was at home[171].

Volatile organic compounds having their origin in building materials were discussed in Chapter 1.5 of *Floors and flooring*[23]. From BRE measurements carried out in dwellings there was no significant difference between the mean concentrations in living rooms and bedrooms. However, the mean indoor concentration of the VOCs measured was 16.5 times higher than that outdoors.

Radon is a naturally occurring radioactive gas with no taste, smell or colour. It is formed from the radioactive decay of uranium and is present in small, variable quantities in all soils and rocks. It can seep from the underlying ground into houses through any opening in the structure. The air pressure in the house is often slightly lower than that in the soil and so air is drawn from the ground into the house. Exposure to radon and its decay products over many years increases the risk of developing lung cancer.

Measures to prevent the penetration of radon into buildings are described in *Floors and flooring* from which Figure 3.21 below is

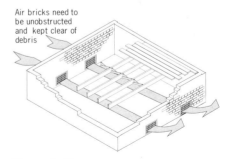

Air bricks need to be unobstructed and kept clear of debris

Figure 3.21
In areas of the UK which are subject to radon, effective remedial measures should be undertaken. Of the five main methods to prevent the gas from seeping into habitable rooms in existing buildings, ventilation of the underfloor space can be very effective. There should be a free flow of air from one side of the subfloor void to the other

taken. Maps of the areas most affected are available in *Radon: guidance on protective measures for new dwellings*[172]. Further information is available for existing dwellings in *Surveying dwellings with high indoor radon levels: a BRE guide to radon remedial measures in existing dwellings*[173], for other building types in *Radon in the workplace*[174] and from the National Radiological Protection Board (NRPB).

Dealing with pollutants
In dwellings the key pollutant has traditionally been water vapour from cooking, bathing, clothes drying etc, which needs to be removed in order to reduce the risk of condensation and mould growth. It was assumed that dealing with water vapour automatically included other indoor pollutants. However, changes in modern lifestyles and using modern domestic appliances (eg pre-prepared foods, dishwashers and tumble driers etc – especially those that are not properly installed) call that assumption into question. Other pollutants, many from modern materials and appliances (eg VOCs and formaldehyde) may be of new significance in dwellings.

In non-domestic buildings there is not usually much of a problem with condensation and mould growth; the key pollutant is metabolic carbon dioxide, which is used as a proxy indicator for many other indoor pollutants.

A common misconception is that dilution ventilation is the only way to remove harmful contaminants from within occupied spaces. The Health and Safety Executive (HSE) gives guidelines[175] for acceptable concentration limits for most common pollutants in the workplace. Dilution ventilation to reduce or eliminate these contaminants to levels below these guidelines may not be better or more economical than other methods.

The UK COSHH Regulations[176] lists the following methods in order of preference to ensure maintenance of good indoor air quality:

- remove contaminating substances
- substitute the substances by others which are less hazardous
- enclose contaminating processes
- partially enclose contaminating processes and provide local extract ventilation
- provide general (dilution) ventilation
- provide personal protection

Air filtration or air cleaning should be considered as a last-resort measure. In the office environment, generation of internal pollution should either be avoided (eg low-emittance furnishings and carpeting) or be controlled locally (eg by local extract ventilation near photocopying machines). If this strategy is carried out, ventilation through fresh air is then only needed to:

- provide sufficient oxygen for breathing
- dilute body odours to acceptable levels
- provide an adequate level of indoor air. An absolute level of less than 1000 ppm metabolic carbon dioxide, monitored internally, is generally used. In the absence of tobacco smoking and with normal occupancy densities, this level translates to a fresh air rate of about 8 l/s per person. The UK requirements may therefore be summarised:
- 0.3 l/s as the minimum necessary to provide oxygen for life
- 5 l/s (or preferably 8 l/s) to satisfy current recommended minimum ventilation rates for background ventilation, not including the dilution of tobacco smoke[169].

Tobacco smoking is an avoidable contaminant. Buildings can have either a no-smoking policy or separate ventilated smoking rooms. If these are not possible, and smoking is allowed, the following fresh air requirements are recommended for each occupant:

- light smoking (25% of occupants smoking): 16 l/s
- heavy smoking (45% of occupants smoking): 24 l/s
- very heavy smoking (70% of occupants smoking): 32 l/s

As indicated, depending on the proportion of occupants smoking, ventilation requirements can not only be doubled, but can be quadrupled. The UK Government's code of practice on smoking in public places contains suggestions on ventilating smoking areas and rooms[177].

Products of combustion
One of the most important matters relating to safety relates to cooker hoods or other types of powered extraction in spaces in close proximity to heating appliances which draw air from the same space and discharge it via a flue or chimney. Fans and cooker hoods remove moisture and smells rapidly from where they are produced, improving the indoor environment. It is important, though, to carry out safety checks when a fan and an open flued appliance are present in the same dwelling because, under some conditions, spillage can occur. Extracting air from a room creates a pressure difference between inside and outside. This causes a drop in pressure at the base of a flue. If the pressure at the base of the flue is lower than at the top then the flue gases will reverse and flow down instead of up. The flue gases will then disperse through the living space where they can be a danger.

Ideally a combustion product spillage test should be carried out with the fan operating so that it generates the maximum achievable pressure difference across the appliance.

Greater risks lie with extract fans rather than with whole house ventilation systems which are covered in Chapter 3.3. Extract fans usually have high air extract rates which are concentrated in a single room, whereas whole house systems, while having similar total extract flow rates, will usually extract air from several rooms. An extract fan, then, is likely to generate a greater depressurisation in a room than a whole house system.

Durability

The durability of windows was considered in *Walls, windows and doors*, Chapter 4.1.

The durability of fans is very variable. There seems to be a dearth of published information, but anecdotal evidence is that manufacturers expect at least 20,000 hours for an ordinary domestic fan with phosphor-bronze bearings in clean air, and would hope for double that life on average. (One year is 8,760 hours). Talcum powder in bathrooms in particular is said to suck the oil out of bearings and lead to early failure. Manufacturers began, in the 1990s, to introduce domestic fans using highly efficient brushless DC motors with ball bearings and seals; these are said to have a very much improved life expectancy – more than 40,000 hours – but time will tell!

On the other hand, feedback from BRE researchers suggests that an ordinary extract fan in a bathroom, which is operated from the light switch, may have a life of as little as two or three years, so improvement is badly needed.

Maintenance

The ability to control natural ventilation rates is one of the key factors in improving the habitability of large blocks of flats, many of which have been plagued by dampness and mould growth. However, BRE investigators have observed very wide differences in how dwelling occupants use their windows in the same weather conditions, with some people keeping windows shut for 24 hours and others keeping them open all day. An important factor is the ease with which the windows can be opened and closed, and the precision of the degree of opening to suit the occupants and the conditions. (See *Walls, Windows and doors*, Chapter 4.1. Maintenance of mechanisms for opening windows is also dealt with in the same chapter.)

Work on site

Measuring techniques

Full information on interzonal air movements can be obtained from techniques using multiple tracer gases. Field measurements, carried out using automated measurement systems, have been developed for this purpose. Theoretical and experimental work has shown that it is probably sufficient to seed part of a complex building with a single tracer gas to measure the approximate overall infiltration rate[178].

Workmanship

Sealing the junctions between different elements in a building is very important in producing an airtight barrier; for example, junctions between

- doors and windows, and their frames
- window and door frames, and the walls
- wall-to-ceiling joints
- wall-to-floor joints.

With the wide variety of materials available for infiltration control, the most appropriate to the application can be chosen[179].

It is difficult to seal a dwelling to achieve a 50 Pa air leakage rate of 4 ach. Just draughtproofing windows and doors is rarely sufficient. Some of the details that need special attention are service entries, wall-to-floor junctions, suspended timber floors, boxed-in pipe runs, behind bath panels, around window and door frames and plasterboard dry-lining (which must be perimeter sealed and not just dab fixed) (Figure 3.22). Flexible or rigid sealing materials should be used according to the circumstances and conditions. It can be quite impracticable to seal some potential leakage paths in an existing dwelling; for instance, where intermediate floor joists penetrate the inner leaf of a cavity wall (see BRE Digest 398[180] and BR 359[149]). Achieving a high level of airtightness in new construction is much easier if details are carefully worked out at the design stage, and satisfactorily complied with on site.

Supervision of critical features

Inspecting a building during its construction may be the only way of ensuring that a continuous airtight layer is incorporated correctly. It also gives an opportunity to ensure that the construction team has a clear understanding of the importance of the airtightness layer. A rigorous approach is even more important where the airtightness of existing buildings is being improved.

A suitable measure used to quantify the overall air-leakiness of a building envelope is the leakage index, Q_{25}/S, where Q_{25} is the airflow rate at the imposed pressure differential of 25 Pa and S is the total permeable external envelope area. A typical airtightness target for a 'tight' UK building might be set at a value less than 5 m³/h per m², for an average building less than 10 m³/h per m², and, for a leaky building in the region of 20 m³/h per m² or more[156].

Figure 3.22
Gaps caused by unfilled mortar joints of lightweight blockwork in an external wall under construction. Airtightness will depend entirely on the integrity of the surface finishes, and junctions with other elements and components

Air leakage in a large industrial building
A model which made the link between leakage measurements and ventilation characteristics was applied to a large, industrial building. Air leakage measurements with the building 'as found' and then with its loading doors sealed showed a 14% reduction at an inside-to-outside pressure differential of 25 Pa. Using these leakage characteristics, the model predicted ventilation rates which corresponded well with measured values. Meteorological data at the site for the heating season were combined with the ventilation characteristics of the building (given by the model) to predict the ventilation performance of the building over that period. The results indicated that the building, as found, would have an average air change rate of 0.5 h–1 during the heating season. Sealing the loading doors would reduce this rate by 24% (ie to 0.38 h–1 ach).

The space heating energy requirements for the building over the heating season were assessed. The results showed that ventilation heat losses accounted for 44% of the total required energy. Calculations also showed that replacing the loading doors with ones which were more airtight and better insulated would reduce by 25% the energy required to cover ventilation losses and by 5% the losses through the building fabric – an overall reduction of 14 % of the total requirement[181].

Testing
Envelope leakiness can be quantified by measuring the whole-building air leakage rate at appropriate applied pressure differentials between the inside and outside. The fan pressurisation technique is the most convenient method of carrying this out. This technique involves sealing a portable fan, such as the BREFAN system, into an outside doorway and measuring the airflow rates from the fans required to maintain a series of pressure differentials across the building envelope.

A single whole-building pressurisation test, using robust and practical equipment, can quantify in a very short time – say 2 hours in a dwelling, perhaps a day in a large non-domestic building – the air-leakiness of the building envelope. However, these measurements do not give a direct measure of the ventilation characteristics of the building which normally requires time-consuming and specialist tracer gas tests.

Ventilation is a significant factor in providing a comfortable and healthy indoor environment and represents a rising proportion of the heat loss from buildings as insulation levels increase. However, in large and complex buildings such as offices, measuring ventilation rates still requires some expertise and not a little understanding of virtually invisible processes. Consequently there is very little available information on actual performance. Nevertheless, the passive tracer gas technique is a simple and inexpensive development which can provide a single value measurement of ventilation rate over a given period, whether a day, month or whole heating season. It can be applied equally to naturally ventilated or air conditioned buildings, large or small, simple or multi-roomed[182].

Some of the problems foreseen in the preceding sections are very difficult to locate on site, and this is one of the reasons why an air leakage test may be thought necessary.

The problems to look for are:
◊ inadequacy of existing opening lights
◊ wrong types of opening lights
◊ draughts
◊ dry-lining by plasterboard dabs instead of ribbons
◊ draughtstripping missing
◊ noisy fans
◊ fans installed the wrong way round (ie pressurising rather than extracting)

Chapter 3.2　　　**Natural ventilation, passive stack**

Natural ventilation is not just about openable windows. It is, rather, a holistic design concept that has been widely used in the last years of the twentieth century in the design of large offices and other building types. Design for natural ventilation ideally encompasses passive ventilation, based on the 'stack' (temperature) effect and wind pressure differentials, to supply fresh air to building interiors even when the windows are closed. As part of this process, many designs incorporate atria or internal stairwells which, in some instances, use low energy fans to provide assisted natural ventilation (ie 'low energy ventilation').

Passive stack ventilation (PSV) systems work by a combination of the natural stack effect (ie air movement that results from the difference in temperature between indoors and outdoors) and the effect of wind passing over the roof of the dwelling. Hot air rises through its buoyancy. Natural ventilation can be effected by channelling, through a duct, this warmer indoor air, using as motive power the pressure differential created by its buoyancy. During the heating season, and under normal UK weather conditions, the ducts will allow warm moist air from the 'wet' rooms to be vented directly outside. The airflow rate in each duct will depend on several factors, the most important at low wind speeds being the temperature difference between indoors and the outside air. As the temperature indoors increases, because of cooking, heating, washing or bathing, the airflow rate

will also increase. Wind speed and direction also influence the rate of airflow but in a relatively complex way that depends on the interaction between factors such as the airtightness of the dwelling, the type and position of air leakage paths and the position of the duct outlet terminals.

There are two distinct types of stack ventilation systems, small and simple systems for dwellings (BRE Information Paper IP 13/94[183]), and larger, more elaborate systems for non-domestic buildings. Where necessary these two types are distinguished in the following sections of this chapter.

Domestic

The principle of ventilation by stack effect has been known for centuries, and has been exploited in a number of Victorian buildings. BRS first became involved in experimental work in the 1950s, measuring rates achieved in blocks of flats in the London area and actually building some experimental houses incorporating PSV to windowless bathrooms and WCs as early as 1960 (Figure 3.23). The changes made to Approved Document F, foreshadowed in the late 1980s and actually made in 1995, describing ways of satisfying the Building Regulations (England and Wales) governing ventilation, allowed the use of PSV ducts as an alternative to mechanical extract fans. It is since 1995 that PSV has grown in use.

PSV is likely to be an attractive alternative to mechanical ventilation in some situations: for example, in cold houses with high moisture production where fans would run for long periods; where occupants are unwilling to run fans; or where occupants are particularly sensitive to noise. Nevertheless, the number of dwellings equipped with passive stack ventilation, although several thousands in total, is a very small proportion of the total stock. Of the

Figure 3.23
Passive stack ventilation ducting being installed in windowless bathrooms and WCs in two-storey housing in 1960. The flow of air through the ducts was controlled by hit or miss shutters. The ducts are larger than would now be considered necessary

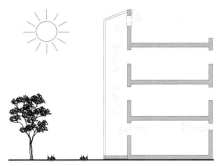

Figure 3.24
Solar chimney principle

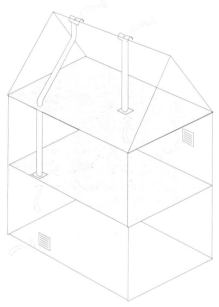

Figure 3.25
PSV layout suitable for most two-storey dwellings

new homes built between January 1992 and November 1993 by British local authorities, it was estimated that less than 1% had a PSV system installed.

In dwellings, BRE investigators found little difference in the performance between PSV systems serving downstairs rooms and those serving upstairs rooms, because the whole dwelling tends to act as a large stack duct. This would not necessarily be the case in tall multi-celled buildings.

Non-domestic

Stack ventilation is essentially cross-ventilation as far as the individual spaces are concerned in that air enters from one side of the space and leaves from the opposite side. The air may flow across the whole width of the building and be exhausted via a stack or chimney, or it may flow from the edges to the middle to be exhausted via a central stack, chimney or atrium. Airflow which is temperature driven is thought to work best via an atrium (ie ductless ventilation) serving open-plan offices on at least three floors. Buoyancy is improved as the stack height increases. This requires critical dimensions in the total height of the stack, and the height from the ceiling of the top floor to the top of the stack.

A vertical pressure gradient is created by thermal buoyancy which depends on the temperature difference between the column of warm air and the ambient temperature, and the height of the column of warm air. For each degree of temperature difference between inside and outside, a pressure difference is set up of about 0.04 Pa per metre of building height.

Passive stack elements can also take the form of 'solar chimneys' (Figure 3.24) which create a column of air at temperatures higher than ambient. This, in turn, generates higher pressure differences and so further enhances the stack effect. Enclosed spaces around atria in which passive stack ventilation operates can also act as buffers to reduce fabric heat losses.

Characteristic details

Description of the system: dwellings

Passive stack ventilation in a house normally comprises a system of vertical or near-vertical ducts running from the ceilings of the kitchen and bathroom to terminals on the roof (Figure 3.25). The ducts serve to extract moist air from the 'wet' rooms of the house, venting it directly outdoors. It does this by the stack effect – by the movement of the air due to the difference in temperature between inside and outside, and by the effect of wind moving over the roof of the house.

The system works in conjunction with fresh air entering the house via trickle ventilators, cracks and other openings, and may reduce the need for extract fans (though these may still have to be installed to comply with the Building Regulations for England and Wales). A passive system will provide almost continuous ventilation since the temperature within the house is usually higher than that outside. Higher ventilation rates will occur when temperature differences are higher, as in winter, and when room temperatures rise (eg during cooking). Therefore the system should be most effective when it is most needed. The wind also has a great effect on the airflow rate of passive ventilation systems, and is especially important when temperature differences are small.

The layout shown in Figure 3.25 is considered to be suitable for the majority of two-storey dwellings. The system should be designed to:
● avoid cross-flow ventilation between kitchen, bathroom and WC
● prevent, as far as possible, airflow in the ducts being adversely affected by the prevailing wind speed and direction, or by sudden changes in these
● minimise resistance to airflow by having ducts that are as near vertical as possible

A separate duct is taken from the ceiling of each of the kitchen, bathroom, utility room and WC to a separate terminal on the roof. Ducts should not be joined together; they must be as near vertical as possible to maximise the stack effect and be well supported. The preferred place for the ducts to terminate is at outlets on the roof ridge so that wind gusts and certain wind directions are less likely to affect performance adversely.

The proper design and installation of these PSV systems is vital to their successful performance, and BRE researchers undertook considerable work in support of the 1995 edition of Approved Document F of the Building Regulations.

Duct runs
To achieve an adequate but not excessive airflow rate, the diameter of the ducting normally should be 100–150 mm. Off-the-shelf PVC-U pipes and fittings, of the type used for drainpipes, are suitable. Flexible ducting is more expensive but has the advantage of being easy to install, particularly in the roof space where access may be restricted. It is also available in insulated form.

Grilles or terminals on the inlet ends of the ducts are required for aesthetic purposes only. The grilles should present the least possible resistance to the flow of air and be easy to clean. Most concentric-ring and egg-crate types are suitable.

For duct systems which terminate on the ridge of the roof, standard ridge terminals with appropriate adaptors are suitable. Gas-flue ridge terminals may offer less resistance to airflow up the duct than other types.

Table 3.2 Recommended sizes for circular ducts†	
Room	**Duct diameter (mm)**
Kitchen	125
Utility room	100
Bathroom with or without WC	100
WC	80

† Non-circular ducts of equivalent cross-sectional area and flow resistance may be used.

Where ducts do not terminate on the ridge, flue or soil pipe terminals should be used. The design of the outlet terminal should prevent rain entering the duct.

Description of the system: non-domestic
Although the physical principles of PSV systems are the same for both domestic and non-domestic buildings, in practice the non-domestic situation is much more complex.

Solar chimneys, in which glazed elements are incorporated into the chimney structure, can enhance stack pressures (Figure 3.26). Wind pressures can also be utilised by placing the outlet in a negative pressure zone relative to the inlet. In these designs, airflows into the building are at low level and exhausted at high level; therefore some care has to be exercised in determining the different sizes of ventilation openings for each floor of the building if equal ventilation rates are required. A relatively simple procedure to do this is now available[184].

Main performance requirements and defects

Outputs required
By the very nature of its design, passive ventilation will not be at the same rate for the whole of the time it is required, and under certain circumstances may be subject to reversal, just as happens with ventilation through windows when the wind blows in the wrong direction. Precise wind effects for natural stacks depend on terminal design, wind speed and wind direction, and the nature of the surroundings and roof design.

Dwellings
In dwellings, PSV systems must be designed to avoid (as far as is possible) the reversal of airflow. In BRE test house investigations it rarely occurred, but in field tests it occurred where the outlet on the roof was poorly positioned.

Figure 3.26
Solar chimneys in a BRE office building. The sun warms the external duct to enhance buoyancy, which in turn increases airflow

PSV ducting should be sized according to the room it serves (Table 3.2). Sharp bends will cause a resistance to the flow of air and so reduce the effectiveness of the system. Any bends should therefore be of the 'sweep' rather than the 'sharp' type, with no more than two in a single duct. In addition, no section of duct should be at an angle of more than 45° to the vertical and outlets as close to the ridge as is practicable.

If the PSV system is to work effectively as an extract system, there must be an air supply to the rooms it serves. This can be achieved by using controllable, purpose-provided ventilators (eg trickle vents) as described in Chapter 3.1. Ventilation rates given in Chapter 3.1 should all be achievable.

Case study

PSV systems in occupied dwellings

BRE researchers carried out a programme of experiments in a test house on the BRE Garston site to assess how well various passive stack ventilation (PSV) systems performed under controlled conditions. To determine how well these systems work under normal conditions of occupation, BRE also monitored four dwellings for approximately three weeks during the heating season.

Particular objectives of the study were to:

● measure the flow rates obtained in the stack ducts and relate these to the humidity within the dwellings
● determine whether the PSV systems would keep the relative humidity at sufficiently low levels to minimise the risk of condensation
● assess the design and installation of each system, and to give advice on possible improvements

The four dwellings which were monitored (A–D in Tables 3.3 and 3.4) were all two-storey, end-of-terrace maisonettes on the second and third storeys of three-storey blocks. The dwellings were built in the mid-1970s, originally with flat roofs and wooden framed, single glazed windows. They were refurbished approximately two years before testing, with the addition of pitched roofs, double glazed windows and PSV systems for the kitchen and bathroom. The position of the duct inlet serving the kitchen was somewhat unusual in that it was situated on the landing at the head of the stairs, the kitchen being at the foot of the stairs. In two of the dwellings the kitchen was open plan with the hall and in the other two it was separated from the hall by a door.

The ducting used for the PSV systems was an insulated, flexible type: 155 mm diameter used for the landing and 100 mm diameter for the bathroom. The landing duct terminated at a ridge ventilator while the bathroom one had been designed to terminate at a tile vent, situated low down on the roof and little higher than the bathroom ceiling. The ceiling outlets were of the circular valve type with a central adjusting section to regulate airflow. On most of the systems tested these inlets were initially fairly tightly shut, thus restricting the airflow, most probably because this is the way they had been delivered from the manufacturer and had not been opened properly when fitted.

On inspection, the PSV systems were found to be badly installed with tight bends, excess ducting and no supporting framework. In one dwelling the landing duct had become detached from the ridge ventilator and was lying on the floor of the loft. The bathroom ducts were of very poor design, with little difference in height between the ceiling outlets and the tile vent terminals on the roof slope.

Airflow rates within the ventilation ducts were measured, together with humidities, temperatures and climatological data. When monitoring had been completed, the air leakage of the dwelling was measured using the fan pressurisation technique described in Chapter 3.1.

Table 3.3 Percentage of time that relative humidity (RH) was greater than 70% before and after modifications[†]				
Dwelling	**Bathroom**		**Landing**	
	before mods (time)	after mods (% time)	before mods (% time)	after mods (% time)
A	16.6	9.3	0.9	0
B	2.7	2.1	0	0.2
C	0.6	0	0.9	0.7
D	24.0	8.5	7.9	0.1

Table 3.4 Results of air leakage measurements with PSV vents open and air pressure at 50 Pa			
Dwelling	**Positive pressure (ach)**	**Negative pressure (ach)**	**Mean air pressure (ach)**
A	11.5	10.8	11.2
B	12.1	11.7	11.9
C	11.2	10.5	11.9
D	10.4	9.8	10.1

† If 70% RH is exceeded for lengthy periods, mould growth may occur.

BS 5250 recommends ventilation rates of 0.5–1.5 ach for the control of condensation. By applying the one-twentieth rule (natural ventilation rates are approximately one-twentieth of the air leakage rate measured at 50 Pa) to the air leakage rates of the test dwellings, it can be seen that they equate to natural ventilation rates of just over 0.5 ach. The dwellings should, therefore, have sufficient ventilation for controlling condensation.

The PSV systems in the bathrooms had originally been designed so that the ductwork ran parallel to the loft floor and had sharp bends, restricting the airflow. The majority of the systems monitored had been badly installed with considerable excess ducting, too many bends and no support; they gave better airflow performance when the excess ducting had been removed and the bends straightened out as much as possible. In spite of poor design and installation the systems coped well with removing moisture and kept the relative humidity levels below 70% for all but a small percentage of the time[185].

Non-domestic buildings

For non-domestic buildings there will be a wide range of solutions, and the systems tend to be designed for each individual application. taking into account the type of accommodation and the air change rates required. Ventilation rates given in Chapter 3.1 should all be achievable. As with the domestic situation, some degree of control may be incorporated to avoid excessive ventilation rates in winter.

Terminals

Separate ducts are taken from the ceilings of the kitchen, bathroom, utility room and WC to separate terminals on the roof. A common outlet terminal or branched ducts between these rooms should be avoided as they could (usually in high wind speed conditions) result in air from one room being routed to another.

In all installations, upper terminals and lower grilles (and ducting) should be specified and fixed in such a way that the cross-sectional area of the system is consistent throughout its length.

Placing the outlet terminal at the ridge of the roof is the preferred option for reducing the adverse affects of wind gusts and certain wind directions. This layout usually, however, introduces bends into the ducts because of individual room configurations. Wind tunnel tests carried out by BRE on various terminal types have shown that those least adversely affected by wind are the 'H' pot terminal and the multi-vane type that revolves with the wind.

An alternative layout is to have the ducts running vertically and penetrating the roof away from the ridge with the centre line of the PSV within 1.5 m of the ridge. They should extend above the roof level to at least ridge height to ensure that the outlets are in the negative pressure region above the roof. While this removes the problem of bends in the ducts, there are disadvantages in extending a duct above roof level because of the considerable visual impact, the difficulty of insulating this part of the duct and the susceptibility of the duct to wind damage.

Ducts that terminate on the slope of the roof (tile vents) should not be used because, under certain wind conditions, airflow will almost certainly be reversed. This would result in moist air being transferred to other rooms in the dwelling and perhaps cause discomfort. Also, the positioning of the outlet on the roof slope leads, particularly with ducting from a room immediately below the roof, to a significant loss of stack height which is one of the main driving forces of the system.

The proximity of tall buildings to the building in which the PSV system is installed will affect its performance. If a dwelling, in which PSV is proposed, is situated near a significantly taller building (ie more than 50% taller), it should be at least five times the difference in height away from the taller building. For example, if the difference in height is 10 m, PSV should not be installed in a dwelling within 50 m of the taller building.

Noise

When a simple PSV system was installed in the kitchen of the test house at BRE, it was noticed that the duct admitted some noise from a motorway (about 250 m away) and from aircraft. There did not appear to be a noise nuisance in the bedroom through which the exposed duct passed. Tests have shown that rigid ducts transmit more noise from outside than flexible ducts.

In situations where external noise is likely to be intrusive (eg near busy roads and airports) some sound attenuation in the duct may therefore be sought. Where external noise is likely to be intrusive, fitting an off-the-shelf sound attenuator (essentially a straight length of highly perforated duct wrapped in mineral wool) where the PSV duct passes through the roof space is likely to be effective.

Condensation

The ducts should be insulated in the loft, or other unheated spaces, with the equivalent of at least 25 mm of a material having a thermal conductivity of 0.04 W/mK. Where the duct extends above roof level, the section above the roof should also be insulated. This helps to maintain the stack effect and reduces the risk of condensation forming inside the duct and running back down into the dwelling.

Commissioning and performance testing

In practice, provided the design rules have been followed, PSV systems perform well. It is not considered practical to proof test a passive stack ventilation system – for one reason there is no suitable test regime in existence.

Health and safety

In dwellings, a duct from the kitchen could, in the event of a fire, act as an easy path for the transfer of smoke and fire to a first floor living space.

In a house of one or two storeys, PSV may be treated like any other service, with no special fire precautions required. However, where the dwelling extends to three storeys or more, additional precautions (eg encasing the duct in half-hour fire retardant board) should be provided to ensure that the escape route is not prejudiced. Where PSV systems are installed in blocks of flats (up to four storeys), the PSV duct should run in a service duct that complies with the guidance given in the Building Regulations. The reduced standard enclosure for drainage or water supply pipes in the Building Regulations will not be

Day centre in an urban location

A four-year-old, two-storey building with a gross floor area of 1350 m², situated in an urban area adjacent to a very busy high street, was examined by BRE investigators. The construction was steel frame with brick external walls. Thermal mass was provided by the exposed areas of brickwork and a concrete ceiling to the ground floor.

Due to budget restrictions it was decided not to incorporate any mechanical ventilation in the building but to provide a purely natural ventilation system. However, openable windows were not really an option due to the projected loss of privacy and to the noise and poor air quality from traffic pollution from the high street. The solution was to create diaphragm walls where each room is connected to a vertical outlet chimney to exploit the stack and wind effects, and so achieve a variable level of exhaust ventilation. The rooms on the noisy high street side of the building had air inlets at low level, fed through a supply duct routed within the footings to allow clean air to enter silently from the courtyard side, displacing carbon dioxide and contaminants which were carried through to the PSV system.

At the top of each chimney, but within the insulated envelope of the building, an opposed-blade, volume-control damper, controlled by the occupants and by a central control system, allowed the passive effects to be optimised without compromising energy efficiency in winter when the stack effect is most pronounced.

appropriate for PSV ducts.

It is also recommended that a quick-acting fire damper, such as the fusible link type, should be properly fitted close to the inlet of the kitchen duct, either below the ceiling or within the floor.

Durability

There is little to go wrong with passive stack systems, and provided the duct insulation does not deteriorate, lives in excess of 30 years ought to be achieved. However, some sources in the industry suggest that shorter lives of around 15 to 20 years may be more appropriate.

Maintenance

Grilles need to be kept clean, since the ducts will virtually act as vacuum cleaners to the ventilated spaces. Depending on the design of the grille and its location, dust, and sometimes grease, will be deposited.

Work on site

Workmanship

The performance of a well designed system can be ruined by poor installation practices. Field studies by BRE investigators on installed PSV systems using flexible ducting have shown that some of the most common faults are:

- that ducting is too long (ie longer than necessary to join the exhaust grille to the roof terminal) causing the duct to have too many bends
- the ducting is not properly supported causing it to sag or become detached at joints, or at inlets or outlets
- the supports are too tight around the duct causing restrictions to the airflow

To enable a system to work effectively the ducts must not have excessive bends or restrictions to flow. This is not usually a problem when rigid ducting is used.

All ducts should be insulated where they pass through the roof and sealed with mastic (or equivalent) where they pass from one room to another.

Inspection

The problems to look for are:
◊ duct diameters too large or too small
◊ excessive bends leading to reduced airflow levels
◊ ducts sagging
◊ terminals inappropriately sited (eg on the roof slope)
◊ ducts not insulated, leading to condensation risk
◊ grilles blocked

Chapter 3.3   **Forced air mechanical systems**

This chapter deals with whole-building ventilation systems which are designed to run continuously. Although manually controlled extract fans are obviously mechanical devices, they were nevertheless described in Chapter 3.1: they do not operate on a continuous basis, as do the systems covered in this chapter, and so are treated for the purposes of this book as auxiliary to natural ventilation systems. Air conditioning was dealt with in Chapter 2.4, although clearly there are ventilation as well as heating and cooling aspects.

Forced mechanical heating systems in the school building programme of the late 1950s and 1960s have already been mentioned in the introductory chapter. Heating systems of this kind often take advantage of the circulation of warmed air to provide appropriate levels of ventilation in buildings, but there are also many systems which provide ventilation alone, without the heating. Forced air systems may also be needed for particular types of buildings (Figure 3.27).

In the UK domestic sector, the use of continuously operated mechanical ventilation (MV) is becoming more common. There are two main types:
● balanced supply and extract mechanical ventilation with heat recovery (MVHR)
● mechanical extract ventilation (MEV)

In both, the air is collected or distributed through a duct network. Figure 3.28 (on the next page) shows a typical installation of an MVHR system. An MEV system consists essentially of the extract components of an MVHR system and is cheaper to buy and install.

Another, very different system is mechanical supply ventilation in which a fan takes air from the roof space (or in the case of flats directly from outside) and delivers it to a stairway or central hallway. For many years the fan units came from only one manufacturer, but there are now several different examples on the market from other manufacturers.

The effectiveness of these units in ventilating dwellings is still not well understood, although more studies are currently under way, but anecdotally they have a good track record for dealing with condensation and mould. *Positive pressurisation: a BRE guide to radon remedial measures* gives more details and describes some potential problems[186].

The main reason for installing some form of MV (with or without heat recovery) is to control condensation and indoor pollutants such as tobacco smoke, metabolic carbon dioxide and odours which are difficult to control at source.

Figure 3.27
Air handling equipment occupies practically the whole of the roof surface of this laboratory building

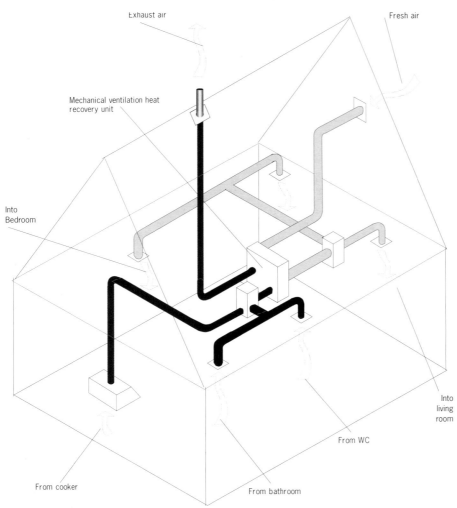

Figure 3.28
A typical system for mechanical ventilation with heat recovery

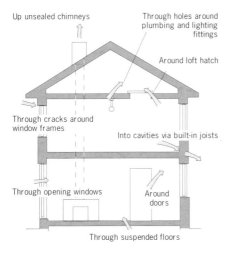

Figure 3.29
Measures may need to be taken to improve the airtightness of a dwelling before it is worthwhile introducing mechanical ventilation

Mechanical ventilation systems are usually designed to extract air from bathrooms, kitchens and utility rooms, so that water vapour is removed from as close as possible to its source, and from WCs to remove odours. When weather conditions are such that natural ventilation rates would be low, an MV system can provide adequate ventilation, even in an airtight dwelling.

In recent years, MVHR systems have been combined with warm air heating systems. This allows the same duct system to be used for both heating and ventilating, and may give the opportunity to recover energy from the products of combustion at the heat source. Although these systems are more complex than ordinary MVHR, design guidelines for ventilation still generally apply. The dwelling should be as airtight as practicable for economic operation of an MEV or MVHR system (Figure 3.29). Airtightness was also dealt with in Chapter 3.1.

Heat recovered from vitiated air before it is expelled from forced ventilation systems is a practice developed largely in the 1960s after the development of the thermal wheel. The mode of operation, in its simplest form, is that ducted hot air blown or sucked over the wheel at one part of its cycle heats those segments in its path. When the heated portion of the wheel rotates through 180°, fresh, and by implication also cold, air is then blown or sucked through it, and in the process is heated before being input into the building.

An unwelcome feature of this type of device is the risk of significant cross-contamination between the two airstreams. Purging the system to avoid cross-contamination results in lower efficiency.

Another form of heat exchanger with no moving parts which avoids this risk of cross-contamination is the flat plate heat exchanger – a collection of simple thin diaphragms forming partitions dividing a duct into several portions, with the warm vitiated outgoing air on one side of each partition heating the cooler fresh incoming air on the other side. Although this type can match the efficiency of thermal wheels, it is less common.

Characteristic details

Descriptions of systems
Domestic
In MVHR systems the fresh air is distributed to habitable rooms by simple fan-and-duct systems which are comparatively small in size. Guidelines on the design and installation of mechanical ventilation systems in typical UK housing are given in *Domestic mechanical ventilation: guidelines for designers and installers*[187].

Non-domestic
Air handling plant for forced air ventilation systems in non-domestic buildings is usually fairly bulky. It is best sited near to zones of maximum demand. Plant room sizes typically can be around 1 m² per 100 m² of floor area served, subject of course to a minimum smallest area for items of basic equipment, and some pro rata reduction for the very largest installations. Additional clearances may be needed round electrical equipment for safety reasons.

As with hot air heating systems and air conditioning ducts covered in Chapter 2.4, generally speaking, circular ducting for forced ventilation will be used for the higher velocities, and rectangular for the lower, but this rule is not universal. It is not only the space taken up by the duct itself that is important, but also the space around the duct necessary for the provision of access to its contents. BS 8313 gives the necessary clear space requirements[112].

Duct sizes
Ducting of 100 mm diameter circular cross-section is normally adequate for air distribution and collection in small dwellings. Some sections of the duct system may need to be 125 mm or 150 mm diameter to keep air speeds below 4 m/s to limit noise generation and excessive airflow resistance. Non-circular ducting of equivalent cross-section may also be used.

Ducting for non-domestic buildings obviously needs to be sized according to the circumstances and the system to be used (Figure 3.30). It is impossible to give simple guidance.

Figure 3.30
A forced-air ventilation system retrofitted in a weaving shed during the 1970s

Grilles and duct terminals
Air transfer grilles to allow the passage of air between rooms within a building are essential only if MVHR is combined with a warm air heating system. If the bottom edges of internal doors clear the floor surface by 5 to 8 mm, there is likely to be a sufficient opening for air movement. Transfer grilles are usually positioned not more than 450 mm above the floor. If positioned higher, grilles could allow, in the event of fire, the rapid movement of toxic combustion products and the spread of fire throughout the building.

The duct terminal fittings must be capable of passing the required airflow rate, at the available pressure drop, without generating excessive noise.

Supply grilles should give good mixing with air in the room without causing draughts. They should be sited as far as practicable from internal doors, at high level and directed across an area of unobstructed ceiling.

Extract grilles should be positioned so that they clear air from as much of the room as possible, ideally as far as possible from the internal door and as high as possible; above a shower cubicle or bath is a good place in the bathroom.

Main performance requirements and defects

Choice of solution for major items of plant

Ensuring a clean environment has become especially important in recent years in a number of manufacturing processes, not to mention hospital and other clinical situations.

Although the first cooling systems used ice, chilled water became the common medium in the early years of the twentieth century, and spray washers began to be used as a means of cleaning the air as well as adjusting its humidity. Filtration was by fine gauze, sometimes pleated to increase the surface area, which needed to be regularly washed, or by disposable cellulose filters.

High efficiency electrostatic filters, in which resin carrying an electrostatic charge is incorporated into a filter medium, may also be found. Odour removal by these filters is not usual though.

The cost effectiveness of MVHR systems with respect to energy savings is not encouraging. However, a good reason for installing full mechanical ventilation in buildings (with or without heat recovery) is control of indoor pollutants, particularly water vapour.

All things being equal, the greater the distance the outlet is from the inlet, the greater the loss of pressure, and the smaller the amount of flow, largely because of friction losses from the containing walls of the pipe or duct. This is usually compensated for in one of two principal ways: either by increasing the pipe or duct diameter, or by increasing the pressure, provided the fan is big enough.

Pressure drop is also influenced by the number of bends, by bore smoothness, and by the shape of the outlet. Furthermore, the rougher the bore, the sharper the bends and the greater the speed of flow, the greater the noise, not only in the plant but also in the pipe or duct.

Computational fluid dynamics (CFD) can be used to predict natural ventilation rates in new buildings or when substantial modifications are to be carried out to existing buildings.

Outputs required
Whole-dwelling mechanical ventilation

Ideally, the air supply and extract rates for MV should be tailored to suit the pollutant production rates in the dwelling, but pollutant production depends on the number of occupants and their life-styles. It is, therefore, difficult to predict with accuracy. A whole-house ventilation rate of 0.5 ach, with some additional extract ventilation in the wet rooms during moisture production, will generally keep water vapour and other pollutant concentrations below accepted maxima. The total extract airflow rate during normal operation of the MV system should be equivalent to between 0.5 and 0.7 ach based on the whole dwelling volume, less an allowance for background natural infiltration. Individual room air change rates will be significantly higher, perhaps 2–5 ach in rooms with an extract terminal.

A facility to boost the air extract rate from the kitchen during periods of cooking, and perhaps from the bathroom during washing, is highly desirable. Many systems increase the total extract airflow rate by higher fan speeds but there are several other methods. They all have their merits and it is difficult to give definitive guidance on the amount of boost required. However, it is suggested that an increase in extract airflow rate of 50% in a single room, or 25% for the whole dwelling, would be a reasonable minimum.

With the facility to boost the air extract rate from the bathroom or shower room, care should be taken that the replacement air entering the bathroom does not cause unacceptable draughts to the occupant.

Supply air is normally ducted to all 'dry' (ie habitable) rooms: bedrooms, dining rooms, living rooms, studies etc. There are normally no extracts in these rooms. Total supply airflow rate is usually distributed in proportion to the room volumes.

Air is normally extracted from the 'wet' (ie service) rooms: kitchens, bathrooms, shower rooms, utility

rooms and WCs, to remove odours. There may be an extract terminal in a larger dry room to help avoid excessive extract rates in wet rooms; this is typical where a dwelling has one kitchen and one bathroom but many dry rooms. Recirculation by the system of moist air from the wet rooms to the dry rooms should be avoided in dwellings.

With MEV, outdoor air must be admitted into the dwelling to replace extracted air. With a leakage rate of 4 ach at 50 Pa, it is probably sufficient to rely on adventitious air leakage. In more airtight buildings, small openings, such as trickle ventilators, are needed in each room to avoid excessive depressurisation by the ventilation system. These openings are unnecessary with MVHR systems.

Balanced flue (room sealed) combustion appliances are preferred in dwellings fitted with MV because their operation is not affected by any pressure difference between the inside and the outside of the dwelling, and they do not have any permanently open air leakage paths. The exceptions are some combined warm air heating and ventilating systems which may incorporate an open flued heat source with a fan driven combustion air supply (see BS 5864[189]). Flueless appliances may be used but, apart from cooking appliances, are not recommended because of their high production of water vapour and other pollutants.

In buildings containing several flats or maisonettes, a central extract system can be used which draws air from kitchens and bathrooms in all units via ducts and a roof mounted fan. These systems normally operate on a 24 or 18-hour basis and ensure reasonable levels of background ventilation in all units. The advantages are that the need for individual humidistat controllers is eliminated and the likelihood of any individual householder not using their extract system is greatly reduced. However, the systems require maintenance by the building operator to ensure continuous use. Information on the system should be brought to the attention of the occupants to stop them tampering with the extract grilles.

To reduce the heat loss resulting from mechanical extraction of air from a house, the cold incoming air can be heated, via a heat exchanger, from the warm outgoing air. Heat recovery can be included in a whole-house ventilation system or with an extract or input fan.

Non-domestic mechanical ventilation

Mechanical ventilation systems installed in some industrial buildings may be complex in design and handle large air supply and exhaust rates (Figure 3.31). The very largest installations may deal with tens of thousands of cubic metres per hour. Commissioning procedures, concerned with flow rates, for example, generally are not always capable of detecting how efficient the system is in supplying air to those locations where it is needed. However, this efficiency may be expected to have a significant impact on the energy consumption and indoor air quality, and affect the response of the ventilation system to changes in load.

Modern control and energy management systems promise to improve individual comfort and reduce energy consumption at the same time. However, fully automatic control is only part of the answer: the user interfaces, both for the individual and for building and organisational management, also must be better understood. Studies have revealed that control systems need to be matched more closely to the way in which buildings are

actually used and managed, particularly in multi-tenanted buildings. Otherwise systems tend to be left on, causing considerable energy waste, particularly with air conditioning. Individual occupants require systems not only to provide comfortable conditions but also to respond rapidly to alleviate discomfort when it is experienced.

Assessment of designs

Established methods of control for HVAC applications assume particular kinds of system behaviour in controlling the performance of many plant and environmental variables. However, many systems display marked characteristics that can lead to instabilities and other problems; therefore controllers must be tuned to give a very slow response in order to minimise these problems. Poor tuning in turn leads both to poor comfort levels, to poor energy performance, and to excessive wear in plant items. These problems are exacerbated during the lifetimes of plant. Degradation can be partly compensated for by adaptive control, but with only limited success[191].

The BREEZE computer program to evaluate ventilation airflows in large buildings was published in 1991. BREEZE is a suite of integrated programs for IBM PCs or compatibles operating under MS-DOS. The analysis program applies methods of network flow computation to calculate balanced mass flows and pressure differences throughout a building. Auxiliary results include volume flow rates equivalent to the analysed mass flows and certain purely local volume flows arising from temperature effects[192].

The design of terminals is crucial to satisfactory performance. Terminals attached to the outlet of a ventilation system can affect the airflow in three ways:
- inducing suction due to wind action
- causing blow-back due to wind action
- causing flow resistance due to a blockage effect

Figure 3.31

A wide range of ventilation systems has been provided for these buildings which have different ventilation requirements

As with air conditioning registers, intake and outlet positions must be sited to avoid pollution and be covered in mesh against the entry of birds etc. Inside the building, grilles must be sited to avoid short circuiting and draughts, and to heat the whole space. Extract grilles should incorporate a dust filter and, when fitted in a kitchen, a grease filter. The system should include variable fan speed control or variable damper control, or both, to permit air extract boost during periods of cooking, showering or washing.

Noise and other unwanted side effects

Noise

So far as domestic installations are concerned, there are three main sources of noise from a fan system:
- fan noise
- regenerated noise
- vibration and re-radiated noise

Fan noise is caused by turbulence of air within the fan casing rather than bearing noise from the impeller shaft. Regenerated noise is caused by changes in air velocity as it moves past an obstruction, particularly at bends and junctions. All fans vibrate to some extent, depending largely on

Figure 3.32
Insulated ducts for mechanical air extraction on the roof of a building containing photographic laboratories. If not insulated, condensation would form within the ducts when warm moist air reaches the cold surface

running speed, and this vibration can be passed on to the building through the mounting, and re-radiated from other parts of the building fabric[193].

Although small duct sizes using high velocities may save space, there is the inevitable drawback of increased noise from the system. Noise caused by air flowing in the ductwork (regenerated noise) can be reduced by minimising velocities and pressure drops across duct components, and by having duct terminals, branches and bends well spaced. Ventilation ducts should be sized to give air velocities below 4 m/s during normal operation to minimise noise. Greater air velocities are usually acceptable for boost operation. Whistling from ductwork usually indicates an air leak in the duct walls.

The MVHR unit or MEV fan may be sited anywhere provided there is adequate access for cleaning and maintenance, and that any noise produced by the system will not disturb the occupants or their neighbours. In some cases it may also be necessary to fit flexible couplings between the ductwork and the MVHR unit or MEV fan to prevent transmission of vibration. For small installations, mounting the MVHR unit or MEV fan housing on a 50 mm thick mineral wool slab, laid on a sturdy board, may be sufficient.

Ductwork must be designed and constructed not only to avoid noise generation and transmission, but to be airtight and capable of being cleaned internally. It must be insulated where it passes through unheated spaces. Other things being equal, noise will be lower from circular ducts than from rectangular ducts of the same cross-sectional area, though lengths of lightweight flexible ducting should be kept as short as possible. A sound absorber may be needed in the air supply duct.

Vibration transmission problems can occur, both within the dwelling and with adjoining dwellings, when they are mounted on timber or other lightweight walls. These problems can be difficult to cure because ordinary acoustic isolators are unlikely to be effective. The best fan

mounting positions are on heavy concrete slabs or brickwork.

With respect to non-domestic buildings, many of the same considerations apply as in the domestic field, although in many circumstances, such as industrial premises, a higher ambient noise level may be encountered than can be tolerated in the domestic situation.

Further guidance will be found in the *CIBSE Guides*[162,163] and in manufacturers' literature. See also *Ventilation and acoustics*[194].

Contaminated air supply

BRE has concluded that some methods of protecting homes against radon can indirectly cause problems of their own. In a small number of cases, sumps and mechanical ventilation systems may interact with open flued appliances, causing toxic gases to spill into living spaces. Radon sumps can also cause an unacceptable level of noise for occupants[195].

The possibility of unacceptable internal air pollution levels may cause concern at the design stage given the potential for cross-contamination between building ventilation intakes and exhausts. The complexity of airflows around buildings, however, makes it extremely difficult to predict the contamination levels at the intake locations. A wind tunnel technique has been developed which uses a model of a proposed building to determine the pollutant levels that can be expected at various inlet locations when noxious emissions from two outlet stacks are re-ingested[196].

Condensation

If the MEV fan or MVHR unit is installed in a heated area, the risk of condensation on the cold casing surfaces can be reduced by insulating and then wrapping them with a vapour control layer (many units are factory insulated). Condensation may also form on the internal surfaces of ducts which are routed outside the main building envelope (Figure 3.32).

Heat exchangers should be provided with a condensate drain. This can be run directly to the outside if protected from freezing, but it is often convenient to connect it to a waste outlet. The duct network also needs insulation and vapour control layers[180].

Heat recovery

The principles of recovering heat from extracted air can also be used with individual extract fans. Small wall mounted units incorporate cross-flow heat exchangers and two fans so that air drawn into the room can be heated from the warm extracted air. Up to 70% thermal efficiency, in terms of heat transfer between exhaust and incoming air, is in theory possible, but average efficiency is below 50%. Capital costs are much greater than for extract systems and the heat exchanger requires periodic cleaning.

The air intake and discharge grilles are very close together on single room heat recovery ventilators (Figure 3.33), so they easily suffer from short-circuiting. In other words, the same air keeps recirculating through the unit. There is as yet no standard test regime to mitigate this problem.

Heat exchange units are unlikely to be cost effective (taking into account the extra electricity usage for a double fan system and for overcoming the resistance of the heat exchanger) except where exhaust air temperatures are likely to be high over long periods of time and there is a requirement in a particular room for heating.

Ventilation for tumble driers

Most tumble driers exhaust large quantities of moisture into the atmosphere; this can cause serious condensation problems if the exhaust terminates within the dwelling. All dwellings arguably should have a vent fitted through an outside wall (sleeved through the cavity) in a suitable location for exhausting a tumble drier. Gas fired tumble driers may discharge their products of combustion through the same duct as the discharged moist air.

Health and safety
Products of combustion

MEV and MVHR installations are not recommended in dwellings fitted with open flued combustion appliances (any appliance which draws some or all of its combustion air from the living space) since these systems might interfere with the operation of appliances and combustion products could be drawn back into the living areas. For most types of appliance, regulations require that an air supply is provided by means of an air vent of specified sizes through an outside wall of a dwelling. These vents, and the open chimney or flue, reduce the airtightness of the dwelling and, therefore, its suitability for mechanical ventilation.

Where MV systems are fitted, all extract points should be treated as if they were extract fans, and a combustion product spillage test carried out (see BRE Information Paper IP 21/92[197] and BS 5440-1[89]) with the MV system operating so as to generate the maximum achievable pressure difference across the appliance. In very airtight dwellings the operation of an appliance might be affected by extract points in other rooms.

Fire precautions

Ducting within a kitchen and connected to the cooker hood (if fitted) should be made of steel. A fire damper to close off the cooker hood's air extraction opening is essential in all installations and should be as close as possible to the hood. Quick acting fire dampers, such as a spring loaded or gravity operated flap with thermal release, are preferred.

If metal extract ducting is used then a fire damper or thermal fan cut-out switch should be fitted in the air extract system upstream of the MEV fan or MVHR unit. If plastics ducting is used, for supply or extract, fire dampers should also be fitted wherever the duct passes through any floor and through those ceilings and internal walls which are required to be fire resisting.

Radon

MV systems should not be installed without expert advice in dwellings which may have high radon levels. There is some further information in Chapter 1.5 of Floors and flooring which lists the available references, but other advice may be required.

Durability

Galvanised steel ducting not carrying any corrosive products can last 30–35 years. Plastics ducting may give shorter lives, and is not so robust as steel or able to withstand mechanical damage during maintenance etc. Fan life is also considerably shorter, some 15–20 years.

Unpublished BRE investigations dating from the early 1980s put the expected lives of heat exchangers at around 20 years. It does not appear that things have changed significantly in the intervening period.

The lifespan of fans installed in non-domestic MV systems, of course, depends heavily on their original design, the nature of the environment in which they are installed, and how intensively they are operated, though an indicative figure might be around 15 years. Fan life is discussed in Chapter 3.1.

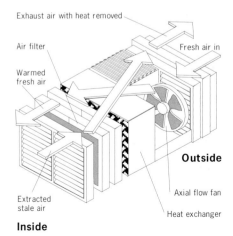

Figure 3.33
A heat exchange ventilation unit

Terminals on the outside of a building will be subject to weathering as well as carrying vitiated air. Unsuitable materials can deteriorate quite quickly – even stainless steel in adverse conditions (Figure 3.34).

Air handling equipment in swimming pools needs careful monitoring for corrosion. For stainless steel units and equipment, a visual inspection twice a year is suggested[198].

Maintenance

A master on–off switch should be mounted on or near the MVHR unit or MEV fan to isolate the system electrically during cleaning and maintenance.

The terminal fittings must be durable and easy to clean. Extract grilles should incorporate a dust filter and, if fitted in a cooker hood, a grease filter. Cooker hood grease filters may need cleaning monthly to prevent contamination of the ductwork and heat exchanger.

Where fans are noisy, it may be possible to remedy noise and vibration problems by enclosing the fan with a quilt lined box, or modifying or boxing in the duct system to reduce noise breakout. Reducing the speed of airflow will also help. In the last resort, it may come to replacing the fan with a quieter model.

Cleaning intervals depend largely upon the location and effectiveness of a system's air filters. Air filters, and air supply and extract grilles, will probably need to be cleaned at least two or three times a year; the heat exchanger in MVHR systems annually. Fan impellers can be inspected, and cleaned if necessary, when the heat exchanger is cleaned. The inside of ductwork rarely needs cleaning.

Work on site

Workmanship

In domestic buildings, in unheated areas, ducts should be insulated with the equivalent of at least 25 mm of insulating material with a thermal conductivity of 0.04 W/(mK). Cold air ducts should be wrapped additionally with a vapour control layer outside the insulation. The same should apply with the air supply duct, between the fresh air intake and the heat exchanger, where it passes through heated areas.

Vertical exhaust ducts should be fitted with condensate traps. Horizontal exhaust ducts should slope away from fans to prevent condensate running back and contacting electrically live parts.

Where ducts or vents pass through an outside wall or attic floor, they must be sealed to the building envelope to maintain airtightness.

The fresh air intake and extract air outlet should always be outside the building (on, say, a roof, wall or soffit) and away from any noise-sensitive areas such as bedroom windows. Intake and outlet ducts must not terminate inside the roof space because this would increase the risk of high humidity levels and condensation.

On exposed sites, it may be better to position the intake and the exhaust on the same side of the building. This helps to reduce the effect of wind generated pressures on the building which can unbalance the system airflow. The extract air outlet should be installed some way from the fresh air intake to minimise

short-circuiting; a separation distance of 2 m should be adequate. Combined intake and outlet fittings may be used if they keep the two airstreams separate.

The fresh air intake fitting must be installed away from boiler and chimney flue outlets, foul and surface water drain vents and contaminated air outlets. Intake and outlet fittings should have a louvre, cowl or similar device to prevent rain, birds, and large insects from entering the system.

So far as non-domestic buildings are concerned, many of the same measures and precautions apply as in the domestic field, though of course the scale of the ventilation plant is greater, and the consequences of defects probably much more severe. In particular, wind induced pressure differences are much more severe on tall buildings than they are on low-rise ones.

Figure 3.34
Baffles corroding on an austenitic stainless steel air handling grille

Inspection
The problems to look for are:
◊   user instruction manuals not available
◊   regular cleaning not carried out
◊   equipotential earth bonding missing from steel duct frames
◊   inlets and outlets too close together
◊   cowls or screens absent from inlets and outlets
◊   low level outlet grilles causing draughts
◊   condensation traps absent
◊   flue spillages when fans operating
◊   carcassing insufficiently airtight
◊   ductwork with too many bends
◊   fans noisy
◊   whistling from air leaks in ducts
◊   fire dampers absent
◊   metal ducting and grilles corroding

Chapter 4   **Piped services**

Figure 4.1
Part of the services installation for a large swimming pool

This fourth chapter deals with all kinds of piped services to buildings and the methods for disposing of waste products via pipes. These services include mains water supply, hot and cold water supplies, soil and waste systems, and sanitary fittings; also gas supply and refuse disposal using piped systems.

Chapter 4.1   **Cold water supply and distribution**

By 1991, ninety-nine out of every one hundred dwellings in England were provided with cold water from a mains supply. However, there was still a small residue of dwellings at that time (around one in a thousand) which had no piped water whatsoever[26]; most of this residue was supplied from wells. The situation is still understood to be much the same. Dwellings without piped supplies tend to be in rural areas above the 200 m contour level, especially those also remote from established communities. The Farm Survey of 1940[199] reveals considerable numbers of relatively prosperous households completely without piped water. However, even though piped water might have been available to some rural dwellings, it was commonly confined to the

kitchen sink, and rarely supplied to the remainder of the house until after the 1939–1945 war. Considerable progress has been achieved since that time.

The specific properties of drinking water supplied for public consumption varies widely in degree of hardness, and the amount of chemical treatment which is necessary for adequate purification. Hard water contains more calcium and magnesium than soft water which is often shown by build-up of 'lime' deposits within pipes and on the surfaces of sanitary ware (Figure 4.2).

Until 1999, water was supplied in England and Wales by statutory water undertakers who were controlled by the provisions of the Water Act 1945. Installations in

buildings were required to comply with relevant byelaws made under the Water Acts. These byelaws, based on the Model Water Byelaws[4] primarily covered prevention of waste, undue consumption, and misuse and contamination of water. Byelaws have been replaced by the Water Regulations 1999[5] which came into force on 1 July 1999. (See feature panel opposite for an outline of the legislation in England and Wales.)

Scotland is covered by regulations for public and private water supplies: the Water Quality (Scotland) Regulations 1990 (as amended) and the Private Water Supply Regulations 1992. Additionally, Scottish local authorities have water byelaws, the Water Authorities Enforcement Regulations, which were updated in April 2000 and are used to implement the European Drinking Water Directive.

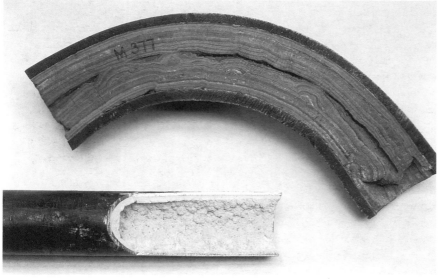

Figure 4.2
Two examples of furring in water pipes . Deposits in once-through hot water systems are generally calcium or other insoluble carbonates (top) but, in systems which recirculate water, different types of scale occur (bottom)

The intention of the Water Regulations is to provide for a less prescriptive approach than the former byelaws while maintaining the primary imperative to prevent contamination of water supplies. The emphasis is on performance standards that must be met rather than by specifying particular devices or technology that must be used. The intention is that manufacturers will be able to introduce innovative products to achieve the same levels of performance but in a variety of ways. The requirements are based on harmonised European standards, or where these do not exist, on national standards.

The main exception to performance standards is in the area of back-flow prevention, where the Regulations remain prescriptive in order to protect public health. Here, water types are covered by five 'fluid categories' ranging from 1 where the water 'is of a quality that is not impaired by any change in taste, colour or odour' (ie mains water) to 5, where the fluid represents a serious health hazard because of the presence of pathogenic organisms, or radioactive or very toxic substances (see Table A of the Water Regulations).

A variety of devices exist for back-flow prevention. Table B of the Water Regulations describes these devices and notes their suitability for protecting against both back-pressure and back-siphonage for a specific fluid category.

Conservation of water is another prime area for the Water Regulations where the intention is to reduce the maximum flush volume of WCs from 7.5 litres to 6 litres for new WC suites; there will, though, be a transition period of two years. Dual flush WCs will be reintroduced and a substantial reduction be required in the maximum volume of water used per cycle in washing machines and dishwashers.

It is expected that consumers will appreciate the savings that these measures will give, and that the environment will benefit from lower abstraction levels leading to restoration of river flows.

Water quality

Little comment is necessary on water quality from public supplies. Water with the equivalent of 50 ppm of calcium carbonate is classed as soft; over 350 ppm is very hard. Hardness removed by boiling is 'temporary' hardness; that remaining after boiling is 'permanent' or 'non-carbonate' hardness, usually due to the presence of calcium and magnesium sulphates[200].

Although unfortunate lapses occur from time to time, and contamination of supplies occurs by human error or from causes unknown, these are outside the control of building industry professionals. However, private water supplies are a different matter entirely. It will be necessary to make individual arrangements to monitor and test private supplies regularly to ensure that vegetable, mineral or bacterial content remains acceptably low. A private supply to a house examined by a BRE investigator in rural Wales in the early 1990s was discovered to have high levels of manganese. This was a comparatively rare case since the prescribed concentration value (PCV) for the year in which the test took place was exceeded in just over 1% of the total number of tests in England and Wales. In another location, in Scotland, at about the same time, the water was only marginally acceptable for bacterial content. In the Scottish case the supply was drawn from a shallow collecting chamber sunk alongside a stream where there was the likelihood of surface run-off getting into the supply; this particular parameter was exceeded in around 7% of the determinations in Scotland for that year[201].

In the past, a wide variety of apparatus has been employed to deliver water supplies to dwellings: gravity, bucket and chain, force pumps, lift pumps, rotary pumps and hydraulic rams. Although these are rarely to be found in working order now, their remains are often identifiable.

In the last years of the twentieth century, the policy was still to supply all water from the public mains of a quality suitable for drinking, even though only a small proportion of this was used for drinking and cooking. Depleted rainfall figures and the consequent reduction of river flows, and reservoir and aquifer levels – most years of the last decade of the twentieth century have been officially classed as drought years – has led to water shortages in many areas of the UK. This policy will not be sustainable indefinitely. Greater use of grey water – or water defined as waste water not containing faecal matter or urine, and which is not of drinkable quality – may become preferred or even obligatory for applications such as flushing WCs, car washing, and garden watering. However, most black water, defined as waste water which contains faecal matter or urine, will still need to be treated off-site for the foreseeable future.

Grey water and rainwater

Grey water is water that has already been used in the household for washing purposes and is subsequently used for another purpose that does not require water of drinkable quality; for example, for flushing a WC.

Recycling of water is possible but may present potential problems, and the need for filtering and disinfecting. Waste water from sinks, baths and basins can be used for irrigation purposes but, if it is used for WC flushing, soap and other contaminants may be deposited in the cistern's mechanism. Other problems associated with recycling water are storage, contamination and separation from the drinking water supply[†].

---

† BRE is currently assessing the implications to health and hygiene of using grey water in domestic properties[202].

Bacteria can proliferate in a grey water collection tank if the water is not filtered and chemically or otherwise treated, causing odours and presenting a hazard to health. The grey water needs to be collected in a tank at low level; it is then pumped to a header tank at a higher level where it is treated and stored for use in the dwelling†.

One of the most popular techniques for saving water is to use rainwater for tasks where wholesome water has normally been used, though large storage capacities are required. Potential uses include WC flushing, garden and window box watering.

Rainwater, collected into roof tanks and piped to the kitchen sink or wash boiler, used to be a familiar feature of houses in a number of areas of the UK in the 1920s and 1930s. In at least one West Midlands town, many new houses built during

† Further information on suitable installations is available from the BRE Water Centre.

these years were equipped with an open topped, rectangular galvanised steel tank of around 2 m × 3 m × 0.3 m containing upwards of 1,500 litres of water (Figure 4.3). The considerable loads from these tanks were usually carried on reinforced concrete slabs over attached external WCs and fuel stores built of loadbearing brickwork. Of course, the tanks needed to be drained and cleaned regularly, but there is no doubt that the water was especially popular for the weekly laundry, as the soft water was found to need less soap than the local mains-supplied hard water. These were the days before detergents.

Water conservation

Standard charges based on the old rateable values of domestic properties has been the common method of paying for water. One possible means of reducing overall consumption is to install water meters (an option available to existing customers of water authorities but compulsory for

new properties) which measure the volume of water consumed: it is presumed that knowledge by users of rates of water usage should help to reduce profligacy and so conserve supplies. There are, however, many social, political and financial implications. Metering of supplies to non-domestic buildings is usual.

BRE has been involved in the search for acceptable water conservation measures since the early 1960s. A number of developments were identified at that time; for example, WCs which incorporate dual-flush cisterns, user controlled flush cisterns and low-volume flush WCs (approximately 4.5 litres compared with the previously normal 9 litres). On the same basis, waterless urinals, conventional spray showers, atomised spray showers, spray taps and handwash units, and low-water-use automatic washing machines and dishwashers have been identified as promising ways of saving water. Water savings of about 50 litres per person per day would be possible. Many of these economy measures could be installed in existing dwellings as well as in new or refurbished ones[203].

In more recent years, the efforts to reduce consumption even further have been stepped up. Ways of conserving water now include measures to:
- reduce the demand for water
- improve the efficiency of water using appliances
- reduce the loss and waste of water

A reduction in water usage increases the availability of limited supplies during a period of low rainfall. This may enable the continuity of water supply during a time when drought orders would normally have to be issued and avoid the inconvenience to consumers of restrictions in water use (eg by introducing hosepipe bans and standpipes).

Local water companies now actively pursue schemes for installing suitable water saving devices: dual flushing for WCs, low capacity cisterns, and showers instead of baths, for instance – all designed to promote economy in use.

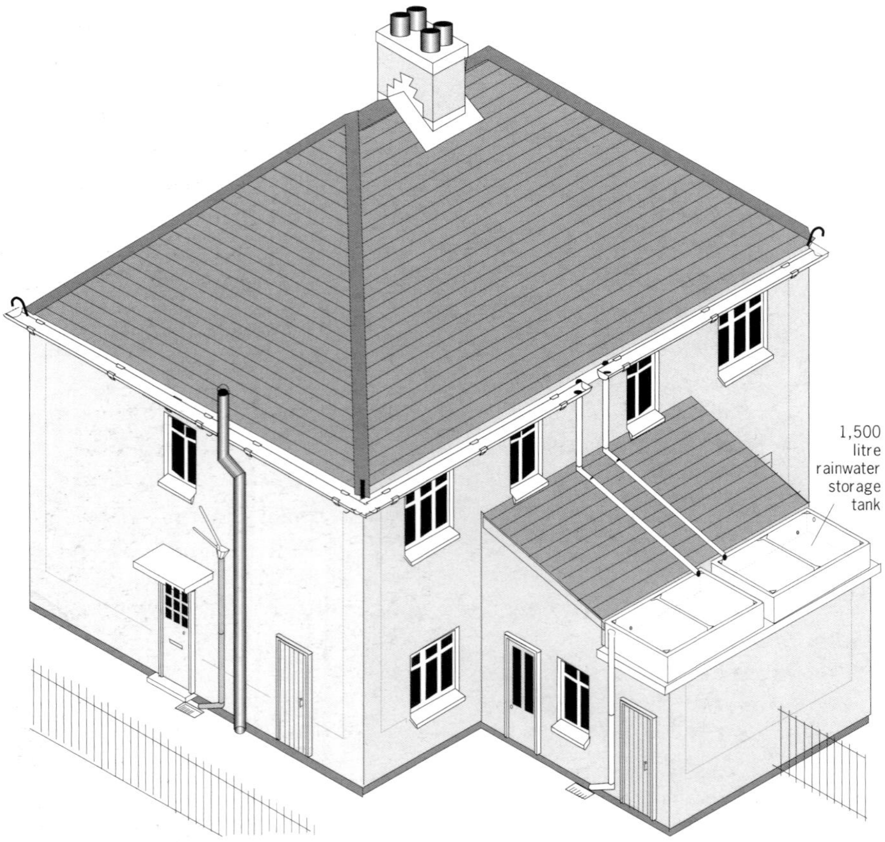

1,500 litre rainwater storage tank

Figure 4.3
Rainwater storage tanks on the roofs of 1938-built semi-detached houses in a West Midlands town. These installations were common in this part of the UK, with whole estates being equipped in this way, although the practice certainly was not universal

Increasingly, automatic leak detectors are becoming available in the UK. These devices, which are fitted into the incoming mains, close when a leak is detected and so are able to prevent wastage of water and damage to property. Some operate by sensing a high flow rate and others use conductivity detectors to activate valves[202].

The reduction in water use has implications for energy consumption, both directly and indirectly. There are energy requirements in supplying water and treating the subsequent wastewater; these include energy use for pumping and ancillary equipment. For example, a family of four in the UK can use 220,000 litres of water a year. This requires 120 kWh of energy to pump it to the point of use and 100 kWh to treat the subsequent wastewater. Generating this amount of energy releases 200 kg of carbon dioxide into the atmosphere. Water conservation measures must be considered in the context of the total amount of resources used; that is, the total water and energy consumed. Water savings might be offset sometimes by an increased energy use; for example, domestic composting toilets use little or no water but may require electrical power (see Chapter 4.3).

A reduction in the amounts of water consumed can alleviate damage to the environment by:
● maintaining water flows in rivers and streams
● maintaining groundwater levels
● eliminating or postponing the need for the construction of new reservoirs required to cope with a rising demand or to meet shortfalls in existing demand

Water storage

Water which is stored, even though it might have originally been of drinking quality, may deteriorate. For many building types other than housing, in which large quantities of water are used for sanitary and washing purposes, supplies may be prone to interruption making it necessary to store large quantities of water which then becomes unfit for drinking. In those circumstances, drinking water is normally supplied separately, directly from the mains.

There has been extensive debate over the need to store both drinking water and water required to service appliances, including WCs. In the mid-twentieth century, local water undertakings required storage arrangements of certain minimum capacities as a precaution against failure of the public supply. It is arguably the case now that interruptions in the public supply are so infrequent that all outlets can be supplied off the mains with no storage. This practice is already widespread in, for example, France. Common practice in the UK seems to be in the process of changing from the storage of these comparatively large quantities of cold water to systems fed directly from the mains. Where cold water storage for drinking purposes is employed, it may be necessary to disinfect the system before use. (See the Water Regulations 1999[5] and BS 6700[204]).

Many factories using large quantities of water, such as breweries and distilleries, use their own private supplies where available.

Water metering

Metering water supplies is a technique that has social, political and financial implications. Presently, the majority of domestic water used in the UK is charged on a tariff which is based on the previous rateable value of a building, though this practice is due to be phased out. A charge based on the volume of water consumed is an alternative to the rateable value method but requires some form of metering (Figure 4.4). Commercial and industrial buildings are already commonly metered.

Metering trials have been conducted in twelve areas of England, covering a range of geographical areas and social groups: a total of 56,570 households. A survey carried out by OFWAT indicated a high level of acceptance of water metering, 72% of those questioned. It was concluded that the installation of water meters, and their use for charging for water on the volume used rather than on a flat rate basis, could be used to encourage water conservation. The actual tariff system adopted for metered dwellings will affect the potential water savings that can be achieved. New metering technology allows for rising block and other types of tariff. The relative contributions to the total water bill from standing charges and charges related to the volume of water used are important for creating financial incentives for water conservation.

The majority of metered householders in the study, 59%, attempted to reduce their water consumption. This was achieved by a number of methods including:
● less plant watering
● less toilet flushing
● taking showers instead of baths
● using washing machines less frequently
● sharing baths, bath water, or showers

Even though installing meters may be costly in existing buildings, there are hidden benefits such as detecting existing leaks. Many buildings built in recent years have been fitted with a water meter or provided with suitable connections for installation later. Although current trends indicate an increase in unmetered water consumption, the costs of reading and maintaining the meters will have to be considered in making final judgements on their suitability.

Figure 4.4
A typical water meter installation

Characteristic details

Pipe runs

Most dwellings in the UK have a single supply, although BRE investigators have occasionally seen dwellings with more than one supply. In certain areas it was also common in the period between the wars to supply more than one dwelling from a single pipe, although more recently each dwelling has tended to have its own supply. On one estate built in the 1930s examined by BRE, groups of at least four houses were served off a single supply, and when any of them had any plumbing work done, all four houses temporarily lost their supply.

BRS Digest 15[205] describes common practice for domestic applications during the 1960s; there has been little change in intervening years. It is normal to find that pipe runs have been concealed wherever possible, though easy access to all taps and joints was required at the time of construction.

Pipe runs in non-domestic buildings built in the 1990s will probably have been installed in ducts to the requirements of BS 8313[112], giving the necessary clear space requirements to accommodate flanges and supporting brackets. Most small single pipes need ducts of about 100–150 mm × 100–150 mm cross-section (for ducts without removable covers), with pro rata increases where more than one pipe is installed. Where removable covers are installed, spaces are often reduced. However, for earlier buildings, a very wide range of practices will be found.

Taps and valves

Leaking, float operated inlet (ball) valves were at one time commonplace, leading to considerable wastage of water and the need for frequent replacement. These valves were invariably of a type (called the Portsmouth valve) in which a sleeve supported a plunger carrying a small rubber disc which was pressed against a nozzle, so sealing off the incoming water. Friction between the plunger and its sleeve, together with the accumulation of scale deposits, impeded full closure of the valve. The end of the nozzle also became pitted due to erosion and cavitation, and the rubber disc no longer sealed effectively. Furthermore, the design tended to be very noisy in operation. The BRS or Garston ball valve was perfected by Sobolev in 1956 (Figure 4.5). By substituting a diaphragm for the rubber disc the BRS valve overcame the defects of the Portsmouth type and also operated with greatly reduced noise. Diaphragm ball valves are now in common use, as are the so-called quiet equilibrium, float operated valves (Torbec type).

More recent developments for taps have included the ceramic disc and single hole taps which are not suited to supplies at differential pressures. The increased use of unvented systems, with their higher operating pressures, have enabled taps of European design to be used in the UK.

Storage tanks

Cold water storage tanks are usually positioned at high level within buildings so that draw-off points can be fed by gravity. However, in some very tall buildings, where the loadings from water storage would have a significant effect on the design of the structure, storage may be at low levels with pumped supplies to high levels.

The fear of interruption of mains supplies, and possible contamination of mains supply by back-siphonage from mains-fed sanitary appliances, has led to this almost universal use of large storage tanks in all building types. Of course, the provision of a storage tank can be a useful measure to accommodate any interruption in the supply. The only exception normally allowed was one tap at the kitchen sink to draw water of drinking quality directly from the main.

Since cold water from wash hand basins is often used for drinking, there would seem to be case for requiring these to be on mains and not the stored supply.

Water softening or conditioning

Water softening is the process in which calcium and magnesium salts are removed from cold water supplies. It may need to be done to reduce the deposition of calcareous deposits on the internal surfaces of systems, and sometimes even for health reasons.

There are two basic processes: the first is separately dosing with various chemicals before filtration; the second is by incorporating chemicals into the process of filtration itself, and activating from time to time by the addition of common salt. The former tends to be used on the larger installations, and the latter for domestic. Water conditioning operates on a less-well understood principle of preventing limescale precipitation by magnetic means.

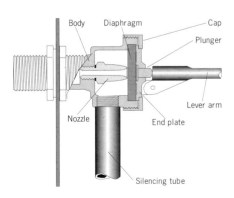

Figure 4.5
Prototype BRS Garston ball valve

Table 4.1 Minimum rates of flow at point of delivery	
Appliance	**Minimum rate at mains pressure (l/s)**
Sink, 15 mm pipe dia	0.20
Bath, 22 mm pipe dia	0.30
Wash handbasin, 15 mm pipe dia	0.15
WC flushing cistern	0.10
Bidet	0.15

Main performance requirements and defects

Strength and stability

Where storage tanks are sited in the roof space they need to be properly supported. The adequacy of support to these storage tanks has been found, in BRE site investigations of quality in housing, to be defective in a large number of newly built dwellings (Figure 4.6). The topic was covered in Roofs and roofing, Chapter 2.1.

If loads are concentrated on too few trussed rafters, or on cross-members set at some distance from node points, they could be at risk. In addition, many supports are of chipboard which deforms and deteriorates, when leakage occurs or when condensation drips from uninsulated tanks. Tanks of 230 litre capacity should be supported over at least three trussed rafters, and those of 330 litre capacity on four; and both tank sizes on substantial bearers on the rafter joists[206].

Outputs and system requirements

Although it is difficult to generalise, some idea of the relative amounts of water consumed in domestic and office buildings in the UK can be gained from Figure 4.7, even though there is a wide variation in actual use depending on such factors as socio-economic group and region. In the Figure, external domestic use includes water for gardens and car washing.

Cold water taps and stop valves need to be leakproof when closed, and should control and deliver water in adequate quantities, at not more than a specified noise level, over their design lives. The taps should be comfortable to use and easy to operate. There may be additional requirements, such as elbow operated taps in healthcare buildings.

The cold water system should supply water at mains pressure to the kitchen sink for drinking purposes and to any cold water storage tank. Stored cold water at sufficient rates of flow needs to be supplied at all times to all non-mains outlets while providing a small reserve. Complete

drainage of the system, including the rising main, should be possible without dismantling.

When preparing performance specifications for the selection of fittings, the flow rates shown in Table 4.1 (see page opposite) may prove helpful.

BS 6700 gives flow rates for other installations. These flow rates are easily checked on site by timing the filling of graduated containers. Other criteria are:
- stored water pressures to be not greater than 300 kPa at point of delivery
- mains pressure not to exceed 600 kPa. If supply is in excess of this value, a pressure reducing valve should be installed
- cold water storage capacity in relation to household consumption; for a household of four people, say 450 litres
- the system to be sterilised before use (eg chlorine dioxide of 50 ppm)
- inlet valves to cisterns to be of a type appropriate to the pressure
- warning pipes are to have a cross-sectional area not less than twice that of the supply pipe
- cold water pipes at high levels wherever possible to be run beneath hot water pipes, or the hot pipes to be insulated

- mainly in relation to future maintenance work, cold water tanks and system pipework to be installed so that they are not obstructed by, and do not obstruct, other equipment and pipework

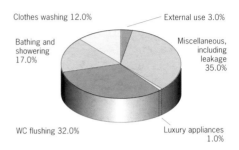

Typical use of water in UK households

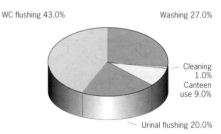

Typical use of water in offices in the UK

Figure 4.7
Typical use of water in domestic and office buildings in the UK

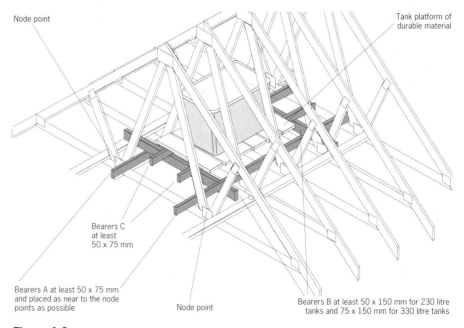

Figure 4.6
Recommended supporting arrangements for cold water storage tanks (bracing and binders are not shown). Tanks have been often found to be inadequately supported

A bath uses about 80–90 litres, a shower 30 litres (but a powered shower may use more), a dishwasher 20–50 litres, and a washing machine 100 litres. Old standard flushes for WCs took up to 13 litres whereas newer models use as little as 6 litres. A hosepipe can use 600–1000 l/h. Total household consumption averages 140–190 litres per day per person.

Unwanted side effects
Back-siphonage
The system must not prejudice the supply; that is to say, there should be no possibility of contamination of the supply. The risk which has given most concern over the years is that of back-siphonage. This is where stored water is siphoned back into the system against the normal pressure, thereby creating a risk to health. Pressures in public supplies can fall below atmospheric pressure at times. Under these conditions, if the supply inlet to the tank is well clear of the surface, air would be drawn into the supply pipe and no harm would result. If, however, the supply pipe is very near to, or, worse still, actually below the surface of the stored

Figure 4.8
Demonstration of back-siphonage with a 40 mm air gap

water, contaminated water might easily reach the mains (Figure 4.8). A major study of the risks of, and prevention of, back-siphonage was published in 1974[207]. Since that time, preventive measures have been strictly observed, and the risk of contamination is now thought to be very low.

Condensation and risk of freezing
Condensation on storage tanks is probably inevitable, even where insulated, because of gaps in the insulation and migration of moisture laden air into the roof spaces in which tanks are usually situated. It is common to see signs of condensation having run off onto the supporting timbers. Arguably, trays of sufficient depth to contain any condensation should be placed beneath tanks so that condensation water can be removed by hand or by natural evaporation in suitable conditions; or the insulation should be fitted with a vapour control layer. Lids must, in any case, be fitted; not just to reduce evaporation, but to keep out birds and insects which penetrate or bypass the screens at eaves. Condensation on cold rising mains is common, and can only be prevented by meticulous attention to pipework thermal insulation. The incidence of condensation can, of course, be reduced by ensuring ventilation of roof spaces; particularly by providing airways at the eaves and ridge vents. Blocking eaves with insulation material will encourage condensation.

It could be argued that thermal insulation should be provided for cold water pipes even those well out of reach of frost action - not for the purpose of fuel conservation but to give a measure of protection against condensation which could lead to wood rot, even insect attack, if excessive. If a vapour control layer is fitted against any insulation, it should have a resistance of not less than 250 MN s/g.

Noise
From measurements taken in dwellings while water was being drawn from the supply system, BRE investigators found noise levels ranging generally between 35 dB(A) and 55 dB(A), and occasionally up to 60 dB(A). Not all of this noise emanated from flows of water in the pipes. Taps can be noisy, though spray taps are quieter than other taps. Attenuation values measured for plastics pipework suggest that a short length of sound absorbent pipe could prove useful for isolating a noisy fitting from the main pipe system. Bellows-type pipe vibration absorbers gave attenuation values of 5–15 dB(A) according to circumstances[208]. For an explanation of noise descriptors, see Chapter 1.9 of *Walls, windows and doors*.

Noise from a cold water storage tank can be caused by the inlet valve and by water running into the tank, and be particularly disturbing at night if the tank is just above the bedroom ceiling. The remedy is to fit a quiet type of inlet valve, but it must discharge above water level to prevent back-siphonage.

Vibration in any part of a plumbing system can pass readily along the water pipes to other parts of the house. The pipes themselves may not radiate much sound but, if they come into contact with or are fixed to partitions or ceilings, they may, in turn, cause vibration in these components; the resulting noise can be appreciable. As a general rule, heavy walls will radiate less sound than light ones. If the pipes are firmly fixed at short intervals to walls (or any other well-founded structural element), sound transmission will be limited; more noise will be heard, though, from the walls nearest the source, especially if they are of light construction. Noise transmission through walls was dealt with in *Walls, windows and doors*, Chapter 1.9.

Water hammer

When water flowing in a pipe is stopped suddenly by rapid closure of a valve or tap, a high impact pressure is produced leading to a surge or wave which rebounds from the valve and passes back down the pipe. In domestic installations this can occur when the water is turned off at a sink or basin. The pressure wave travels to the stop valve, which may be of the screw-down pattern with a loose washer plate or jumper; the jumper, being free, oscillates until the wave action is damped out by friction. This results in the knocking sound known as water hammer or impulsive noise. The pressures created by this effect in small rigid pipes can rise to several times the normal pressure, possibly leading to damage to the pipe system.

Water hammer may also occur when certain kinds of innovatory valves or self-closing taps are fitted.

The stop valve is not the only source of water hammer. Water hammer might occur at an inlet valve when a tap elsewhere is closed, the sudden increase in pressure in the pipe causing the float to oscillate; or an inlet valve discharging into a cistern may set up waves on the surface of the water and so develop a self-induced oscillation which results in water hammer. These effects are liable to occur when the float exerts insufficient force, through the lever-arm, on to the inlet valve piston. A possible remedy is to increase the size of the float.

Water hammer should not arise in taps on the hot or cold circuits from the storage cistern because of the low pressure and the absence of restriction on securing the tap jumper. Water hammer in taps is sometimes attributed to worn washers. There is no clear evidence on this point, but replacing worn washers is always desirable in the interests of water economy.

Renewing a tap washer in another material may stop or lessen water hammer, but, in difficult cases, inserting an air reservoir to absorb the maximum pressure or using short lengths of flexible pipework in the system is required; in the latter case the material must be chosen carefully.

Commissioning and performance testing

Pipes and fittings should be to the appropriate British Standard for the material concerned.

Where copper tubing is specified for pipework, it should be clean and any carbon deposits caused by drawing lubricants be removed; otherwise there is a risk of pinhole corrosion.

Where it is deemed necessary to carry out a proof test of a cold water supply installation on completion, one and a half times the working pressure for 30 minutes may be specified.

Flow restrictors are available readily and can be fitted to many appliances, but their use must be appropriate. They are a cheap way of reducing water wastage when taps are left open by careless users and where items are washed under running water. A more effective, but more expensive, solution would be to install taps operated by proximity sensors.

Health and safety

The temperature of cold water in storage tanks should be maintained below 20 °C, and if possible below 15 °C, to prevent the growth of the legionella bacteria. (See Chapter 2.4 for a full description of Legionnaires' disease. Also see the next chapter, Chapter 4.2, for a description of BRE studies of microbiological conditions in hot water storage installations).

Although the risk to health from bacteria is probably greater in non-domestic installations, even in some domestic systems, particularly large ones, it may be necessary to consider precautionary measures against bacterial growth.

Because rats and birds require easily accessible supplies of water with their diets, water in tanks should not be allowed to be uncovered; tanks must be covered with close fitting, vermin-proof lids. In roofs, this will both deny water to pest species and protect the supply from contamination with droppings if the roof should become infested. While the provision of tank covers is mandatory in new installations, they

should be fitted to existing installations too. Screened vents are now required for ventilation purposes in cold roof spaces.

Pests may gain access to parts of buildings where pipes pass through walls, floors, ceilings or foundations. Polyurethane foam, flexible silicone and acrylic sealants can be effective for sealing around small pipes. Where pipes or cables pass through fire barriers, the continued integrity of sealants and barriers must be regularly checked. If service pipes are subject to significant movement due to expansion or contraction, some form of tightly fitting sliding seal should be considered.

Waters containing organic acids or free carbon dioxide are plumbo-solvent. A danger to health is probable if water flowing through a lead pipe contains more than 0.1 mg/cm^3 of lead.

Durability

The properties of water that most affect the durability of plumbing installations are:

● the nature and concentration of impurities
● the ability to form a protective scale

Of the mineral impurities, the chlorides and sulphates are the most aggressive to metals. In addition to mineral impurities, natural waters also contain dissolved oxygen and carbon dioxide which influence their corrosive behaviour. Oxygen is dissolved in varying amounts in natural waters, usually within the range of 2–8 mg/cm^3.

The composition of natural waters is governed by the nature of the ground from which they originate. Surface waters from lakes, rivers or reservoirs are maintained by rainfall in which the main dissolved constituent is carbon dioxide – up to 5 ppm. If the ground within the catchment area is relatively insoluble, the water will be soft and its acidity (low pH) will be controlled by the amount of decayed vegetation or peat within the area. Alternatively, if the water drains over or through chalk or limestone, it is likely to be hard.

Figure 4.9
A corroded galvanised steel domestic cold water storage cistern removed after 35 years' use. There is no sign of rusting on the outside surface, but rusting inside shows that the cistern is at imminent risk of perforation. The vertical pipe in the cistern is the warning (overflow) pipe, almost completely blocked by corrosion

The chemical composition of deep well waters depends on the ground in which they are located, but the composition (and temperature) are much more constant than for surface supplies.

Iron and steel

For supplying fresh water, bare iron or steel is unacceptable owing to the severe rusting that occurs. Galvanizing gives effective protection against almost all hard waters but not against acid waters or soft waters with a free carbon dioxide content of 30 mg/cm^3 or more (Figure 4.9). The ability of the water to put down a protective, inert calcareous scale is of prime importance in considering the suitability of galvanized steel.

Pitting corrosion of galvanized components may arise in mixed metal systems from the deposition of copper from the water on to the walls of a cold water storage tank: as little as 0.1 ppm of copper is sufficient to encourage this process.

Most waters will pick up this quantity of copper from new pipework; however, continued pitting of cold water cisterns from this cause is more likely to occur by back-flow of copper-bearing hot water through the feed pipe or drips from the vent tube.

If copper is added to an existing galvanised system, the bore of the galvanised pipe will already be coated with a film of scale, and copper bearing water will not therefore come into contact with the galvanised steel. In cases where the water flow is from the galvanised steel to copper pipes (eg copper tails at taps), there is not a problem. It is not uncommon for galvanised systems to be successfully repaired with copper pipe.

Lead

Although lead was used widely for the supply of wholesome water until the mid-twentieth century, belated recognition of its harmful effects means that it is not now used. Unfortunately, lead as a material is quite durable if fatigue and creep failure are avoided. Some dwellings are still served by lead piping.

Copper

Almost all waters react with copper but the action ceases as a protective scale forms on the metal. A cupro-solvent action takes place in water with a high free carbon dioxide content. The amount of corrosion is too slight to have any detrimental effect on durability and, since copper, unlike lead, is not a cumulative poison, amounts of up to 1.5 mg/cm^3 can be tolerated. Amounts in excess of this react with soap, producing green staining on bathroom fittings. Lower concentrations greatly accelerate pitting corrosion of galvanised steel tanks and aluminium components.

Cupro-solvency can be a problem in some areas where it can lead to the thinning of the walls of copper pipes. In these areas it is not advisable to use copper pipework.

Copper hot water storage cylinders have sometimes developed leaks through pitting corrosion after comparatively short periods of service. This has occurred only with hard, or moderately hard, deep well waters and outbreaks have been unpredictable and sporadic, with most of the installed tanks and cylinders not being affected. The apparent cause of the trouble lies in the type of oxide film which forms on the surface of the metal when it is first put into service. In most waters, copper develops a thin oxide film which is protective but, in the troublesome areas, it occasionally produces a film which breaks down and allows pitting to occur. The formation of a protective film can be ensured by controlling the electrochemical potential of the copper while the film is forming. Fitting an aluminium rod inside the tank or cylinder and connecting it to the copper produces the right result.

A problem recently identified, called rosette corrosion of copper, occurs primarily in copper cylinders and is associated with using aluminium electrodes and nitrate bearing waters.

Under certain conditions copper tube may suffer intense localised pitting corrosion. This has occurred when the tube contains an almost continuous film that is cathodic to the underlying metal in the pipe. The copper corrodes at breaks or fissures in the film. Since 1960, mechanical cleaning of tubes after manufacture has dramatically reduced, though not entirely eliminated, the incidence of attack.

In soft moorland waters containing manganese salts, a cathodic scale has been found to be deposited slowly in the hotter parts of the system leading to pitting corrosion.

Plastics

Since the 1980s, there has been a significant growth in the use of tubing manufactured in various kinds of plastics. The general requirements for plastics pipework installations, so far as the materials are concerned, are to be found in BS 7291-1[209]. The particular materials are covered in the other Standards in the same series: polybutylene[210], polyethylene[211], and chlorinated polyvinyl[212].

Brass

Many components used in water services consist of hot-pressed (alpha-beta) brass. While satisfactory in very many waters, zinc may be lost from the alloy in alkaline waters, producing a porous copper layer which may extend throughout the wall of the fitting and cause water seepage. When this occurs, the zinc dissolved out of the brass forms a bulky corrosion product and eventually blocks the fittings. While high tensile (alpha) brasses may be made immune from dezincification by incorporating 0.03% arsenic, no simple remedy will prevent attack on hot-pressed brass, and alternative copper or gunmetal fittings must be used.

Cavitation erosion

Cavitation erosion occurs mainly at seatings of taps and ball valves where water speeds are high and where the direction of flow changes. Cavities or bubbles filled with water vapour may form in the water stream, and these collapse when reaching a region of higher pressure. The sudden compressive stress causes erosion of adjacent solids, and generates a considerable amount of noise (Figure 4.10). Cavitation appears to be a relatively rare phenomenon.

There is, though, another form of pressure-related erosion-corrosion which is more common. This occurs when calcium carbonate deposits on the inside of copper plumbing systems erode, leaving the bare metal to corrode[208].

Cavitation can also occur in parts of pumped systems in which water is turbulent due to water speed being too high; or it can be due to changes in the direction of the pipework or to a reduction in pipe diameter, thereby increasing water speed.

Frost

Freezing pipes in homes were relatively common events until the central heating revolution in the 1960s. Nevertheless, it does happen occasionally when houses are left unheated for long periods in winter conditions, and plumbing systems are not drained.

Thawing frozen pipes by passing heavy electric currents through them, although possible, can be hazardous since the current may pass along paths other than those intended. The consequences are serious risks of fire and electrocution. Expert handling of equipment is required[213].

Warning pipes which drip and are left unattended can lead to large icicles forming in frost conditions.

Maintenance

Changes in the layout of cold water systems, together with provision of additional or improved features, may result in substantial, perhaps total, replacement of piped water services and fittings. However, wholesale replacement may not always be essential, and careful appraisal should be undertaken to decide what parts may be capable of economic reuse, especially where only relatively minor changes are envisaged. External appearances alone, however, are not an adequate guide to the suitability of pipes and fittings for reuse. All retained water

Figure 4.10
Mechanical erosion of a ball valve seating due to cavitation. Collapsing bubbles generating sudden compressive stresses can erode pipes and valve seatings

supply fittings should be approved by the local water authority or the National Water Council (NWC), and fitted to avoid back-siphonage and consequent health risks. Water authorities do not allow 'dead legs' to remain where remote appliances have been permanently removed. Existing systems may already contain such defects.

Ageing taps, siphons and ball valves will probably be worn or damaged and not easily maintainable. In hard water areas, pipes, fittings, cylinders, tanks, radiators and immersion elements may be furred up. In areas with aggressive water, sacrificial anodes may be exhausted and brass fittings, in particular, may be corroded and furred. Washers or discs need to be easily replaceable.

Early pipework will be in imperial sizes and replacement fittings and adaptors are becoming increasingly rare. Rehabilitation provides a good opportunity to trace and overhaul or replace external and internal stopcocks, and to renew (and re-route if required) the water main, particularly if it is of lead.

Work on site

Workmanship

Workmanship should be to the requirements of BS 8000-15[214].

If fittings are of current British Standard pattern and good quality (non-dezincifiable perhaps), it may be better to overhaul rather than replace with, possibly, inferior components. Pipework should be flushed through and pressure tested where reuse is anticipated. The incoming water main should be traced, and stopcocks and fittings operated to check that they still function and have a reasonable lifetime ahead of them.

Pipe runs should be arranged to avoid, where possible, notching of timber joists. Where notching has to be used, it is to be confined to the end third of the length of each joist with the pipes protected against puncturing by nails by suitable means (eg 2 mm steel plate). This topic was dealt with in Floors and flooring, Chapter 2.1.

Cold water service entry pipes must be suitably protected from frost action (see *Floors and flooring*, Chapter 1.3). Pipes in unheated areas (eg in outside toilets) can be protected by electric heat trace cables.

All pipes concealed in hollow floors arguably should have the routes marked on the sub floor surface.

Copper alloy holderbats, strap clips of copper, copper alloy or plastics, and purpose made straps and hangers may be used to fix copper tubes to the structure.

Inspection

The problems to look for are:

◊ diameter or capacity of pipes not adequate
◊ pipework liable to airlock
◊ pipes inadequately clipped
◊ stop valves leaking or ineffective
◊ pipes in unsuitable position to isolate, drain down or flush out
◊ pipework obtrusive
◊ water pipes vulnerable to damage or in unknown locations
◊ water pipes or cisterns liable to freeze (Figure 4.11)
◊ hot and cold water pipes uninsulated and touching or too close, leading to heat transfer
◊ pipes in contact with outside walls
◊ insulation on cisterns insufficient
◊ pipe entering buildings at insufficient depths below ground
◊ cold water storage tanks in roof spaces inadequately supported
◊ platforms for storage tanks deteriorating due to condensation

Figure 4.11
An uninsulated cold water storage cistern in a loft. In spite of all the publicity, systems vulnerable to frost action can still be found

Chapter 4.2   Domestic hot water services

The introduction to this book (Chapter 0) referred to the provisions for heating domestic hot water alongside open fires in many cast iron ranges of the late nineteenth century. Few buildings then had piped supplies. Before that time, domestic hot water had to be heated in pans over the fires used for cooking.

Water can be heated for domestic use either directly by means of what is known as a primary circuit, or indirectly by a secondary circuit. With the primary circuit, all the water to be heated passes through the heating appliance. With the secondary circuit, water recirculates through a closed circuit comprising, basically, the heating appliance and a heat exchanger; in the exchange unit, the heat is transferred to water in another circuit to become the domestic hot water. There are considerable advantages with secondary circuit systems, particularly in hard water areas where scale deposition within systems is reduced.

In the mid-twentieth century it was sometimes the practice to equip dwellings with multi-point, open flued, instantaneous gas heaters, and not to have any storage capacity. The very largest multi-point heaters had the capability of supplying only around 10 litres per minute. This was not really a sufficient flow for filling a bath in a reasonable time. Nevertheless many households made do with this equipment.

Problems with maintenance, and in particular accidents involving deaths from carbon monoxide poisoning have made them less popular. Where there is no central storage of hot water, or where the storage is a considerable distance from the outlets, it is now usual to install a small gas powered balanced flue boiler on the inside of an outside wall (Figure 4.12). Electrically powered installations are also feasible. The need for improved designs has now been realised in the increasing use of the combi-boiler, which heats both domestic hot water and hot water for space heating purposes from the same installation. (For a description of combi-boilers, see the section, 'Main performance requirements and defects, Combi-boilers', later in this chapter).

The years since the turn of the nineteenth century have witnessed a sea-change in common practice in the supply of hot water. Just over two-thirds of dwellings in England now have domestic hot water supplied from a central heating boiler, with the backup of an electric immersion heater. However, 1 in 7 dwellings still rely solely on an immersion heater. Half of these immersion heaters use off-peak electricity[7].

A proportion of dwellings in the UK still have inadequate or no thermal insulation to tanks and pipes in loft spaces, and to hot water storage cylinders.

Figure 4.12
A small gas fired boiler dedicated to instantaneous supply of domestic hot water. This example was about to be replaced after 10 years' service (Photograph by permission of B T Harrison)

Characteristic details

Space requirements for major items of plant (dimensions etc)

Vented systems

Until the 1980s, storage-type domestic hot water systems in the UK typically used a storage vessel supplied with cold water from a cistern and vented to atmosphere through an open vent pipe (Figure 4.14). The other common domestic hot water system used an instantaneous water heater taking cold water either from a cistern or directly from an unvented mains supply (Figure 4.15).

In houses, domestic hot water (DHW) is usually provided by the same boiler that meets space heating requirements. Summer operation, supplying hot water only, will be met by either the boiler or an immersion heater. In non-domestic buildings, DHW is usually taken from a separate boiler to improve summer efficiency.

Unvented systems

Unvented DHW systems have been permitted since 1986. A former typical, though now less commonly found, directly heated, storage-type, domestic hot water system is shown in Figure 4.16; and one heated indirectly by means of a primary coil and secondary circuit is shown in Figure 4.17. Figure 4.18 illustrates a system that includes a primary coil for use in conjunction with hot water space heating, and an electric immersion heater that heats the water directly for use in summer when space heating is not required.

Expansion vessels are required to take up changes in water volume; cold water filling points with pressure gauges are also needed. Normally components are bought as a package with all the safety devices in place. There are regulations on the safety aspects, not only on the risk of explosion but also on back-siphonage into the mains supply. This is definitely not a situation for DIY installation.

The advantages of unvented systems are that there is no risk of cold water freezing in the roof space,

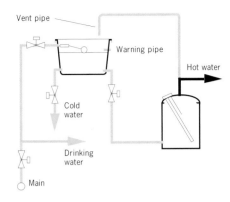

Figure 4.14

A typical storage-type domestic hot water system

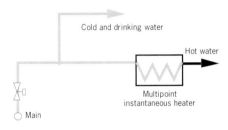

Figure 4.15

A system with a multi-point instantaneous heater

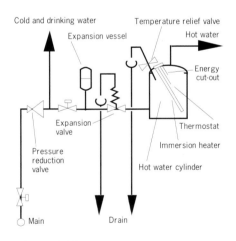

Figure 4.16

A directly heated unvented system

no airlocks, and, with combi-boilers, hot water is available at mains pressure.

With each installation, the source of heat is controlled by a thermostat, supplemented by a temperature operated energy cut-out that interrupts the heat input should the thermostat fail; the storage vessel is protected by a temperature operated relief valve. A separate vessel

Figure 4.13

A domestic installation for storing hot water heated mainly by solar power accommodates expansion of heated water where back-flow is prevented. In case this fails, and to prevent expansion of trapped water splitting the hot water storage vessel, an expansion relief valve is fitted. The pressure of the water supply to the system is controlled by a pressure reducing valve and, to achieve the advantages of balanced hot and cold water pressures, the cold water outlets in the building are also fed from this same control valve.

An unvented hot water system requires the same back-siphonage protection as a cold water system that is connected directly to the mains. In a single dwelling this is usually provided by air gaps or other protection at each outlet or point of use. It is also important to ensure that, even under fault conditions, hot water cannot enter the pipework to cold water outlets; this is achieved by including a check valve in the pipe feeding cold water to the hot water storage vessel. It is also necessary to ensure that the water level in the storage vessel does not fall to such an extent that the functioning of the temperature operated safety devices is impaired. In most cases the hot water storage vessel will be designed to withstand negative (vacuum) pressures as well as positive pressures, but where this is not the case, provision for vacuum relief should be provided; temperature relief valves that incorporate this function are available[215].

Hot water storage capacity

In the early 1950s, where the output from a boiler was very small (eg from a back boiler of an open fire, or from a small solid fuel space heating or cooking appliance), it used to be recommended that the capacity should not exceed 30 gallons (136 litres). If these figures were exceeded, an unsatisfactory service could result from difficulty in reheating the water within a reasonable time after a heavy draw off. Storage capacities up to 40 gallons (182 litres) were recommended with independent boilers if justified by the boiler rating[20].

In practice, although these conditions no longer apply and virtually any reasonable capacity is possible, cylinder sizes remain much the same as they ever were (Figure 4.19). BS 853[216] may be relevant. Although 120 litres capacity is usual for small dwellings, larger dwellings with more than one bathroom will require larger storage. BS 6700[204] recommends a minimum capacity of 100 litres for a solid fuel installation, 200 litres for an electrical off-peak, or 45 litres per occupant, whichever of these produces the largest figure unless the design of the system justifies a smaller capacity. A five-person house would therefore need 225 litres. High performance cylinders with rapid heating coils are now available which reduce the time

Figure 4.19
A factory insulated, domestic hot water storage cylinder photographed during the 1985 BRE quality in housing site studies

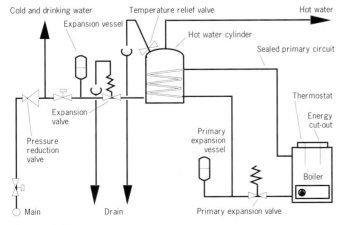

Figure 4.17
An indirectly heated unvented system. Current practice has changed, and installations will often have motorised valves on primary circuits together with cylinder thermostats and energy cut-outs

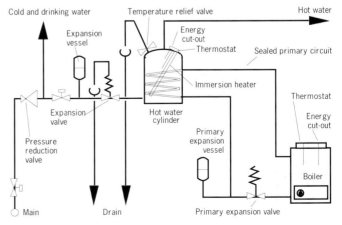

Figure 4.18
A directly and indirectly unvented heating system This is similar to Figure 4.17, but with an added immersion heater

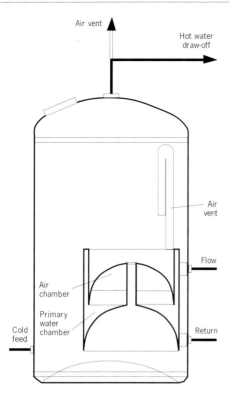

Figure 4.20
A 'primatic' cylinder

taken to reheat a cold cylinder, and may also reduce boiler cycling.

Modern practice is to pump both primary and secondary circuits with priority for hot water if required. Recovery times can be more rapid than for gravity systems.

Some cylinders may be found in which the water for the primary circuit is obtained from inside the cylinder instead of from a separate header tank. These units were known as 'primatic' cylinders (Figure 4.20). They operate on the principle of air separation of the primary and secondary heating circuits, with expansion and any necessary air venting taking place within the unit, and with the primary system being automatically replenished from the secondary whenever required.

Boiling water heaters

A boiling water heater adjacent to the sink can replace the use of kettles. These devices are not in common domestic use, but may be found, for example, in offices and where it is difficult or hazardous for people to lift kettles.

Pipe and trunking runs

Hot water pipes should be designed and supported so that they do not make noise when expanding and contracting.

In hard water areas, indirect hot water cylinders reduce the amount of scale deposited in boilers by avoiding continually heating fresh water.

For sizes of ducts, see the appropriate section in Chapter 4.1.

When choosing sizes for pipework for unvented systems, it should be borne in mind that the pressure head available is often greater than the head provided by a storage cistern; significant economies can sometimes be made by using smaller pipe sizes: 15 mm instead of 22 mm and 22 mm instead of 28 mm. However, pipework upstream of the separation into hot and cold distribution systems, including the service pipe, may need to be larger than usual.

Whenever cold water is delivered via the feed pipe to the hot water storage vessel and heated to the controlled storage temperature, it expands in volume by up to 4%. In most unvented DHW installations, expansion will be accommodated in an expansion vessel. This contains a diaphragm with air or nitrogen on the other side, compressed to the system operating pressure (ie the outlet pressure of the pressure reducing valve). As the water heats up, it expands and enters the expansion vessel and the air is compressed, the system pressure rising with it. If the expansion vessel is too small for its duty, the pressure will rise until the expansion valve opens and releases water to waste. The vessel must, therefore, be of sufficient capacity for the maximum expected expansion to be accommodated without increasing the system pressure to the point where the expansion valve opens[215].

The discharge pipe for unvented DHW systems should be of a metal suitable for the temperature conditions – at least of the same bore as the valve outlet, throughout its length. Discharge should be via an air break to a tun-dish or equivalent, the pipe laid to a continuous fall and be not more than 9 m long unless its bore is increased to compensate.

Main performance requirements and defects

Outputs required

The water heater is the second largest user of energy in the home after space heating. Therefore, by reducing the amount of hot water used, some of the energy needed to heat it can be saved. In the UK, domestic water heating consumes about 75 PJ each year. Also, a reduction in the delivered water temperature of 5 °C will produce an energy saving of about 10%[202]. Energy is saved by insulating against heat losses from the hot water system – not always done in practice (Figure 4.21), by reducing water temperatures, by using controls to improve appliance efficiencies, and by taking showers instead of baths. Insulation is extremely cost effective and controls are well worthwhile, but measures which employ heat exchangers and extra plumbing may not be economic. In new construction, recent changes to building regulations have been introduced to improve tank and pipe insulation.

A further economy measure to consider is heat recovery from waste water, but equipment is not usually installed for this purpose unless the volumes of waste hot water are considerable.

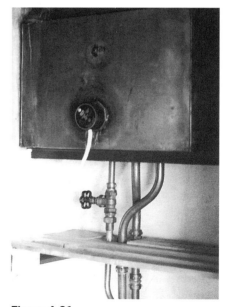

Figure 4.21
An uninsulated domestic hot water system found by BRE investigators in 1985. These can still be found in most parts of the UK

Hot water systems as a whole

Technical Appendix F of the *Housing Manual 1949*[20] recommended that facilities should be provided in all new four-person houses for supplying each week at least 250 gallons (1140 litres) of hot water at 140 °F. This was then said to provide for 7 hot baths per week, 10 personal washes per day, normal washing up requirements, 10 gallons (45 litres) per week for house cleaning and 50 gallons (227 litres) per week for laundry. More recent requirements are contained in BS 6700[204], Clause 6.11.4.3.

The hot water system should be capable of supplying domestic hot water at sufficient rates of flow and at all reasonable times, at temperatures to suit user needs. Complete drainage of the system ought to be possible from dedicated drain-off taps and without dismantling. The installation must not prejudice the water supply; that is to say, there should be no possibility of contamination of the supply. Flow rates required at taps etc should be similar to those for cold water given in the previous chapter.

When preparing performance specifications for the design of a hot water system, the following guide lines may prove useful.

- The system ought to be capable of supplying water at a temperature adjustable between 50 °C and 70 °C, and at a flow rate of not less than 0.45 l/s for a period of up to 15 minutes in any one hour

- Minimum flow rates at each of the individual outlets should be not less than those given in the previous chapter for cold water

- Thermostats should be subject to a tolerance of ±2 °C in both switch-on and switch-off modes

- The predetermined temperature, ±3 °C, at any outlet, should be reached not later than 10 seconds after flow commences

- Expansion loops and resilient packing should be used in fixing the pipes to the building carcass

If the delivery rate is specified on an hourly basis, this will automatically include recovery time. The maximum draw-off rate given above is approximately equivalent to a conventionally sized cylinder with a full recovery time of half an hour.

The overall economic benefit of DHW production by a common boiler, which also serves the space heating load, may be marginal when compared with separate installations, unless there is an appreciable DHW load. There will, of course, be very different conditions between summer and winter conditions; the following guidelines are suggested.

- If the average DHW load is large (with a similar order of magnitude to the space heating load), a separate appliance is recommended. In these cases the load in the summer months will still be significant. High DHW loads are likely in hotels and catering establishments, for example

- Smaller DHW loads are probably better met by separate appliances, especially if these can be decentralised, though there is little experimental evidence one way or the other

- Very large DHW loads (much larger than the space heating load) will usually benefit from a dedicated condensing hot water appliance. For industrial processes in particular, 'direct contact' hot water heaters should be considered; but for DHW, direct contact heaters are not currently recommended

Unvented domestic hot water storage systems

Mains pressurised DHW systems were first introduced where there was insufficient headroom to provide an adequate flow of hot water from a tank feed; for example, in a flat or where it was deemed preferable not to have a water storage tank in a loft space. The design and installation of unvented domestic hot water systems is covered in the Building

Regulations Part G3. This system was illustrated in Figures 4.16 to 4.18.

The conventional feed and expansion tank can be replaced by direct connection to the mains water supply and by providing an expansion vessel if not already accommodated in the cold feed pipework.

The Approved Document recommends:

- temperature and pressure relief valves, and a non self-resetting thermal cut-out in addition to the control thermostat

- for a system with a boiler, that the non self-resetting thermal cut-out is connected to a motorised valve so that, if the temperature of water from the boiler becomes excessive, it is diverted from the primary heating coil

- all safety devices to be factory fitted and all functional controls to be supplied either factory fitted to the cylinder or in a kit form supplied by the manufacturer

- the system is serviced regularly by a competent person

The British Board of Agrément has published a Method of Assessment and Test for unvented hot water storage systems[217] †.

Temperature of stored hot water

The thermostat in unvented systems should be set to control water temperature at the lowest acceptable value so that energy wastage is minimised. For domestic use, 60 °C is recommended, though some users may prefer a higher temperature; having a higher temperature, though, increases the risk of scalding, especially where children or elderly people are concerned. An absolute upper limit is imposed by the requirement that the water temperature must be controlled to a value that will not operate the energy cut-out unless a malfunction has occurred. The cut-out operates at

† At the time of writing, 16 Agrément certificates for domestic hot water systems were current. This scheme was the only one in operation until 1991, but since that time third party certificates for products may now be issued by other bodies such as the Water Research Council and Wimlas.

Combi-boiler replacing existing installation

A 1970s built bungalow was originally equipped with an open flued gas boiler which served a radiator system for space heating, and domestic hot water via a calorifier and storage cylinder. When the dwelling was upgraded in 1987, with additional thermal insulation and replacement double glazed windows, the opportunity was taken to install a more efficient boiler of smaller capacity than the original. At the same time, the occupants were persuaded that they could release more usable space in the rather cramped accommodation by doing away with the hot water storage cylinder, and installing a combi-boiler which would deliver both hot water and space heating.

When the BRE investigator visited the site in 1999, the occupants complained about the quality of service obtained from the new system. With the old installation, they had been used to drawing domestic hot water for more than one appliance simultaneously, but they could no longer do so. When the boiler was supplying one appliance, the other appliances would lose their instant supply. In particular, the quantity of hot water to the bath was considerably reduced, and filling the bath took much longer than before.

The occupants were advised that these symptoms were characteristic of the system they had installed. Although it was possible for a larger capacity boiler and larger diameter supply pipes to be fitted, this was not an economic proposition. If they were to maintain the advantage of lower gas bills, they would have to get used to the lower quality of service delivered by this type of boiler and system.

about 85 °C and, allowing for tolerances and overshoot, this implies that the thermostat should not be set above 75 °C, and preferably not above 70 °C.

In traditional vented systems, the vent pipe releases any pressures generated in the system.

DHSS recommends 43 °C for health buildings which is based on evidence from the Medical Research Council. See also the section on microbial growth and legionella later in this chapter.

Electric immersion heaters

There are around 13 million electric immersion heaters in the UK. In a study of electric immersion heaters carried out in 1991 by the Electricity Research and Development Centre, it was found that nearly three quarters of the heater thermostats examined were set at temperatures greater than 65 °C. They could usefully be set to a lower temperature.

Combi-boilers

Both the domestic hot water storage cylinder and associated cold water storage tank can be eliminated by using the combination boiler – the so-called combi-boiler (Figure 4.22). The boiler incorporates both central heating and domestic water heating direct from the mains via separate heating coils, the central heating coil being cut off automatically when domestic hot water is required. Combi-boilers will always give a supply of (nearly) mains pressure hot

water while storage cylinder systems can run cold when hot water demand is high. Combi-boilers are particularly well suited to smaller dwellings, where hot water demand is limited.

A combi-boiler system is particularly advantageous where space is at a premium – but see the case study on this page.

Combi-boilers are frequently used with sealed space heating systems, though models are available which use sealed or open vented circuits.

All instantaneous water heaters may suffer from pipe scaling in hard water areas as mains water passes directly over the heated surfaces in the heat exchanger. Precautions may therefore be needed in hard water areas to prevent furring in pipes. Water softeners or conditioners should be fitted if scaling is a problem in the area.

Thermal store

In a thermal store system (Figure 4.23), the boiler maintains the temperature of hot water in a tank or thermal store. The hot water is pumped from the store to the radiators which are under thermostatic control. The space heating part of the system can be vented or unvented. Domestic hot water is provided by directly heating mains cold water fed through a heat exchange coil in the thermal store. Mixing the hot water in the store with the incoming cold water controls the tap water temperature. The advantages are that there is no

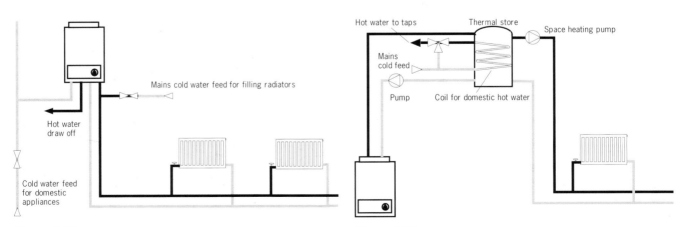

Figure 4.22
A combi-boiler system

Figure 4.23
A thermal store system

risk of the system freezing, hot water is supplied at mains pressure, and there is rapid response to demand for space heating. Because the store is constantly held at a high temperature, it is essential that it, and its associated pipework, is very well insulated to reduce heat losses. As with the combi-boiler, users in hard water areas must not set tank temperatures too high if they are to reduce calcium deposits and furring-up.

Taps

Hot water taps and stop valves must be leakproof when closed, and they should control and deliver water in adequate quantities, at not more than a specified noise level, over their full design lives. The taps should be comfortable to use and easy to operate. In most respects, taps with ceramic disc are superior to the traditional designs, being far less prone to leakage. Their operating characteristics are different, requiring, for instance, less rotation of the spindle from open to closed positions than the screw-down type with conventional washers.

Appropriate standards which incorporate test criteria include BS 1010-2[218] and BS 6864[219,220].

When preparing performance specifications for making a choice between the taps available on the market, the following guidelines may prove useful:

- minimum rates of flow from sink and wash hand basin taps are 0.16 l/s, and from bath taps 0.32 l/s, when fully open and at pressures of 100 kPa, ±1 kPa
- the taps are leakproof (from both washer and glands) at 2 MPa hydraulic pressure after 70 000 cycles using water at a temperature of 75 °C ±5 °C. This criterion gives a factor of safety of about 20 over normal running pressure. There is nothing more annoying than a dripping tap, even disregarding the waste of precious water
- there is no wear which would impair performance after 250 000 cycles
- the handle or crutch not to exceed 40 °C when the water temperature is at 60 °C

- the operating member of a tap should have a shape that permits a torque of up to 3 Nm to be applied to the spindle without discomfort to the wet, soapy hand of an operator
- the jet or spray from a tap with a flow rate of 0.16 l/s should be contained within a 300 mm diameter cylinder and that from a 0.32 l/s tap within a 400 mm diameter cylinder
- the temperature of a hot water tap in a health care building should not exceed 43 °C

Noise and other unwanted side effects

Noise

Water hammer is unlikely to be generated by DHW systems. Most noise, at least from low pressure systems, will come from the float operated valve when the supply tank is refilling.

Explosions

Different means of controlling the upper temperature and pressure limits in water heaters are used in different countries. Where temperature control is applied, the maximum setting for a temperature activated relief valve is generally 99 °C. Where pressure activated relief valves are used to protect a water heater, a situation can arise where superheated water is contained within the vessel under fault conditions. If the storage vessel ruptures before the relief valve opens, flash steam may be produced with an accompanying explosion[221] (Figure 4.24).

As far back as the late 1940s it was noted that, where the temperature of stored water was higher than 140 °F (60 °C), there was an increased risk of scale deposition in hard water areas[20]. The scale is generally calcium carbonate, but other types of scale will be found[222]. Blocked pipes will increase the risk of explosion.

Figure 4.24
The aftermath of an explosion of an unvented domestic hot water installation. This was a test carried out at BRE

Water for domestic purposes is only needed at temperatures below 100 °C; an explosion cannot result when water is released at these temperatures, however high the pressure. Accordingly, requirement G3 of the Building Regulations specifies that there shall be adequate precautions to prevent the temperature of stored water exceeding 100 °C in an unvented hot water supply system of over 1 5 litres capacity. These precautions usually include:

- a thermostat
- a temperature operated cut-out acting on the energy supply
- a temperature operated relief valve

These are designed to operate in sequence as the temperature rises. For the water temperature to exceed 100 °C, all three means of protection must have failed.

Commissioning and performance testing

For systems generally, the following may be considered:

- proof test on completion to one and a half times the selected working pressure for 30 minutes
- proof test of draining the system

Work was carried out at BRE in the mid-1980s to determine potential modes of failure and assess the durability of the three types of temperature control devices which were fitted to unvented hot water systems. Thermostats for electric heaters were shown to be liable to fail with their contacts closed after repeated operation and there was considerable variation in durability between the different products available in UK and other European countries. Energy cut-outs were shown to be liable to fail with their contacts welded closed during short circuits. Improved designs now ensure that unvented hot water systems have energy cut-outs that survive short circuit currents when the electric heater elements fail[223].

Health and safety
Microbial growth and legionella

Poor design and maintenance of water services can lead to conditions where bacteria multiply and become suspended in an aerosol leading to a health risk for a building users. Hot water storage should be maintained at least above 46 °C to prevent the growth of the legionella bacteria. (See Chapter 2.4 for a description of Legionnaires' disease.)

BRE investigators have discovered that levels of bacteria may be higher in expansion vessels than in the incoming water supply. Examination of 35 expansion vessels revealed one installation which contained legionella bacteria and another 30 vessels which had positive bacteria counts. The presence of a biofilm in the majority of expansion vessels is contrary to Clause 3.2(b) of BS 6144[224] which states that materials used in the construction of expansion vessels that are in contact with water should not support microbial growth.

Specifiers and installers need to consider the choice of materials used in expansion vessels and their ability to support a biofilm. British Standard BS 6920-2.4[225], contains a test for determining whether there is microbiological growth. Approval given under the Water Fittings Byelaw Scheme is subject to the membrane proving satisfactory when tested in accordance with the appropriate parts of this British Standard. Even approved expansion vessels have been shown to support microbial growth.

All the expansion vessels examined had a single connection to the unvented hot water storage system; in other words, the connection acted as the inlet and the outlet. Alternative types of expansion vessels are available. One is a modification of the single connection expansion vessel which has integral inlet and outlet flow-ways, allowing water to flow into the vessel each time water is drawn from the system. A second alternative is a flow-through expansion vessel with separate inlet and outlet connections. This device allows a replacement of water contained in the vessel each time water is drawn from the system. These alternative types of expansion vessels do not allow a quiescent zone to occur, even under fault conditions (ie depletion of gas charge), so there is less likelihood of stagnation.

Overall, in specific applications the use of expansion vessels could present a health risk, primarily through the presence of bacteria. Risk sites include healthcare facilities, such as hospitals and nursing homes, where occupants may be particularly vulnerable. Further guidance is available on installing and maintaining expansion vessels for using with drinking water systems to reduce this risk[226].

It is important to avoid water being stored at, and forming aerosols at, temperatures of 20–46 °C.

Risk of scalding

The temperature of water from a hot water cylinder should be controlled thermostatically so that the risk of injury by scalding is minimised. There is a balance to be struck between high temperatures for activities such as dish washing, and lower temperatures for personal washing. Thermostatic valves can be provided to individual fittings at points of delivery but it is preferred that temperatures are controlled at the points of supply. Depending on the length of draw-offs, the temperatures at the supply points should be 43–50 °C, giving temperatures at the points of delivery of approximately 40 °C. The lower temperature for points of supply should be used for health buildings.

Hot water at 60 °C from a tap can cause a partial thickness burn in 5 seconds, whereas water at 54 °C will take around 30 seconds. Lowering the temperature of hot tap water considerably reduces the risk of thermal injury[80].

Where a gas water heater is to be enclosed, provision should be made for an ample supply of combustion air; the free area of the air inlet should be not less, and preferably greater, than the area of the flue socket on the heater.

BRE has seen many instances of blocked or reduced air supplies to open flued appliances supplying space heating, and the same can happen to those appliances supplying domestic hot water.

The air inlet and overflow pipe should be screened against pests using a corrosion resistant mesh. Some lagging material for hot water systems running in the loft space may provide a source of nesting material for birds and rodents. Where this is the case, the lagging should be sealed in a stout or well jointed casing to prevent its removal.

Durability

Excessive scale in pipework, fittings and appliances leads to inefficiencies in plumbing systems because it reduces the flow capacity of pipes, reduces their heat transfer characteristics, and may produce almost complete blockage of heating pipes and impairment of waterways.

In hard or moderately hard areas, galvanized steel cisterns and hot water storage tanks, and galvanized pipework and cast iron boilers, were used for many years without any serious defects. Occasionally, pitting was experienced which may simply have been due to iron filings or rubbish left at the bottom of the tank impeding the formation of a protective scale, or, with hot water cylinders, excessively high temperatures during the early life of the system.

Where metal pipes are still in use, 40 years seems to be an average life expectation, with 25 years as a minimum. If plastics pipes have been installed which are not suitable for the temperatures experienced, they will distort and prematurely fail in use.

Examples of the principal designs of electric thermostats have been subjected to performance and durability tests at BRE, and thermostats removed from service have been examined and tested. Rod-type thermostats used in the UK in electric immersion heaters (Figure 4.25) have an average service life of about five years, although individual units show a great variation about the mean, indicating a significant risk of early failure. Hydraulic (or bulb) type thermostats used in the UK have an average service life of about seven years, but similar units used in Europe, where unvented hot water systems are common, have an average service life of nearly 14 years.

Clearly, durability at least equal to that of the storage vessel is desirable, since it may reasonably be assumed that when either vessel or thermostat fails, both will be replaced.

Maintenance

Every hot water installation needs repair, maintenance and replacement of components at stages during its working life. To ensure safety, it is essential that all work on unvented systems is carried out by personnel who are at least as well qualified and competent as those who installed the equipment.

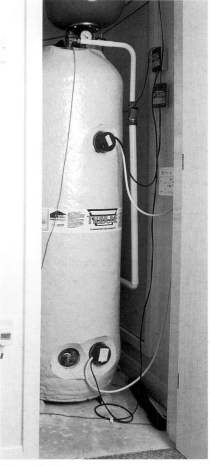

Figure 4.25
Rod-type immersion heaters have an average service life of around five years. This installation has two immersion heaters, one low down for use during off-peak hours, and one higher up for top-up purposes at other times

Work on site

Workmanship
Workmanship should be to the requirements of BS 8000-15[214].

All pipes concealed in hollow floors arguably should have the routes marked on the sub floor surface. Pipes carrying hot and cold supplies should not be run in close proximity to each other. On a number of occasions BRE site investigators have found them touching.

Copper alloy holderbats, strap clips of copper, copper alloy or plastics, and purpose made straps and hangers may be used to fix copper tubes to the structure. Site inspections have often revealed inadequacies in fixings.

It is essential that the installer of unvented DHW systems has a very good understanding of the principles as well as the practice of these systems. The Approved Document requires that installers should themselves be competent. They should be properly qualified to do the work, having successfully completed an additional training course and examination in unvented systems.

Supervision of critical features
Where electric immersion heaters are fitted in hot water cylinders, they should, of course, be thermostatically controlled. It is important that hot water storage cylinders are insulated with the recommended thickness of approved insulation . Vent or expansion pipes should run horizontally for a distance of not less than 500 mm before rising vertically. This will prevent internal circulation of hot water within the expansion pipes.

Unvented systems, being comparatively recent, may present a number of problems, though there are in general no problems with installations which meet the requirements of building regulations. In BRE site studies of installations that did not meet the requirements of Building Regulation G3, the most serious deficiency was the omission of temperature relief valves which occurred with several systems. There were also instances where discharge pipework from combined temperature and pressure relief valves was seen to be unsuitable; these included incorrectly sized pipework and problems with the terminal points of discharge.

Inspection
Because domestic unvented hot water systems represent a departure from traditional UK experience, using third party certification of the equipment and competent installers is vital for safety. However, in addition to factory inspection of the components and equipment, inspection of the completed installation is also essential. This should be carried out by the installer as a final check on the installation process, and should be followed by an inspection by a building control officer (or an Approved Inspector if this form of procedure is being followed), checking for compliance with building regulations and third party certification. The user also may wish to check that he has a safe and adequate system. Third party certification schemes include unannounced checks on work in progress and completed installations.

The issue of a third party certificate gives an assurance of factory inspection of newly supplied units but a further inspection should be made after repair or maintenance work, at least by the approved installer responsible for the work.

The problems to look for in all hot water systems are:
◊ diameter or capacity of pipes not adequate
◊ pipework liable to airlock
◊ pipes inadequately clipped
◊ gate (stop) valves leaking or ineffective
◊ pipes in unsuitable positions to isolate, drain down or flush out
◊ pipework obtrusive
◊ water pipes vulnerable to damage or in unknown locations
◊ boilers or controls obsolescent or inefficient
◊ hot and cold water pipes not insulated and touching or too close, leading to heat transfer
◊ user instruction manuals not provided

Also, in vented systems:
◊ hot water cylinders having inadequate storage capacity, being poorly positioned or with long (uneconomic) legs to taps

In unvented systems:
◊ installers not third party approved
◊ temperature relief valve discharge pipes not easily visible
◊ thermostats not fitted
◊ temperature relief valves absent
◊ discharge pipework from combined temperature and pressure relief valves not suitable

Chapter 4.3

Sanitary fittings

Although the Romans were keen bathers, their facilities tended to be of the communal kind and disappeared when the buildings containing them were destroyed. It was the religious communities in the monasteries who to some extent revived the lavatorium during the Middle Ages. In much later times, public slipper and Turkish baths were built in quite a few cities; many were retained up to the 1960s and a few still survive.

The small private bath used by the aristocracy up to mid-Victorian times had, of course, to be filled by water carried by hand from the kitchen ranges – a labour intensive operation.

Some monastic sanitary installations featured water-borne sewage removal, though it was left to nature to effect disposal or, strictly speaking, dispersal. For other large buildings such as castles, the discharge from the privy or garderobe was frequently to a midden in the open air via a dry chute which regularly had to be cleaned. No wonder cholera was rife!

For ordinary dwellings, from medieval times, of course, chamber pots were used indoors, and the 'night soil' disposed of into middens in the public thoroughfare. Soil was collected by a public carrier with his horsedrawn container. Such practices were common in suburban areas up to the 1914–18 war, before the installation of mains drainage, and were still to be found in rural areas right up to the 1950s – a group of several inhabited houses in one remote area of the country seen by a BRE investigator in 1967 still had no sanitary facilities whatsoever, not even an earth closet, though they each had a single Belfast sink in the scullery. There are still dwellings relying on cess pits which need regular evacuation by suction tanker.

Earth closets, cleaned out annually, were also in common use in rural areas until mains water supplies became available, largely during the 1950s. Composting toilets are installed occasionally by ecologically sensitive householders, and they are referred to later in this chapter.

The water closet came into use in the early part of the nineteenth century, and its development seems to have been more closely related to the supply of piped water than to facilities for sewage disposal. An enormous number of different designs were produced, each with different claimed advantages, but most using a water sealed trap. Some early bowls were made of metal, but most were of glazed earthenware, and in later years of vitreous china[18]. Before the 1880s they were enclosed in wooden frameworks and some even had upholstered seats. The development of the freestanding unenclosed appliance dates from around this time.

This chapter does not include information on, or activity space allocation for, sanitary fittings for use by the disabled, or for specialised appliances and facilities (eg the very large range of surgeons' basins, physiotherapy and birthing pools etc used in hospitals) for which specialised literature should be consulted.

Basic amenities

Basic amenities, defined as kitchen sinks, baths and showers, indoor WCs, and hot and cold running water, are lacking in only around 1% of dwellings in England; of the quarter million or so dwellings involved, around half were vacant at the time of the survey. Only 44,000 dwellings lack an internal WC[7].

Nearly 90% of dwellings in England now have a modern bathroom, defined as constructed or modernised after 1964, and around 1 in 8 dwellings have more than one bath or shower. Around 1 in 3 dwellings have a second WC, with, as is to be expected, the proportion rising to nearly half for those households of five or more people. Most en-suite bathrooms date from 1980 onwards.

In Scotland, around 1 in 6 dwellings have an additional WC, with the greatest proportion, as might be expected, in detached houses[9].

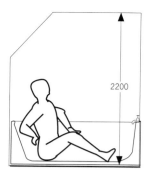

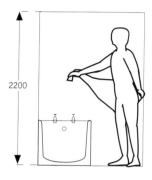

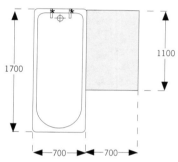

Figure 4.26

Space needed to use a bath

Figure 4.27

A modern shower installation with the waste water outlet in the floor

Characteristic details

The sizes of normally installed appliances are given in the following sections, together with the size of the space necessary for using each appliance – called the activity space. Further information is available in the *BRE Housing Design Handbook*[14] and in Part 2 of BS 6465[227].

Baths

Until the end of the nineteenth century, and even up to the 1939–45 war, where there were no piped supplies of water, baths tended to be portable tubs of galvanised light gauge steel. Before the Miner's Welfare Commission began a programme of pithead bath construction in the 1930s, a collier would come home to a portable bath in front of the living room fire. There were some baths also of glazed ceramics. Plumbed-in baths of cast iron came into almost universal use in the early years of the twentieth century, and were finished in enamel, sometimes highly decorative. It was only during the mid-twentieth century that casing a bath became popular – until then the bath was usually freestanding. Enamelled steel baths were introduced in the interwar years. Showers only became popular after the 1939–45 war.

The current tendency is to baths with lower rims, and for elderly and disabled people this may prove easier and safer for getting in and out. British Standards size baths are 1700 mm × 700 mm with coordinating heights in multiples of 50 mm and other plan sizes increasing in multiples of 100 mm.

Corner baths have become popular in recent years, but their dimensions vary considerably and so does the space needed to use them. They also use much more water than a conventional bath.

European Standards, in addition to the British Standards, are applicable. They include EN 198[228], BS EN 232[229] and EN 263[230].

Activity space

A clear floor space of 1100 mm × 700 mm is needed, with the longer dimension adjacent to one side of the bath (Figure 4.26). This allows for getting in and out of the bath, for drying, and for an adult beside the bath bathing a child. The defined area does not extend for the whole length of the bath and can be positioned at any point along it, although where there is choice it will preferably be at the tap end. The area beside the bath beyond the activity space should be included within the floor area of the bathroom. This will enable bath taps to be reached and the tub to be cleaned from the activity space whatever its position, although it can be occupied by a second appliance or by other items such as a stool or laundry box.

Full floor-to-ceiling height is required over the activity space and the length of bath related to it. Beyond this area a change in ceiling plane is possible and could be convenient where, for example, a bathroom is being installed in an existing building.

An area 1100 mm × 900 mm is recommended for dressing and drying, and for activities not fully covered by the appliance activity space. It can also be a useful check on bathroom plan area and shape. Usually, however, when the bath is combined with other appliances, space will be available within the resulting plan without need for this special provision.

Showers

Plan sizes of 800 mm × 800 mm and 900 mm × 900 mm are the most common. A tray height of 150 mm is considered to be a practical minimum. There are many different materials and designs for shower surrounds; they include plastics curtains, tiling of the wall surfaces, glazed screens and folding doors.

These sizes apply to all types of shower: an unenclosed but drained corner of the bathroom, as more commonly found abroad (Figure 4.27); basic floor trays, either set into or standing on the floor plane, and which can be fitted with various types of screening; and the packaged shower unit supplied as an integral floor tray and enclosing cabinet.

European Standards, in addition to the British Standards, are applicable. They include EN 249[231], BS EN 251[232], EN 263[233] and EN 329[234].

Activity space

If enclosed on one or two sides, an area is needed on plan of 400 mm × 900 mm (or the width of tray used) adjacent to one open side of the shower (Figure 4.28). This space is for access and allows for drying, partially within the shower. If enclosed on three sides, an area 700 mm × 900 mm. is needed This space is for access and towelling. Full floor-to-ceiling height is needed over the shower and its activity space.

As with the bath, an area 1100 mm × 900 mm is recommended clear of the shower for dressing and as an alternative area for drying. This may overlap or include the defined activity spaces.

Wash hand basins

Until the later years of the nineteenth century, wash hand basins tended to be portable and made of highly decorated ceramics. They were later inserted into marble surfaced wash stands, and served by taps screwed to the wall above.

It was from these early models that the elaborately shaped wash hand basin made of fireclay or vitreous china, and including waste and tap holes, fascias, special rims and soap housings was developed. Wash hand basins are now 600 mm × 400 mm on plan, sometimes larger. Front rim height is 800 mm for pedestal sets.

Ideally, the basin needs to be fixed at a lower height for washing face and hair than for hand rinsing. The 800 mm height is really a compromise for family dwellings. Where the basin is more likely to be used by adults, a height of 900 mm may be acceptable. For small children a height of 800 mm will be difficult to reach. This inconvenience is normally solved either by adult assistance or by the child standing on a firm support.

European Standards, additional to British Standards, are applicable for products. See, for example, EN 31[235], EN 32[236], EN 37[237] and EN 111[238].

Activity space

An area 1000 mm × 700 mm is recommended for accommodating the various activities normally carried out at a bathroom basin (Figure 4.29). The critical posture is for hair rinsing with the user bending low over the appliance and with elbows extended and spine in a near-horizontal position. In this posture the user needs the full width of the activity space only at and above the upper plane of the basin. It is therefore possible for the basin activity space to overlap adjacent appliances and their activity spaces which are below this plane without obstructing the use of the basin.

Full standing height is needed over the front half of the basin and all its activity space. Where storage and fittings are located above the basin they should be clear of the arc described by the user when bending down over the appliances.

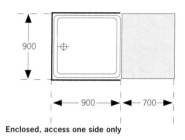

900

|← 900 →|← 700 →|

Enclosed, access one side only

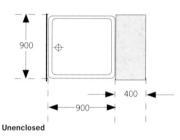

900

→| 400 |←

|← 900 →|

Unenclosed

Figure 4.28
Space needed to use a shower

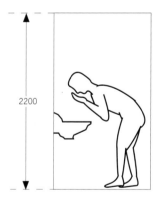

2200

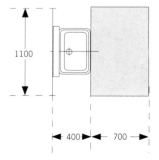

1100

|← 400 →|← 700 →|

Figure 4.29
Space needed to use a wash hand basin

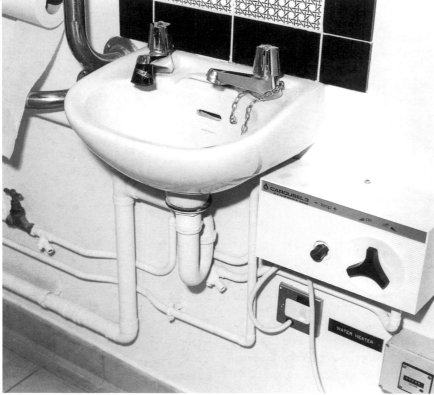

Figure 4.30
A wash hand basin in a disabled person's room with an adjacent point-of-use water heater

Figure 4.31
A hot air hand drier. Automatic sensing rather than manual switching with timed cut-off is now preferred

Cloakroom basins

Where a small basin intended primarily for hand washing is installed in a cloakroom or WC compartment, the basin activity space specified for a WHB can be reduced to 800 mm × 600 mm. This would not apply, of course, to one designed for use by disabled people for which special provision must be made (Figure 4.30).

A 300 mm × 500 mm basin would be suitable for most WC compartments. Many sizes and variations are available, including types which can be set in a corner location or into the thickness of a wall. These will affect the size of the compartment required, but the activity space will, in principle, always be the same.

Hot air hand drying units

These are normally mounted on walls, and are usually relatively small in size (Figure 4.31). However, they need an activity space which is similar in size to that needed for a wash hand basin (see BS 6465-2).

Sinks

Glazed fireclay (or earthenware) sinks could be found in most dwellings built between the last years of the eighteenth century and the mid-1950s (Figures 4.32 and 4.33). Before that time they might have been made of stone or slate, or of wood which may or may not have been lined with sheet lead. Since the 1950s, sinks invariably have been manufactured in stainless steel.

An enormous variety of glazed fireclay sink sizes and shapes have been made: from shallow (around 150 mm deep) to those with sloping fronts and parallel sides (around 500 mm deep) used for clothes washing, some of them with wood inserts on the rims. Sinks without overflows were usually known as London pattern, and those with integral overflows as Belfast pattern. Various other names have been used from time to time, for example 'butler', 'drip' and 'housemaid', but these names have never been universal.

Slop sinks are to be found in hospitals and health buildings where it is necessary to dispose of excreta from bedpans etc. They operate in exactly the same way as a WC, but the pan is sink shaped rather than WC shaped. Otherwise the traps and flushing arrangements are similar.

A single sink in a dwelling should be large enough to take an oven shelf, the largest object commonly washed there, and therefore be at least 500 mm × 350 mm × 175–200 mm deep. If the depth is, say, 250 mm or more, it will also be suitable for washing clothes and household linen.

The type of double sink now most usually found in the UK has two equal bowls about 400 mm × 400 mm, but a more useful combination might be one large bowl (say 500 mm × 350 mm and 175 mm deep) big enough for cleaning oven trays and, if necessary, washing clothes and household linen, and a smaller rinsing bowl alongside (Figure 4.34). A removable rack may be provided over the rinsing bowl so that dishes can be stacked on it and sprayed with hot water, but if this is to be done the bowl should be at least 400 mm × 350 mm.

Whichever type is chosen, it will be useful if there is a waste outlet large enough for a waste disposal unit to be fitted later without major alterations. A basket strainer waste will trap food scraps before they clog up the pipe. Where unequal bowls are being used, this outlet should be fitted to the smaller bowl.

British Standards are BS 1206[239] and BS 1244[240]; European Standards are EN 411[241] and EN 695[242].

Drinking fountains

Drinking fountains for wholesome water are normally of vitreous china or stainless steel fed by water jets designed to eliminate contamination.

Figure 4.32
A survival from Victorian times: a glazed earthenware sink still in use in 2000

Figure 4.33
Belfast sink cantilevered from the wall on cast iron brackets, typical of the kitchens of the 1920s and 1930s. This one was photographed in the mid-1980s in a house about to be refurbished

Water closets (WCs)

Three types of WC appliances may be found in existing buildings:
- wash out
- wash down
- siphonic

The wash out pan leaves the excreta largely exposed in the shallow bowl and away from the trap until the flush operates to wash it out (Figure 4.36 on the next page). This type is fairly rare, and was formerly used to a considerable extent in health care buildings. The wash down is the most common type encountered, with the excreta largely submerged in the front part of the trap until the flush removes it (Figure 4.37. The siphonic type is similar in some respects to the wash down, but in this case the siphon uses atmospheric pressure to force the evacuation of the bowl when the vacuum is created in the discharge pipe (Figure 4.38). The siphon may be of the single trap or double trap kind.

Low level WC suites are around 800 mm wide × 700 mm deep with a 50 mm seal. High level suites may be slightly less deep. Rim height is normally around 400 mm, but may be lower for use by children in schools.

Where vandalism is a problem in public lavatories, WCs may be equipped with inserts, usually of wood, on the rims instead of separate hinged seats.

WCs cantilevered from walls are designed to allow easier cleaning of surrounding floor surfaces, and are commonly specified in health buildings.

WCs fitted with macerators can also be found. These need an electrical supply, and are covered by BS EN 60335-2-84[243].

Activity space

An activity space of 800 mm wide × 600 mm deep is needed, and to full ceiling height to allow male standing use where there are no urinals (Figure 4.39a).

Where WCs are installed in buildings used by the public – especially at airports, railway and bus stations, where passengers may be

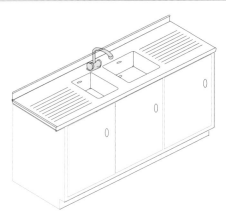

Figure 4.34
Typical sink unit with unequal bowl sizes, fabricated from stainless steel

Figure 4.35
A modern kitchen with a sink unit integrated into the design

accompanied by luggage – the size of the activity space in WC compartments needs to be larger than normal to accommodate the luggage (BS 6465-2).

There is no significant difference in the space requirements of siphon flushed WCs and valve flushed WCs, although the cistern shapes may be different. Valve flushing has not been permitted by the Water Byelaws, but, under the Water Regulations 1999, they will be permitted from January 2001.

Space requirements for installations for disabled persons are given in Building Regulations Approved Document M.

European Standards also apply; for example, EN 33[244] EN 34[245] and EN 37[246]).

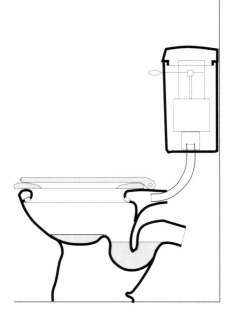

Figure 4.36
A wash out pan: the excreta is exposed in the shallow pan until the flush is operated

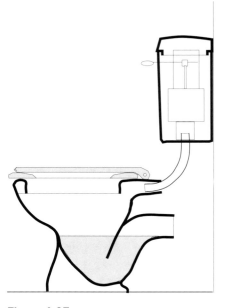

Figure 4.37
A wash down pan: horizontal outlet. Old types were available with either P or S-trap outlets

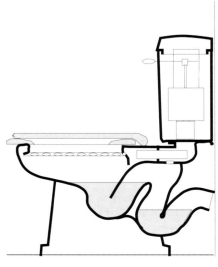

Figure 4.38
A siphonic pan: there are several different designs of siphons, but the principle is to create a partial vacuum on the discharge side of the first water seal to suck out the bowl contents when the handle is operated and before the bowl is recharged with water

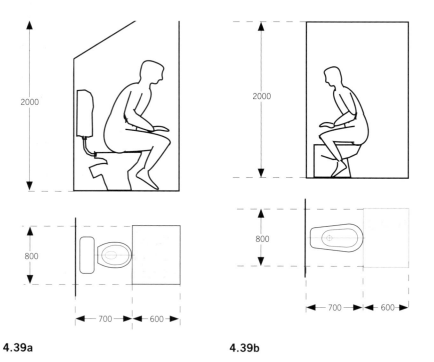

4.39a   4.39b

Figure 4.39
The activity spaces needed to use a WC (4.39a) and a bidet (4.39b)

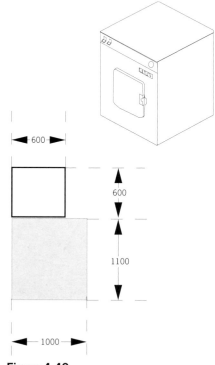

Figure 4.40
The activity space needed for a washing machine

Incinerators

Incinerators are powered either by gas or electricity. They require flues. There is no current standard for small appliances. Hospital incinerators are covered in the various parts of BS 3316[247].

Bidets

Development of the bidet dates from around 1900[18]. Appliance dimensions are 700 mm × 400 mm on plan.

An activity space of 800 mm × 600 mm on plan is needed to use this appliance. Full standing height is also needed (Figure 4.39b).

Most bidets can be filled from taps fitted to the rim, but some are fitted with a spray to which special regulations apply; that is to say, hot and cold supplies must be drawn from tanks and not direct from the mains, and there must be no tees serving other appliances.

British and European Standards are applicable. (See, for example, EN 35[248] and EN 36[249]).

Washing machines

Presently, 85% of households own an automatic washing machine and 10% a dishwasher. They consume about 12% of all domestic water. Ownership of these goods, previously considered a luxury, is increasing.

From the point of view of economy in the use of energy, it is better to use machines that draw water from the main domestic hot supply, rather than heating it in the machine; this requires a permanent connection to the DHW supply.

Most machines are around 600 mm × 600 mm on plan, and need an activity space of 1000 mm × 1100 mm (Figure 4.40).

Urinals

Urinals are basically of two types: the continuous trough and the bank of individual stalls. The latter type can be found in two forms: the full height (1.2 m) stall and the bowl type. Widths vary slightly, but most of those with thin dividers are around 600 mm, centre to centre. Victorian urinals of stoneware or slate tend to be wider (Figure 4.41).

With multiple bowl installations, it is increasingly common practice to consider the needs of young children in fixing at least one bowl at a lower height.

British Standards BS 6465-2[227], BS 4880-1[250] and BS 5520[251] are applicable. A European standard, EN 80[252], also applies.

Dishwashers

Front loading machines are convenient to load and can stand permanently in a position under a worktop. According to the old British Standard, now replaced by BS EN 1116[253], a plan area of 600 mm × 600 mm will accommodate most machines.

Activity space requirements are similar to those for a washing machine.

Main performance requirements and defects

Quality of appliances and their outputs

All sanitary appliances using hot and cold water will have particular input and output requirements. The input needs have been covered in the previous two chapters. This chapter deals mainly with outputs – the foul and waste effluents discharged from appliances, and the appliances themselves.

The requirements for discharge rates from appliances should be a primary consideration in the design and subsequent checking of waste systems. Typical discharge rates for the UK are given in relation to each appliance in the following sections. The sizes of outlets, traps and pipework should be such that the discharges from sanitary appliances are not restricted. Pipes serving more than one appliance should be sized taking account of simultaneous discharge by using the concept of a discharge unit value. This technique of a discharge unit value is not described here but more information is available in BS 5572[254]. A new European Standard, EN 12056[255], uses a different technique, and buildings built in future years can be expected to conform with this Standard. There are differences between the new European Standard and the British Standard, particularly with respect to the calculation of discharge units measured in litres per second. Three systems are described in EN 12056-2, of which system III is closest to UK common practice. The discharge rates shown in the following sections are those given in the British Standard; those from the European Standard are shown in brackets.

Figure 4.41
Many of the best quality Victorian urinal installations have proved to be very robust, and still perform acceptably

Baths

There are three main materials from which baths are manufactured:
● acrylic, to BS 4305-1[256]
● cast iron, to BS 1189[257]
● vitreous enamelled sheet steel, to BS 1390[258]

With an 80 litre capacity and 40 mm diameter branch pipe, the maximum discharge rate will be around 1.1 l/s (1.3 l/s).

Panels for covering the ends or sides of baths will be found in various materials such as framed sheeting of cellulose fibre on timber battens, or moulded plastics of acrylic or glassfibre reinforced plastics (GRP).

Showers

Showers are covered by the various parts of BS 6340-1[259]. Materials from which the trays are commonly made include acrylics to BS 7015[260] and BS 6340-5[261], porcelain enamelled cast iron to BS 6340-6[262], vitreous enamelled sheet steel to BS 6340-7[263], and glazed ceramics to BS 6340-8[264].

Even though not required by legislation, shower fittings ideally should have thermostatic mixing valves to avoid scalding and to allow accurate control of temperature.

Figure 4.42
An electrically heated, instantaneous hot water shower heater and head (Photograph by permission of B T Harrison)

An alternative to a thermostatic mixing valve is an electric shower heater which is fed from the cold supply and is thermostatically controlled (Figure 4.42). They can, though, be more expensive to use than showers fed from the DHW supply.

Flow rates from single head showers are small so that the 40 mm discharge pipes usually fitted do not require venting. However, difficulties may arise in achieving a self-cleaning velocity and adequate provision should be made for cleaning the outlet. A shower unit with more than one shower head may produce considerably greater flow rates than through a single outlet.

For an electric shower of 7–8 kW, maximum discharge rate will be around 0.07 l/s. A low pressure or low volume shower will have a maximum discharge rate of around 0.15 l/s, but a high pressure shower will have a maximum discharge rate of 0.35 l/s (without a plug 0.4 l/s, with a plug 1.3 l/s).

The average amount of water used for a conventional shower is about 30 litres (a bath requires about 80 litres). Initially, it appears that showering is more energy and water efficient than bathing, but the fact is that householders with showers use them more frequently than non-shower householders use their baths. Also, pumped and multi-head showers are not as water efficient as conventional showers. By using suitable shower equipment, all householders could save water; and those with water meters could save money.

Showerheads with low volume flows can save about 27 litres per day for each person taking showers rather than baths. This equates to an energy saving in hot water of 444 kWh (1.6 × 109 J) for each person, each year, for water heated by gas, or 388 kWh for water heated by electricity. The cheaper alternative to low volume showerheads is to fit a flow restrictor to the supply to an standard showerhead, though this may increase the showering time or result in unsatisfactory showering.

Figure 4.43
A Belfast sink on cast iron legs, with lead waste and trap, newly installed during refurbishment of a cottage in 1968

Wash hand basins

With a 6.1 litre capacity and 32 mm diameter branch pipe, the maximum discharge rate will be around 0.6 l/s (0.3 l/s).

Wash hand basins are covered by BS 5506-3[265] – for both pedestal and wall hung basins.

Sinks

Most sinks in dwellings built before the mid-twentieth century would have been of glazed fireclay (Figure 4.43). Fireclay sinks are still covered in BS 1206[266].

Stainless steel sinks with integral draining boards became popular from the middle 1950s, although increasingly sinks are being fabricated in various kinds of plastics.

With a 23 litre capacity and 40 mm diameter branch pipe, the maximum discharge rate will be around 0.9 l/s (1.3 l/s).

Metal sinks are covered in BS 1244-1[267].

Water closets

Water closets are commonly of the vitreous china wash down type. They were formerly available with one of two trap configurations, 'S' and 'P', but since the 1970s the horizontal outlet type has been preferred. With suitable connections, a wide variety of layouts is possible. Floor mounted pedestals are covered in BS 5503[268] and wall hung models in BS 5504[269].

Typical discharge rates are, for:
- washdown with high or low level cistern and 7.5 litre capacity: maximum discharge rate of 1.8 l/s
- washdown with close-coupled cistern and 7.5 litre capacity: maximum discharge rate of 1.2 l/s

(EN 12056-2 suggests for a 6 litre cistern 1.2–1.7 l/s, for a 7.5 litre cistern 1.4–1.8 l/s, and for a 9 litre cistern 1.6–2.0 l/s.)

WCs can be flushed with water using compressed air assistance[270]. Some use the pressure of water in the mains supply to compress a volume of air above the stored water. When the water is released into the pan, it does so at a much greater velocity than it would from a conventional gravity operated cistern. These products are not readily available in the UK but can be found in France and the USA. To work effectively, these cisterns must be matched to WC pans that accept the water at the intended velocity. Another type of water-and-compressed air WC uses the water to rinse the pan and compressed air to evacuate the contents. This type is used in many buildings in the USA.

WC cistern water displacement devices, such as bottles, dams and bags, are inserted into cisterns to reduce the volume of water flushed. Although relatively inexpensive, they can affect the correct and efficient operation of the flush. Some types do not fit easily into UK cisterns with a siphon flush mechanism because they are designed primarily for use in cisterns fitted with flap valves. If all the volume of water in the cistern is necessary to clear the WC pan, a reduced flush volume may not be effective: the user may have to flush the cistern again, increasing the use of water instead of reducing it. Putting a brick in the cistern – advice frequently offered to consumers who wish to save water – will often produce the same result.

Macerators and wastewater pumps

Macerators enable WCs to be connected to internal drains using comparatively small diameter pipes – down to 22 mm for very short runs, but 32 mm for longer runs. They are particularly useful where access to foul drains by normal means is restricted or perhaps impossible, and they may take as little as 3 litres of water to flush. Heavy duty models are available which can pump the slurry discharge up to two storeys high and for considerable horizontal distances. Power for all models can be taken from normal 240 V AC supplies.

Different designs of macerator have different capabilities for dealing with bowl contents, and users can overload them with fibrous material leading to blockages. Systems using large radius bends tend to be more trouble-free than those with sharp bends. Discharge rates vary according to model: generally about 0.4–1.5 l/s.

Electrically powered wastewater pumps are also available which have no macerating capability. They too can be fitted inside the building, and are used where there are problems with restricted falls to sewers.

Composting toilets

Of the toilets that use no water for flushing, the most common type in the UK is the composting toilet. Usually, in its domestic form, it is electrically powered, heating the waste material to promote the composting action. Size is its main problem; the smallest domestic model is about twice the size of a conventional WC suite. Large models (greater than 15 m³ capacity) usually do not require external energy to start and sustain the process as the aerobic decomposition is sufficiently exothermic to be self-sustaining. Large composting toilets may be environmentally acceptable as they consume only a small volume of water, require no drainage pipework and produce compost that can be used in the garden. However, the questions of adequate hand cleansing facilities, if there is no available water supply, and the safety of children using toilets with open chutes, must be considered.

Urinals

Urinal flushing cistern controllers have been used widely in the UK for some time. Water Byelaws used to state the maximum rate at which cisterns may be filled. Since 1989, new cisterns have been required to be refilled only after the urinal has been used. There are various methods of sensing use and operation: changes in water pressure to identify operation of taps and, therefore by association, the use of urinals; passive infrared (PIR) and proximity detectors to determine movement of people in the toilet, and devices to sense the temperature of urine in the urinal traps. The essence of these devices is that they all obviate the flushing of urinals when the premises are not being

The overflowing urinal

The cistern for a urinal was fitted at a very high level which resulted in water for flushing flowing into the urinal at too-high pressure and cascading onto the floor and wetting users. A solution was attempted by applying a silicone sealant to act as a weir to deflect the water away from the sides of the bowl. A better remedy would have been to introduce a flow restrictor (eg a washer with a smaller diameter hole) in the pipe union to the bowl, or simply to lower the cistern (Figure 4.44).

Figure 4.44
Attempting to deflect water (at above normal pressure because of the height of the cistern) away from the rim of this urinal by applying a silicone 'weir' did not succeed. A better solution lay in reducing the water pressure

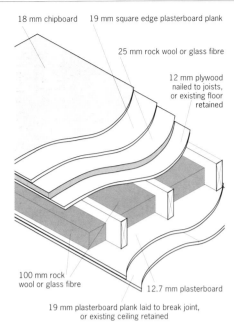

18 mm chipboard   19 mm square edge plasterboard plank

25 mm rock wool or glass fibre

12 mm plywood
nailed to joists,
or existing floor
retained

100 mm rock
wool or glass fibre

12.7 mm plasterboard

19 mm plasterboard plank laid to break joint,
or existing ceiling retained

Figure 4.45
A floating floor installed over an existing
timber joisted floor

used and are usually an improvement
over the use of the traditional 'pet-
cock' that drips water at a required
rate into the cistern until it reaches a
level at which the flushing operation
is activated.

Urinals are covered by BS 4880[271]
and BS 5520[272]. (EN 12056-2
suggests discharge rates for a single
urinal with cistern of 0.4 l/s, and for a
slab urinal of 0.2 l/s per person.)

Using an occupancy detector to
turn off the water supply to a toilet
when unoccupied is another
application of passive infrared (PIR)
technology. This can minimise the
waste in urinal flushing caused by
taps being left open.

Bidets

With a 23 litre capacity and 40 mm
diameter branch, the maximum
discharge rate will be around 0.9 l/s.

Bidets are covered by BS 5505[273].

Washing machines

Water Byelaws used to require that
the maximum permissible volume of
water used for a washing machine
was 3.6 litres of water for every litre
of drum or tub volume; the current
maximum volume of water per wash
programme is around 100 litres.

With a 4.5 kg capacity, the
maximum discharge rate will be
around 0.6 l/s. (EN 12056-2 suggests
for a capacity of up to 6 kg, 0.6 l/s,
and for up to 12 kg, 1.2 l/s.)

Dishwashers

Water Byelaws used to require that
the maximum permissible volume of
water used for a dishwasher 7 litres
of water for every place setting; the
current maximum volume of water
per wash programme is around
25 litres.

With a 12–14 place setting
capacity, the maximum discharge
rate will be around 0.25 l/s (0.2 l/s).

Noise and other unwanted side effects

Avoiding nuisance from bathroom
and plumbing noises is mainly a
matter of correct planning and the
choice of quiet apparatus. Some
noise from the plumbing in
bathrooms and WCs is inevitable;
preferably these rooms should not be
placed adjacent to main living rooms,
and the WC and cistern should not
be fixed to intervening walls. The
door to the WC or bathroom should
be as heavy as practicable and well
fitting, especially if it faces another
door directly across a landing.

In existing houses, where the
bathroom and toilet may have been
placed already over one of the main
rooms, and where the floors are of
wood joist construction, the only
treatment likely to give any useful
improvement is to increase the
insulation of the floor by as heavy a
pugging treatment as the ceiling will
sustain and to reconstruct the floor
boards as a floating floor (Figure
4.45). (See the relevant discussion in
Floors and flooring, Chapters 1.8 and
2.1). If a floating floor is fitted in a
toilet, the connection to the soil stack
from the WC must be made flexible,
otherwise movements in the floor
could cause the WC pan to crack. A
patent rubber ring joint (as for
drainpipes) could be used for this
flexible connection, filling the joint
flush with a mastic jointing
compound. Flexible WC connectors
are now available, though, that
would now be more appropriate.
The cistern should be isolated from
the wall by resilient pads (eg thick
cork or rubber) and the screws or
bolts securing it to the wall should
have resilient sleeves and washers.

Noise from the flow of water
through the float operated valve is
likely to be worse where the cistern
fills directly from the mains rather
than from a storage tank, because of
the higher pressure of the incoming
water, but essentially the noise is due
to the passage of water through the
inlet valve jet. The remedy is to
install a quieter valve.

Another annoying noise from WC
systems is that of the discharge
through the pan. The common
wash-down type of closet is by its
nature, noisy; but there are various
siphonic types which can be very
quiet in operation. For wash-down
types the velocity of the water
passing through the pan determines
the noise, and consequently high
level cisterns create more noise than
low level types. As a rule the quietest
WC suites are those with double
siphonic traps and a close-coupled
cistern.

Commissioning and performance testing

The most onerous of commissioning requirements relate to WCs, particularly low flush WCs. Points to be checked include:

- supply pipes are connected to water inlet valves through isolation valves and filters where needed
- water pressure is adequate
- inlet valves are of correct type (eg high pressure if connected to the mains)
- WCs are securely fixed
- operating and filling mechanisms are free to operate
- there are no leaks
- a single flush clears 12 sheets lightly crumpled toilet paper
- siphonage problems are detected by testing to BS 5572[274] or EN 12056
- all appliance wrappings are removed

Health and safety

Any shower thermostat should fail safe.

Metal baths and sinks must have provision for, or must be capable of being provided with an effective and approved means of connecting an earth continuity conductor. The means of connection is to be sited in an inconspicuous position and where it cannot be easily tampered with.

Durability

Most appliances, unless subjected to impact damage (Figure 4.46), are inherently durable once installed. The replacement of sound baths, sinks, WC pans and cisterns can be justified only if the replacements are of equal or better quality, or if the basic layout of the bathroom or WC is to be altered. BRE unpublished investigations dating from the early 1980s put the expected lives of most sanitary fittings at around 30 years, and it does not appear that things have changed much in the intervening period. Of course, lack of attention to dripping taps, and use of abrasive cleaners can take their toll of even the most durable of appliances (Figure 4.47).

Glazed earthenware and fireclay sinks eventually suffer from crazing of the glaze, and makes satisfactory cleaning almost impossible. If the crazing is felt to be too unsightly, replacement is the only satisfactory solution. WC pans in hard water areas suffer from deposits of calcareous substances at water level. Excessive use of abrasive cleaning agents can damage the glaze which again prompts a decision on replacement.

Maintenance

Some of the early thin (3 mm) plastics baths deflected when being used, particularly if they were not supported and fixed according to manufacturer's instructions. Consequently the joint between a bath and a wall could be broken leading to leakage of water into the floor structure and further consequences of damage to the floors themselves, structural timber, ceilings and electrical equipment. Any signs, then, of fracture or impairment of a sealant at the edges of a bath should be remedied quickly.

Scratches in acrylic appliances may be removed by metal polish. Abrasive powders should be avoided though, even for removing scale in hard water areas and blue-green copper deposits in soft water areas. Minor blemishes on appliances with enamelled surfaces may be removed by specialised polishing techniques.

Regular maintenance of valve-flushed WCs is required to maintain their efficiency. Points to observe include:

- cleaning in-line filters monthly
- checking and replacing defective parts
- checking weekly to detect leaks

A useful tip to detect leaks from the valve of valve-flushed WCs is to spread talcum powder on the surface of the water. Leaks will readily show as disturbance of the surface. A dye test, using for example instant coffee in the cistern, and inspecting for dye coloured leakage into the bowl after half an hour, should be carried out every six months.

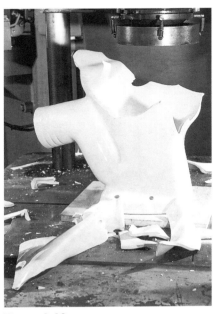

Figure 4.46
A laboratory test on a WC pan. Properly buffered seats should not crack the rim if dropped, but even the slightest movements in the soil stack or floor can disrupt seals – a time-consuming process with a potentially expensive outcome

Figure 4.47
Allowing taps to drip continuously can take its toll of even the most durable of appliances. Using an abrasive cleaner to remove the stain only makes matters worse

Figure 4.48
In winter, a drip from a warning pipe (or overflow) due to lack of maintenance of the water supply system has been allowed to build up to become a huge icicle

So far as WCs in domestic premises are concerned, especially where they are used frequently or blockages are common, rodding eyes located in the WC branch pipe before it enters the stack, and in the stack itself, can make remedial work and maintenance of the soil system easier and quicker.

Buildup of scale deposits on appliances is unsightly, and solvents can be obtained for their removal.

Work on site

Storage and handling of components and materials

Sanitary fittings are at their most vulnerable when being transported, stored and manhandled to their final location. The heaviest appliances, such as cast iron baths, will need to be moved by special trolley. Packaging should not be removed until the last possible moment. Vitreous china appliances are, self-evidently, vulnerable to damage, but so too are the lighter weight acrylic appliances. In spite of being of 'plastics', and also having some of the qualities of plasticity, they are not able to withstand rough treatment; and they scratch easily.

Workmanship

Workmanship should be to the requirements of BS 8000-13[275]. Installations should also meet the requirements of BS 6465-1[227].

Chapter 4.4

Above ground drainage

Above ground drainage systems were normally located on the external faces of buildings constructed before the 1960s. On high-rise blocks of flats, arrays of branches, downpipes, and soil and vent pipes could appear very obtrusive and unattractive (Figure 4.49). Installations were usually confined to the rear elevations of buildings, a factor that limited and still may limit internal planning options for refurbishment.

In respect of building types other than domestic, the drainage systems of the 1960s were still based on archaic practices. It was observed, for example, in 1967: *'In several important respects the design of drainage systems in modern hospitals shows little advance over that of a hundred years ago. Admittedly, most piping is now installed internally rather than externally, and there is greater provision for anti-siphonage devices, but there are still many different drainage layouts, including vertical and horizontal main drains within the building, and the pipe systems which take tortuous paths through the structure. The main reason for this seems to be lack of coordinated planning of the different services, both with each other and with the main structure. The result is a network of piping which is spread throughout hospital buildings and which reveals how little thought has been given to the need for adequate access to the pipework'*[276].

Figure 4.49
Multi-storey flats, with soil and rainwater drainage on the external wall. Five access plates are visible in the three storeys of the right hand stack

There are thousands of buildings which still retain these old features, and some of them involve complex pipework, with vent pipes in addition to the drains and wastes. For both aesthetic and planning considerations, rehabilitation will often include provision of a new internal single stack drainage system. If an existing system is likely to be retained and possibly modified, its ability to function adequately and safely must be determined.

Where street level drainage is at a higher level than the lowest part of the drainage system in a building, a storage system will have to be constructed from which the sewage is pumped to the main sewer.

Rainwater collection and disposal was dealt with in *Roofs and roofing*, Chapter 1.3, and is not dealt with here. However, some rainwater pipes may be accommodated internally, and will require checking for both capacity and leaks since the consequences of overflow internally are more serious than for pipes sited externally.

The purpose of a sanitary plumbing and drainage system in a dwelling is to carry away waste products, to a foul or combined sewer, septic tank or settlement tank. This must be achieved to the requirements of the Building Regulations, quietly and with the minimum risk of blockage or leakage, but without permitting foul odours from the drains to enter the building.

To prevent odours escaping from the drainage system, water-sealed traps are used at each sanitary appliance. Large fluctuations in pressure in the pipework system can, under certain conditions, destroy the seals; for this reason the positive and negative pressures in the system should be contained within the limit of 38 mm water gauge in order that at least 25 mm of water seal is retained within the traps.

A comparatively recent innovation has been the waterless trap, a sealing mechanism based on a self-closing elastomeric membrane which is opened by the flow of water and closes when the flow ceases.

This resists any positive pressure within the system so that foul air cannot reach the appliance.

Excessive fluctuations in pressure can be avoided by venting the system. One method of venting which is still sometimes used in large installations employs a system of vent pipes and branches connecting each appliance. Although installations of this type use separate soil and vent stacks, they are referred to as one-pipe systems. Variations on these systems have either the WC branch only vented, or simply a direct cross-connection between the two stacks. These are known as 'modified one-pipe' and 'modified one-pipe – vented stack' respectively. The new European standard has introduced new terms to describe drainage system layouts. The National Annex for the UK includes a section which relates the old terms to the new terms[255].

Venting is usually by an open ventilation stack taken at least 900 mm above any opening into the building which is within 3 m of it. For internal stacks this means penetrating the roof covering. An alternative is to use an air admittance valve (AAV) fitted to terminate the stack above the flood level of the highest appliance. The AAV opens to allow air into the system when pressure in the system becomes negative.

BRS carried out an extensive programme of research directed towards the design of more economic domestic plumbing installations, and to revolutionising plumbing practice in the UK. Single stack plumbing in the USA provided little economy because of its widespread use of anti-siphonage pipework. Laboratory experiments at Garston, begun in 1949 and supplemented by full scale multi-storey experiments in an office building in London, investigated ways of reducing siphonage and of designing sanitary systems so that water siphoned out of traps was replaced by a continuing or 'trailing' discharge from the surfaces of the appliances. Later work established the requirements for systems serving

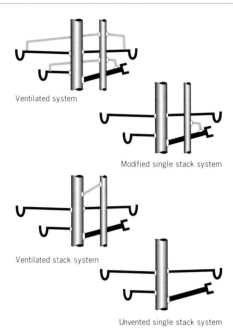

Figure 4.50
Various types of domestic drainage system

offices and other multi-storey buildings. The results were progressively incorporated in the British Standard Code of practice from 1953 onward and subsequently into building regulations.

As laboratory studies carried out by BRS as early as the late 1950s have illustrated (Figure 4.51), the hydraulic and pneumatic conditions in a stack down which water is flowing are complicated[277].

Vertical stacks do not normally run full. Most of the water flows down as an annular sheet round the inside wall, with the core of the flow being formed by air, drawn in from the open vents by the discharge of water. Solids tend to fall down the centre of the stack. Under some circumstances, of course, the suction proves sufficiently strong to draw the water seals of traps connected to the stack if they are not deep enough or if they are not separately vented.

At the foot of the stack, the water flow with its solids follows the outside of the bend, and then occupies the invert of the drain with air flowing above. The whole contents of the flow discharge to the main sewer or septic tank, with the air exhausting via other stacks connected to the system. When there is sufficient foam from detergents to fill the stack, the airflow is interrupted. Occasionally, foam can erupt from appliance seals at the bottom of the stack, though under normal conditions 100 mm-plus diameter stacks can cope with foam without problems. Stacks of less than 100 mm diameter, as sometimes found in older multi-storey buildings, are the main culprits.

The smaller branch pipes connecting to the stack normally flow full, with air bubbles entrained in the flow being sucked in through appliance overflows or the drainage vortex. Large soil branches, on the other hand, usually do not flow full bore unless there is a restriction caused by a sharp bend or by a mass of solids. It is these conditions that give rise to self-siphonage, and the trap seals can only refill by the trickling end of the WC flush or appliance discharge. Plugs of water (ie short lengths of full-bore flow) filling the pipes arise more commonly in the longer lengths – this is the reason for placing limits on lengths in order to reduce the likelihood of the seals being drawn.

Characteristic details

Space requirements
Ducts

BS 8313[112] gives guidance on the design of pipe ducts. Ducts should be constructed appropriately for fire resistance, sound insulation and to limit the spread of vermin; and they should provide ready access for maintenance, cleaning and testing.

Discharge pipes, if located in ducts with high ambient temperatures are likely to dry out between discharges if the flow in the pipe is small and intermittent. This can readily cause a buildup of deposit in the pipe and bring about a pipe blockage. Pipes can also dry out after long periods of non-use. The ambient temperature in the duct should be controlled to prevent this happening. In situations where the discharge pipe is receiving hot water, high ambient temperatures will inhibit heat loss through the pipe wall; consideration should be given, therefore, to insulating discharge pipes. Where pipes pass through walls or solid floors, they should be sleeved; and where appropriate, suitable fire stops should be fitted.

The space required for encasing parts of service areas for ducting depends on the diameter of the pipes or stacks to be cased and the thickness of the casing structure or materials. These ducts can accommodate normal requirements for services, but for unusual requirements the advice of a services engineer should be sought on detailed duct layout.

In simple duct layouts which are to include soil stacks, allowance should be made for the overall sizes of sockets which vary considerably according to the material used. In the case of a 100 mm diameter cast iron soil stack, for example, a space of 160 mm × 160 mm is needed to accommodate a socket. A cast iron stack with a diameter of 150 mm requires 215 mm × 215 mm. Clearance is also necessary around a stack for jointing etc, and, in particular, space should be allowed on the wall behind the stack for water and gas pipe runs, and electric cables. Stacks in other materials require less space because of their smaller overall socket sizes.

A 100 mm diameter stack in cast iron should be given a minimum clear space around it of 50 mm; in other words an overall space of 200 mm diameter minimum, and for a 150 mm diameter stack this would be 250 mm minimum. Where the duct casing is made of thin material and little space is required for framing the casing, the 200 or 250 mm width overall may be adequate to include the casing construction, but in many cases the duct will be formed in heavier constructions which are 50, 75 or even 100 mm thick, and in these cases the overall duct width should be increased to the nearest 50 or 100 mm increment: say 250 or 300 mm overall.

Duct sizes for accommodating internal drains and wastes are given in BS 8313. See also *Space allowances for building services distribution systems – detail design stage*[278].

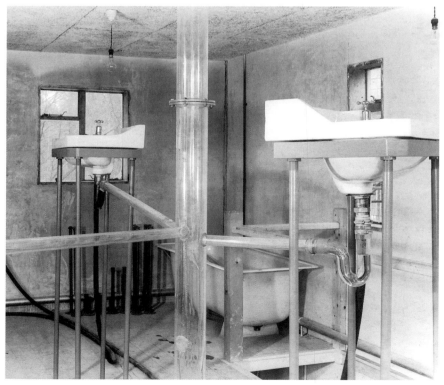

Figure 4.51
BRS experimental installation dating from the 1950s for observing the flow of discharges from wastes and stacks

Pipe runs

Cast iron, lead, copper, asbestos cement, galvanised steel, and occasionally lead drainage systems may be seen in buildings erected before the introduction of plastics. Until the late 1950s, cast or spun iron reigned supreme for soil and waste pipes, though copper can sometimes be found in buildings dating from this time. Since then PVC has taken the major share of the market.

For particularly corrosive industrial effluents, or in hospitals where inspection for blockages is needed, special glass formulations are available for manufacturing pipes.

Stacks and WC connections

It is now established that, provided certain principles are observed with respect to the number of fittings connected, the lengths of waste and the diameter of discharge stack used, a single stack serving as a combined soil-and-vent stack will be adequately self-venting for all domestic work up to 30 storeys. The stack needs to be not less than 75 mm for two storeys, and not less than 100 mm for up to eight or ten storeys, with 150 mm above that. For non-domestic installations, depending on the number of appliances connected, the minimum size is 100 mm, which can be used up to four or eight storeys, depending on the number of connections, and 150 mm above that, again with restrictions depending on the number of appliances connected.

The single stack system has obvious planning and economic advantages when compared to the 'one-pipe' systems. A 150 mm single stack in a multi-storey block for example, requires considerably less space than a 100 mm soil stack and 50 mm vent stack system.

WC branch connections should be swept in the direction of flow with a 50 mm minimum radius sweep. Single WC branches of 75 or 100 mm size do not require venting whatever the length or number of bends included in the run. Bends, however, should have as large a radius as possible to prevent blockage. Falls should be greater than 1°.

For any diameter of stack there is a limitation in height and in the number of appliances that can be connected before risk of seal loss makes it is necessary to add a separate vent stack or a larger soil-and-vent pipe.

The previous chapter gave specifications for activity spaces for appliances in kitchens and bathrooms based on ergonomics and user studies. There are, additionally, specific points of plumbing design which help to achieve simple, economic and efficient plumbing layouts. The soil stack for example should, ideally, be located directly behind the WC in a duct. If this is not possible, due for example to the position of windows or to the WC flushing cistern, the stack should be placed to one side as near to the WC as possible and between the WC and other appliances, particularly the bath (Figure 4.52). This is important in situations where a horizontal outlet from a WC pans enters the stack at a point horizontally opposed to the bath waste. Soil material from the WC should not be allowed to enter the bath waste but this can be difficult to engineer without the use of special fittings which increase the cost of the installation.

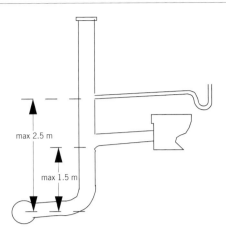

Figure 4.53
A stub stack showing the maximum heights for a WC and the topmost appliance connection above the branch drain to the sewer

Stub stacks

In some areas it has become common practice to connect the domestic sanitary appliances in single-storey buildings to a short straight 100 mm discharge stack with the top closed by an access fitting. This arrangement performs satisfactorily provided the inverts of the WC branch and of the topmost connection are not more than 1.5 m and 2.5 m, respectively, above the centre line of the drain or branch discharge pipe (Figure 4.53). However, where a drain serves a number of dwellings equipped in this way, the upstream end of the underground drain should be vented either by a separate ventilating stack or by a conventional discharge stack, as described earlier in this section.

Air admittance valves (AAVs)

It is now possible to specify AAVs which avoid the need to take ventilation stacks through the roof space (Figure 4.54 on the next page), though these should not be used if the more normal practice of using ventilating pipes to the stacks is possible. AAVs should be installed according to the manufacturer's instructions. AAVs should be removable so that blockages in stacks can be cleared.

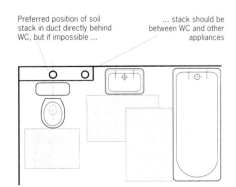

Figure 4.52
Ideal position of soil stack in relation to the WC

Preferred position of soil stack in duct directly behind WC, but if impossible ...

... stack should be between WC and other appliances

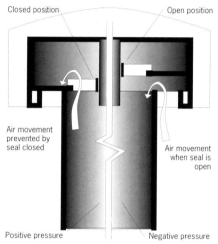

Figure 4.54
An air admittance valve

Approved Document H of the Building Regulations permits AAVs only if they are subject to a current third party certificate, and the conditions of use accord with the terms of that certificate. The Technical Standards for compliance with the Building Standards (Scotland) Regulations 1990 also place restrictions on their use.

By the early 1990s, there were around 3.5 million AAVs installed in the UK. A BRE survey indicated that most AAVs were fitted in individual two-storey houses. A planned European Standard has reached draft stage[279].

Discharge pipes (wastes)

A summary of the requirements for branch discharge pipes is given in Table 4.2 on page 172. In general, the pipework should be kept as short as possible, with uniform gradients and without unnecessary bends; where bends are unavoidable they should be of large radius. Access for clearing debris and blockages from pipes is important, especially for WCs, basins, sinks and urinals.

Installations in new-build can be expected to conform with new European Standards, particularly EN 12056-2[255]. Table 6 in the Standard (for system III unventilated branch discharge pipes) and Table 9 (for system III ventilated branch discharge pipes) contain the appropriate information. (System III is the procedure that most closely corresponding with current UK practice). There are a number of major differences from the old British Standard. For example, minimum seal depths of WC traps are reduced to 50 mm for both greater than 80 mm and less than 80 mm diameter ventilated and unventilated branch discharge pipes; and the former maximum length of 6 m for all discharge pipes from the stack has now been removed for both the ventilated and unventilated cases. Slopes are now given in percentage gradients instead of degrees (°). For other appliances it will be essential to consult the tables in EN 12056-2. For replacement purposes it may be possible to adapt designs in conformance with the new Standards, or more usually it will be a case of replacing like with like. In the following sections, common practice ruling in the 1980s and 1990s is described since most existing buildings built during these years will have been designed to the old British Standards Codes of practice.

The length, diameter and slope of waste pipes from wash basins are particularly important when considering self-siphonage since the 32 mm diameter waste pipes normally run at full bore (Figure 4.55). Where a pipe is not ventilated, the length and slope of the pipe need to be limited (Figure 4.56).

Washbasins, unlike baths and sinks, do not have a beneficial final slow 'tailing off' when emptying, that is to say a progressive reduction in the flow, ending with a trickle. When selecting washbasins it is better not to choose a model which is too funnel shaped, but. at the same time, to bear in mind the type of terminal water fitting to be installed since too-flat a base in some situations could lead to splashing. This applies only to P traps; S traps are liable to severe siphonage which may require additional venting as in one-pipe systems. Risk of seal loss from induced siphonage depends on the number of appliances connected, the profile and diameter of the branch connections, and the height of the stack and its diameter.

Sinks and baths are normally fitted with 40 mm discharge pipes. Self-siphonage is not a problem because the trap seal is replenished by the tailing off at the end of the discharge due to the flat bottom of the appliance. The flat bases of these appliances should still be sufficiently

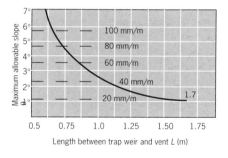

Figure 4.55
Allowable lengths and slopes for discharge pipes

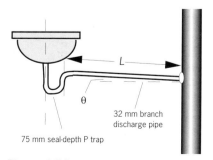

Figure 4.56
An unvented discharge pipe from a wash basin

Table 4.2 Design of branch discharge pipes

Appliance	Number of appliances	Trap size (mm) and type	Branch pipe Max length (m)	Branch pipe Size (mm)	Bends in branch pipe	Design precautions Principle action to be guarded against	Design precautions Trap design	Design precautions Branch gradients
Wash basin	Single	32 P	1.7	32	Not more than 2 (min radius 75 mm to centre line)	Self-siphonage	Tubular	1–2.5° (18–45 mm/m)
	Single	32 P	3.0	40	Not more than 2 (min radius 75 mm to centre line)	Self-siphonage and excessive deposition	Tubular or bottle	1–2.5° (18–45 mm/m)
	Single	32 P or S	3.0	32	Any bends to have min radius of 75 mm to centre line	Self-siphonage and excessive deposition	Resealing	1–2.5° (18–45 mm/m)
	Range of up to 4	32 P	3.0 main 0.74 branch	50 main 32 branch	None	Self and induced siphonage and excessive deposition	Tubular or bottle	1–2.5° (18–45 mm/m)
	More than 4	32 P	10.0 main 1.0 branch	50 main 32 branch	Any bends to have min radius of 75 mm to centre line	Self and induced siphonage and excessive deposition	Resealing	1–2.5° (18–45 mm/m)
	More than 2	32 S	10.0 main 1.0 branch	50 main 32 branch	Any bends to have min radius of 75 mm to centre line	Self and induced siphonage and excessive deposition	Resealing	1–2.5° (18–45 mm/m)
Bath, sink	Single	40 P or S	3.0	40	Any bends to have min radius of 75 mm to centre line	Self-siphonage and excessive deposition	Tubular or bottle	1–5° (18–90 mm/m)
Shower	Single	40 P or S	3.0	40	Any bends to have min radius of 75 mm to centre line	Self-siphonage and excessive deposition	Tubular or bottle	1–5° (18–90 mm/m)
Washing machine	Single	40 P or running trap	3.0	40	Any bends to have min radius of 75 mm to centre line	Siphonage from machine during refill	Tubular or bottle	1–2.5° (18–45 mm/m)
WC	Single	75–100 P or S	6.0	100	Avoid knuckle bends	Excessive deposition	–	1° min (18 mm/m)
	Range of up to 8	75–100 P or S	15.0	100	Avoid knuckle bends	Excessive deposition and induced siphonage	–	0.5–5° (9–90 mm/m)
Urinal	Single bowl	40 P or S	3.0	40	Any bends to have min radius of 75 mm to centre line	Excessive deposition	Tubular or bottle	1–5° (18–90 mm/m)
	Range of up to 5 bowls	40 P or S	As short as possible	50 main 40 branch	Any bends to have min radius of 75 mm to centre line	Excessive deposition and induced siphonage	Tubular or bottle	1–5° (18–90 mm/m)
	Range of stalls	65 or 75 P or S	As short as possible	65 or 75	Large radius bends	Excessive deposition	Tubular	1–5° (18–90 mm/m)

Note: Buildings built before the publication of this book will of course have been equipped with drainage installations which were designed to Standards ruling at the time of construction. For example, during the 1980s and 1990s, the criteria given in this Table can be expected to have been widely used.

inclined that scum and reasonably dense material are carried away, but they should empty slowly enough to ensure that the trap refills. Therefore, the length and slope are not so critical and venting is not normally required, though the maximum length should be restricted to 3 m to reduce the likelihood of blockage from deposits.

Domestic washing machines and dishwashing machines normally have a 40 mm discharge pipe which can be connected directly to a discharge stack or gully, or to a sink branch pipe. Normally a trap should be fitted in the horizontal section of the discharge pipe, but this is not required for connections via a sink branch pipe, where it can be made at the appliance's outlet using a suitable fitting.

Design of stacks and branches

Hydraulic and pneumatic conditions in a stack are complicated, and a complete theoretical analysis of fluid flows has not yet been achieved. Flow characteristics in branch pipes of varying lengths and slopes further complicate the issues; laboratory studies have demonstrated that varying pressures and suctions produce results that can only be described in general terms.

From the experimental work undertaken at BRE and by pipework manufacturers, and based on configurations which have been proved in the laboratory or on site, the following performance information and recommendations for specifications are made.

Branch discharge pipes
Water closets
Pipes of 100 mm and 75 mm diameter, and slope at or greater than 1°. Length is not critical but bends should be of large radius.

Urinals
Pipes of 40 mm diameter. Lengths should be as short as possible but not exceed 3 m. Build-up of deposits in traps may require frequent maintenance of the appliances.

Wash hand basins (with plug waste)
Discharge pipes of 32 mm diameter. Lengths and slopes need to be finely controlled if venting is to be avoided (see Figure 18 in BS 5572[274]).

Bidets
The same arrangements apply as for wash hand basins with plug wastes.

Sinks and baths
Discharge pipes of 40 mm diameter. Trap replenishment at the end of a discharge means that self-siphonage is not a problem. Restricting the maximum lengths of discharge pipes to 3 m will reduce the likelihood of blockages.

Combined branch for bath and wash hand basin
It is not possible to give any general design limits, but 40 mm diameter branches and vertical pipes, with 32 mm connections at basins, have been proved in practice (see Figure 20 in BS 5572).

Showers
Discharge pipes of 40 mm diameter are normal, though with low flow rates it may be difficult to achieve self-clearing velocities and adequate provision will have to be made for cleaning.

Domestic automatic washing machines and dishwashing machines
Discharge pipes of 40 mm diameter should be no longer than 3 m with a slope of 1–2.5°.

Banks of water closets
Discharge pipes are normally 100 mm diameter with no need for branch venting.

Banks of urinals
Branch pipes, normally of 40 mm diameter, from individual bowls should be as short as possible. No venting is needed with 50–75 mm main branch pipes.

Banks of wash hand basins (WHBs)
Venting is often needed, though not for ranges of up to four WHBs and pipe lengths up to 4 m.

Discharge stacks
A bungalow typically includes a WC, bath, WHB and sink, and, where these appliances are closely grouped, a 75 mm stack without separate venting; this arrangement has been shown to perform well. For two storeys with an additional WC and WHB, the stack will need to be 100 mm diameter, again without a separate vent pipe. Other groups and ranges of appliances, and their vent pipe requirements, are illustrated in BS 5572. This British Standard has been superseded by EN 12056[280].

Interceptors
Interceptors, as the name implies, are used to trap fats, oil and grease discharged from appliances connected to drainage systems. They can be fitted either inside the building, or outside. They are rarely necessary in domestic situations, but are invaluable for avoiding drain blockages resulting from the discharges of commercial and industrial kitchens. A variety of sizes and capacities are available. The fats, oils and grease collect in the chamber of the interceptor, and need to be removed regularly, either manually or by bio-chemical dispersion.

Main performance requirements and defects

Outputs required

The basic requirement for any discharge system is that waste material is carried away quickly, quietly, without blockage and without foul air escaping into the building. Water trap seals help to prevent odours escaping, but to be sure that an adequate water seal depth is retained at each appliance, the air pressure fluctuations within the discharge branch and stack pipes must be limited (see 'Unwanted side effects' later in this chapter).

Traps for sinks, basins and baths must have effective seals; pipes and traps should be well designed and manufactured, of durable material, stable at operating temperatures, and of adequate capacity. Traps should resist a pressure or suction of $0.75 \, kN/m^2$ without evacuating, and at no point have an aggregate effective area less than that of the bore of the waste pipe.

Waste and soil pipes also should be well designed and manufactured, of durable material, stable at operating temperatures, of adequate capacity, and installed at appropriate slopes. Design should be to the appropriate self-cleansing velocity requirements.

Waste pipe runs must be kept as short as possible and be adequately supported (Figure 4.57). Kitchen sink wastes, particularly, are liable to build up deposits. The deposits from sink wastes also appear in stacks, but, as perhaps might be expected, the effect of deposit build-up is not as marked in stacks when other appliances are connected as in stacks taking kitchen sink wastes only. The latter tend to have an abnormally high build-up of deposits, making them more susceptible to blockage.

Domestic single stack systems

Offsets on the main stack above the topmost connection have little effect on the performance of a system, but offsets below the topmost connection can produce back pressures and ought to be avoided; otherwise one or more vent pipes may be required to prevent pressure fluctuations in the stack.

The bend at the base of the stack is a critical point in a plumbing system. Unless properly designed it can lead to blockages, back pressures, and build-up of detergent foam. Large-radius bends are essential, preferably a 90° bend composed of two 45° bends; but even when using a one-piece bend, the radius of the bend should be as large as possible, and in no circumstances be less than 200 mm at the centre line.

In single, two, and three-storey houses, the vertical distance from the lowest connection to the invert of the drain or tail of the bend preferably should be 450 mm minimum. For low-rise flats of up to five storeys with ground floor appliances connected into the main stack, this dimension should be increased to 740 mm minimum.

For multi-storey systems the ground floor appliances should be separately connected to their own stack. Above 20 storeys, both the ground and first floor appliances should each be connected to their own stack.

Long horizontal runs of large diameter drains within the building may become blocked if the slope is wrong or if there are offsets. Most stoppages seem to occur where the pipework is complicated by knuckle bends and sharp offsets and 92.5° junctions. In BRE investigations dating from as long ago as the 1960s, no blockages were reported where straight lengths of 100 mm and 150 mm diameter, gradually sloping pipes were used[276].

Stub stacks

Where one or more stub stack connections discharge to a drain, the head of the drain should be ventilated by a ventilating stack or discharge stack that terminates externally to the atmosphere.

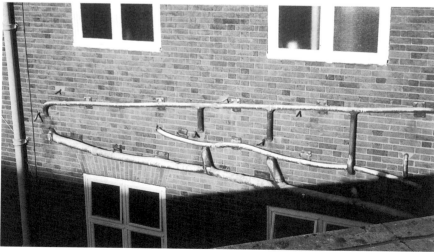

Figure 4.57
Apart from being unsightly, the sag in the vent pipes is not crucial to good performance, but that in the wastes certainly is

Figure 4.58
A PVC soil stack penetrating a pitched tile roof under construction. This example illustrates a reasonable attempt to make a satisfactory joint to prevent water flowing down the sarking from entering the building. Apart from the faults in the battening (joints not staggered), the main problem could result from fine snow blowing under tiles (a risk in roofs of even the best construction quality)

Weathertightness

Until the 1950s, and even later, the majority of waste disposal installations were suspended from the external façade. Although they could hardly be susceptible to problems of rain penetration, the methods of design, fixing, positioning and maintaining drainage systems could jeopardise the weathertightness of a building.

However, the most vulnerable point of drainage systems is where an internal stack penetrates the roof covering (Figure 4.58). Faulty installation was observed on many occasions during the BRE site inspections of the 1980s, and examples were shown in Roofs and roofing, Chapter 2.1. Where the sarking is badly torn there is little protection from rain and fine snow being driven under tiles and flashings.

Noise and other unwanted side effects

With the development of internal plumbing, stacks and waste branches must be sited carefully to minimise nuisance created by noise and by sound transmission where the fitments are adjacent to sleeping or living areas. The discharges from sanitary appliances, and pressure fluctuations in pipework causing loss of seal, are significant sources of noise, but those systems designed to limit pressure fluctuations will tend to be quiet. Some design solutions include using resealing traps with wash basins. These traps have been proved to be effective in practice but they may be noisy at the end of a discharge and less efficient in resealing if they fill with deposits. The noise should not be a problem, though, if their use is restricted to a single wash basin or a range of basins in a single toilet room. In general, noise may be reduced by sound insulation of the pipework from the structure and its containing ductwork; correct fixing of pipework will contribute to noise limitation.

The movement caused by thermal changes in pipework requires special consideration, and therefore, adequate provision for expansion should be made, especially with pipes made of plastics and copper. Where pipes of these materials pass through solid walls and floors, sleeves should be provided – also fire-stopping may also be required to limit the spread of fire between floors and rooms connected by pipework. (For further information on fire-stopping, see Chapter 1.8 of *Walls, windows and doors*.)

In high-rise buildings, suction and turbulence from wind blowing across the top of stacks has caused loss of water seal in traps. Stacks positioned near corners or close to parapets are particularly susceptible: positions which should be avoided. Protective cowls on stacks may help stop this happening in tall buildings.

The Approved Document that supports the Building Regulations for sound insulation recommends that drainage pipework is lagged and boxed where it runs through habitable rooms in dwellings. A manufacturer of steel pipework has suggested that these measures are not necessary for all pipe materials. However, a series of experiments has investigated noise levels in a room containing a length of drainage pipework when a WC is flushed. PVC-U, steel and cast iron drainage pipes, and the effect of lagging and boxing each pipe, were tested. The results supported the Approved Document showing that boxing and lagging pipes reduced the noise levels in the room significantly for all three of the pipe materials. The effect of each pipe on the sound insulation qualities of the floor through which it passed was also investigated, and showed that airborne sound insulation could be compromised by sound transmission down the pipe. This detrimental transmission effect is reduced, though, if the pipes are lagged and boxed. The effects on noise production of various minor changes to the pipework configuration were found to be small compared to the effect of lagging and boxing. Composite and cellular cored plastics pipes that reduce noise transmission are now available for sanitary pipework systems.

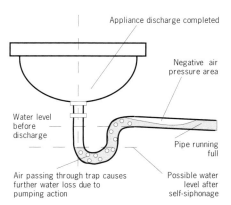

Figure 4.59
Induced siphonage and back pressure due to suction (negative air pressure) in the discharge stack

Self-siphonage

The suction due to full bore flow in a branch pipe may reduce the depth of the water seal in the appliance which is discharging. The incidence of seal loss from self-siphonage need not be a serious problem provided the recommendations on traps, pipe length and diameter, and inclination of waste branches set out in Standards are followed.

Induced siphonage

The suction normally associated with fluctuations in air and water flow pressures down a discharge stack can reduce the depth of water seal in appliances connected to the stack in the vicinity of the fluctuations (Figure 4.59 on page 175). It can also happen due to full bore flow in branch pipes, affecting appliances which are connected to the branch pipe downstream of a discharging appliance (Figure 4.60).

Back pressure

The positive pressure associated with air and water flows down the discharge stack can be exacerbated by offsets and the bend at the base of the stack. This effect may be sufficient to cause foul air to bubble back through a trap water seal or, in extreme circumstances, complete ejection of the water seal.

Avoidance of loss of seal

The traditional way to limit these pressure fluctuations was to use ventilating piping which, if correctly sized, ensured that design limits were not exceeded; it did, however, significantly increase the cost of the discharge system. In the housing field the so-called single stack system, which itself requires no venting, has now become widely used. These simplified systems have also been used in offices and similar buildings where ranges of appliances, each vented to a separate ventilating stack, are typical features in a traditional system. However, it has been found that if the design is correct, much of this pipework is unnecessary or can be reduced in size, saving up to 40% of the cost.

Commissioning and performance testing

In single stack unvented systems, trap water seal depths of 75 mm were formerly required for all appliances except WCs, and were based on the premise that some water could be lost from a seal provided the residual depth was at least 25 mm after discharge. A WC, though, has an initial seal depth of only 50 mm and it was found that suction, or negative pressure, in the stack must be limited to -375 N/m^2 or the seal depth would be reduced unacceptably. Positive pressures are limited, somewhat arbitrarily to 375 N/m^2. The new EN 12056 has now amended the requirement to 50 mm after discharge.

The frequency of use of sanitary appliances needs^c also to be taken into account since it is unreasonable to design any discharge system on the basis that all or most appliances discharge at the same time. Tests can ensure that all appliances, whether discharged singly or in groups, drain quietly and completely. After each test the requirement was for a minimum of 25 mm of water seal to be retained in every trap. To test for self-siphonage, the appliance should be filled to overflowing and discharged by removing the plug. WCs should be flushed. Ranges of appliances connected to a common discharge pipe should be discharged together. In dwellings, where many appliances might be connected to a stack, the WC, WHB and kitchen sink should be discharged simultaneously. Guidance on tests for other domestic, commercial and public use is given in Table 11 of BS 5572, and in the new EN 12056.

Testing for induced siphonage and back pressure in the discharge stacks of multi-storey buildings is handled in a similar way, with the appliances to be discharged being close to the top of the stack and on adjacent floors. Baths are ignored for these tests as they do not add materially to peak flows. Flows from showers are small and can usually be ignored. For non-domestic buildings where the stacks serve other appliances, spray taps and urinals need not be included.

Fire

The risk of spread of fire due to plastics drain pipes passing through fire-resisting elements was investigated by FRS using a gas fired furnace similar to, but much smaller than, that used for fire resistance tests specified in BS 476[281]. The investigation concentrated on pipes of sizes and plastics materials that, when plastics were first introduced for this use in the late 1970s, were commonly being used for above ground drainage installations. The work was divided into two parts: the first dealing with the performance of pipe and wall combinations under fire conditions; the second with pipe systems contained within protective walls or encasements. The results showed that the integrity of the element was usually lost quickly where plastics pipes penetrated simple wall constructions, especially when thermoplastic materials of low melting point were used. Chlorinated PVC gave the most satisfactory results. Other aspects of the investigation were to ascertain the performance in fire of a domestic installation and to assess the risk of spread from one compartment to another through a system contained either within a protected shaft or enclosure. Complete barriers at each storey level within the protected shaft were shown to be necessary together with some control of the shaft walls or enclosure construction[282].

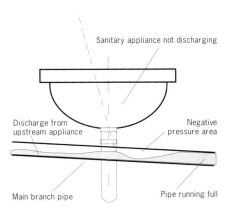

Figure 4.60

Siphonage induced by suction (negative air pressure) in a branch discharge pipe

Health and safety

Although there are several thousand cases of dysentery and many more of gastroenteritis reported in the UK each year, there is no information on the degree to which lack of hygiene in sanitary appliances and drainage systems is implicated.

The risks of infection occurring in single dwellings from these sources are thought to be negligible, provided that appliances and systems are in good repair, and are regularly cleaned. Washing hands with soap and water after using a WC or urinal is particularly important. Washing facilities should be immediately adjacent to WCs and urinals, and notices displayed to encourage users to use the washing facilities.

Where multiple use occurs, in hostels, schools and other such building types, the risks to health are arguably greater, and this places greater emphasis on regular inspection of the facilities and supervision of cleaning procedures.

Drowning in the bath unfortunately accounts for the deaths of a number of small children, for example for the year 1988–89, 25 children under the age of 15 were drowned and 19 survived near-drowning. Four out of five cases occurred when the child was left in the bath alone or with a sibling[80].

Figure 4.61
This inadequately supported plastics pipe cannot possibly function well. In frosty weather it will be vulnerable to freezing of waste water remaining in the pipe

Durability

The 1985 Building Regulations required that materials selected for pipework systems were taken from the following list:
- cast iron to BS 416[283]
- copper to BS 864-2[284] or BS 2871-1[285]
- galvanised steel to BS 3868[286]
- unplasticised polyvinyl chloride (PVC-U) to BS 4514[287]
- polypropylene (PP) to BS 5254[288]
- acrylonitrile butadiene styrene copolymer (ABS), modified unplasticised polyvinyl chloride (MUPVC), polyethylene and polypropylene to BS 5255[289]

and for traps from:
- copper to BS 1184[290] (now obsolescent)
- plastics to BS 3943[291]

European Standards also apply. They include prEN 877[292], EN 1123-1[293] and EN 1124-1[294].

Recommendations regarding choice of materials are also made in BS 5572[274]. Those currently in common use for domestic work are cast iron, copper, galvanised mild steel and plastics, but lead can still be found in older properties. Stainless steel and borosilicate glass may also be seen.

In some pipework systems, more than one material may be used, but using dissimilar metals in the same system should take account of the effects of electrolytic corrosion. The list below shows the effect of combining metals – an earlier-named metal will be attacked by a later-named one; and the closer the metals are in the list, the smaller the severity of attack:
- zinc
- iron
- lead
- brass
- copper and stainless steel

Lead, copper and iron systems have proved to be very durable, often with service lives in excess of 50 years if properly maintained. These systems are frequently replaced because of the high cost of adaptation rather than because of material failure.

Asbestos cement and steel systems are less durable and their replacement will be more routine. Neoprene or a similar non-metallic components should be used for making connections in mixed metal systems.

Relevant British and European Standards include those mentioned above, and BS EN 274[295]; prEN 12763[296] is at draft stage.

Pipes for soil and waste systems are manufactured in a number of thermoplastics materials including PVC-U, MUPVC, high density polyethylene (HDPE), polypropylene (PP) and ABS. All are light in weight, easy to handle and highly resistant to corrosion. Attention should be paid, though, to expansion in plastics drainage systems as the coefficients of expansion of these materials are much higher than for metals (Figure 4.61). Horizontal runs of plastics pipe require more support than metal pipes. Plastics material exposed to direct sunlight may require protection to resist ultra violet degradation. It is advisable to seek guidance from manufacturers of any materials other than PVC-U or MUPVC. Also, some solvents and organic compounds damage plastics materials.

Relevant European Standards include those mentioned above, and EN 1451-1[297], EN 1455-1[298], EN 1519-1[299], EN 1565-1[300] and EN 1566-1[301]; also the drafts prEN 1329-1[302], prEN 1453-1[303] and prEN 1456-1[304].

It should not be assumed that existing plastics systems are automatically suited to reuse. Embrittlement hastened by excessive UV exposure may cause failure within 20 years, or make adaptation of the system difficult after perhaps 10 years. BRE used to receive many reports of sanitary pipework that had failed prematurely, some within a few years of installation. In particular, polypropylene pipes are not suitable for use externally, or even internally when exposed to sunlight; photo-oxidation results in the pipe becoming brittle and the pipe inevitably fails

(Figure 4.62). PVC-U pipes are not affected in the same way[305].

ABS, high and low-density polyethylene, polypropylene and MUPVC are sometimes suitable for use in high temperature conditions but manufacturers should be consulted on the choice of appropriate materials and grades.

PVC-U is the most commonly used plastics material for domestic discharge pipes but should not be used in applications where large volumes of water above 80 °C are discharged; for instance, in dwellings where wash-boil programme washing machines or water heating appliances without thermostatic control are used.

The durability of cast iron pipework to BS 416 depends on its protective coating; each item of these systems should be examined carefully before installation to ensure the protective coating is undamaged. The introduction of flexible joints has considerable advantages in labour costs over the previous hemp and lead joints. Cast iron pipes may be fixed satisfactorily by means of the lugs on the pipe sockets; by cast iron, malleable iron or steel holderbats for building in, nailing or screwing to the structure; or by purpose-made straps or hangers.

Copper tubes to BS 2871-1[285] are available in long lengths, which reduces the number of joints. Copper plumbing assemblies are very suitable for prefabrication techniques. Jointing is by brazing, or compression or capillary fittings. Copper U bends can be problematical, particularly when household bleach from sinks has been passed through them. Copper tubing which cannot be bent or welded is also available to other specifications, and therefore it may prove unsuitable for sanitary plumbing installations.

The component parts of pipework in mild steel tubes to BS 3868[286] should be hot dip galvanised after manufacture.

Maintenance

Suitable access should be provided to all pipework so that it can be tested and maintained effectively during its lifetime. Access covers, plugs and caps should be sited so as to facilitate the insertion of testing apparatus and the use of equipment for cleaning parts of the installation and for removing blockages. Their installation should not be impeded by elements of the structure or by other services. Attention should be paid to the condition of packings,

Figure 4.63
It is not good practice to connect wastes into the WC discharge pipe; if the large diameter pipe to the right of the stack flows full bore, the trap serving the branch pipe connected to it will be vulnerable

ring seals and washers, and any damaged items renewed before replacing covers or boarding.

Access points should not be located where their use can lead to nuisance or danger if spillage occurs. It is preferable that they are above the spill-over level of the pipework likely to be affected by a blockage or are extended to suitable positions at the face of the duct or casing, or at floor level. On discharge stacks, access should be provided to the stack at intervals of not less than three storeys.

All access positions and inspection covers should be readily accessible, and have adequate room around them for using cleaning rods (Figure 4.64). Cleaning access points additional to those at pipe junctions should also be made in long waste branches. Consideration should also be given, during the design of a pipework system, to access for replacing pipe lengths and fittings which may become damaged.

Figure 4.62
Polypropylene pipes that have deteriorated following exposure to UV light

Figure 4.64
In this old cast iron installation, the access plates are virtually inaccessible. Either a trap should have been left in the floor, or the pipe suspended at a lower level

While much can be done to improve the layout and prevent misuse of drainage systems, some risk of blockages will always remain. It is important therefore, particularly in the most highly serviced of buildings such as hospitals, that pipework is easily accessible. BRE investigators have heard many complaints in the past relating to lack of access to and within ducts – too small, blocked by other services, or in the wrong place. On pipework itself cleaning eyes often could not be used; some were partly embedded in concrete or turned towards the wall or ceiling. Access to others was blocked by other services, and in many cases it was not possible to use drain rods. Permanent lighting has been absent in some ducts, and a hand torch does not usually give a plumber adequate light by which to work.

Vertical pipes may become blocked if too many offsets have been used or if the pipes are of inadequate diameter. Horizontal waste pipes may have inadequate falls, be of excessive length or inadequate diameter, connected too close to a WC branch, sag because support brackets are too widely spaced, or simply deteriorate or become damaged in service. Disfiguring stains from leaking pipework can quickly become obvious (Figure 4.65).

Traps should be checked for 'pulling' (ie emptying of the water seals) when other appliances are discharged; proper design should obviate the need for special traps. Falls and length of waste runs should be compared with recommendations in Codes of practice. Of primary importance is the sealing of all branches by water traps and safe venting of the system to the atmosphere. The application of AAVs may simplify this requirement in some difficult situations.

Birds nesting on top of pipe vents may affect performance of traps in WCs, allowing odours in the drains to enter the building.

Blockages may result also from misusing the building while under construction. Builders' rubbish has frequently been found in drainage systems above and below ground. The open ends of soil branch connections are often seen on site stuffed with sacks or cement bags to keep out rubbish (Figure 4.66) – an unreliable method if these materials have to be removed as rubbish themselves from the completed system after handover.

Interceptor traps for fats oils and grease need regular cleaning and either physical, chemical or biological removal of the sludge.

Figure 4.65
Disfiguring stains from leaking waste pipes

Figure 4.66
A floor access point blocked by polyethylene sheeting instead of a drain plug

There is also another kind of intercepting trap, which may be fitted between sewers and drains to deter sewer rats from entering drainage systems. A BRE survey carried out in the early 1990s to establish the level of problems with rats in drainage systems in England and Wales showed that interceptor traps were widely used. However, there was also some evidence that certain designs of interceptor blocked more readily than others, even to the extent of needing their removal to reduce maintenance costs. In BRE tests, cast iron traps performed better than clayware, but are more expensive. Specifiers need to be selective in the choice of patterns.

Building drainage systems can be affected by occasional surcharges. The first sign of surcharge may be an overflowing WC pan. Devices are available in the form of a flap valve or an in-line valve which is manually or automatically operated. A visual or audible warning system is desirable. Where any type of anti-flood device is fitted, it may be necessary to provide additional ventilation to ensure preservation of trap seals.

Work on site

Workmanship

Workmanship should be to the requirements of BS 8000-13[306].

Galvanised steel tubes may be fixed by malleable iron brackets for building into or screwing to the structure, by malleable iron pipe rings with back plates or girder clips, or by purpose made straps or hangers.

Cutting pipework in galvanised mild steel on site should be avoided as this destroys the galvanic protection at critical points in the system; for this reason assemblies are generally prefabricated or purposed designed and galvanised before assembly on site. Painting of galvanised pipes is needed, especially where these are located in service ducts. Where galvanised mild steel pipes pass through floors or walls they should be protected by a suitable tape or waterproof paper wrapping, or be painted with bitumen or primed with one of a number of proprietary products now available (calcium plumbate has been phased out); galvanised products can be attacked by Portland cement, lime, plaster, brickwork and magnesite.

Copper tubes may be fixed using copper alloy holderbats for building in or screwing to the structure, or strap clips of copper, copper alloy or plastics, or purpose made straps or hangers.

Pipes may be fixed with holderbats of metal, plastics coated metal or suitable plastics material. Clips must not bite into the external surfaces of pipes when tightened. Intermediate guide brackets fitted to pipe barrels should allow for thermal movement. Intervals for supporting pipes in various materials are given in Table 4.3.

In multi-storey dwellings, vertical pipes should be supported by metal brackets because of their greater fire resistance.

Plastics pipes in most cases require to be installed in ducts having a fire resistance equal to the building. Guidance on the installation of unplasticised PVC systems is published by the Institute of Plumbing, and set out in the various European Standards referred to in this book.

Solvent welding is used to join PVC-U items, but this requires expansion joints in the system. An alternative form of assembly is to use rubber ring seals. PVC-U becomes brittle at low temperatures, so handling and assembly during cold weather should be done carefully.

Bends of large rather than small radius are always to be recommended but are particularly important with heavy usage of systems. Similarly, oblique junctions are preferred to T-junctions since they allow discharges to flow more readily from branches to main drains.

Access points are advised for all changes of direction except the most gradual. They are also needed at all junctions on horizontal drains. All parts of above ground drainage systems within service spaces and ducting should be easily accessible. Flexible connectors between soil appliances and piping allow for ready removal of the appliances if this is necessary for rodding.

Table 4.3 Maximum distances between pipe supports			
Material	Size (mm)	Vertical (m)	Low horizontal gradient (m)
ABS	32–50	1.2	0.5
Cast iron	all	3.0	3.0
Copper	25	2.4	1.8
	32–40	3.0	2.4
	50	3.0	2.7
	65–100	3.7	3.0
Galvanised steel	25	3.0	2.4
	32	3.0	2.7
	40–50	3.7	3.0
	65–75	4.6	4.0
Polyethylene	32–40	1.2	0.5
	50	1.2	0.6
MUPVC	32–40	1.2	0.5
	50	1.2	0.6
Polypropylene	32–40	1.2	0.5
	50	1.2	0.6
PVC–U	32–40	1.2	0.5
	50	1.2	0.6
	75–100	1.8	0.9
	150	1.8	1.2

Supervision and testing

Tests on installations are described in BS 5572 and EN 12056. These include air tests and other methods of leak detection. The air test provides the more reliable result, possibly used in conjunction with a smoke producing machine. A soap solution can be applied to joints and leakage will be detected by the formation of bubbles. The water test is normally only applied to the lowest part of the pipework and the static head should not exceed 6 m. It should not be applied to the whole plumbing system.

In a BRE survey, nearly a quarter of local authorities reported regular difficulties in carrying out air-tests on drainage systems fitted with AAVs.

Inspection

The problems to look for are:
◊ PVC systems degraded
◊ pipe joints leaking
◊ pipes not supported at sufficient number of points in a run
◊ outlets to vent pipes too near windows
◊ vent pipes and stacks badly sited or modified, and not sufficiently well supported (Figure 4.67)
◊ access covers poorly sited
◊ materials not suited to high temperatures or corrosive discharges

Figure 4.67
This existing soil-and-vent pipe was required to be modified when the dormer was inserted and the plumbing extended! In spite of the stay, it still looks vulnerable to high winds

Chapter 4.5

Fire protection

Nearly 1,000 people die every year as a result of fires in buildings. About 60% of the deaths are due to inhaling smoke and toxic products, and it is necessary, therefore, that particular attention is paid to reducing the spread of smoke in fire as well as to providing means of escape. Some aspects of smoke control involving pressurisation and depressurisation require powered systems, and so come within the definition of building services. For the purposes of this book these matters are dealt with here rather than in Chapter 3.

Passive fire protection measures are those features of the fabric (eg cavity barriers) that are incorporated into building design to ensure an acceptable level of safety. These features have been dealt with in *Roofs and roofing, Floors and flooring* and *Walls, windows and doors*). Measures which are brought into action in the event of a fire – such as fire and smoke detectors, sprinklers and smoke extraction systems – are referred to as active fire protection and are dealt with in this book.

It is widely thought that active and passive systems of fire protection in some way compete. In practice, each system should complement other fire protection systems, working together to enhance the safety of a whole building and its occupants.

The development of fire fighting techniques external to the building is not within the terms of reference of this book. However, those measures which are provided within the building curtilage are.

Portable self-contained fire extinguishers – for example, glass spheres containing water, brine or other dilute chemical solutions, and stored in baskets in the building to be protected – were in common use for fighting fires in the last decades of the nineteenth century. Primitive sprinkler systems were in isolated use in the early years of the nineteenth century, drenching the whole space through permanently open holes in the distribution pipes when the storage tank valve was opened, and regardless of the position of the seat of the fire. The first melting solder automatic sprinkler heads were developed in the 1880s and 1890s when the availability of water at mains pressure became widespread. Smoke control by ventilation was practised from the earliest times, with holes in the roof in primitive dwellings, but deliberate emergency smoke control by ventilation dates only from around the 1950s.

Hosereels connected to the building's water supply, and dry risers for connecting to fire brigade pumps came into widespread use in non-domestic buildings mainly during the twentieth century.

For an overall examination of the main issues relating to fire protection see *Fire safety in buildings*[307].

Fire detection systems, dealt with in Chapter 5.3, based on either heat, ionisation or optical principles will trigger the operation of fans to power pressurisation of escape routes where there is fire in an adjacent room. On the other hand, where the fire is in the same space as the escape route itself, smoke exhaust ventilation is more appropriate. For extinguishing the fire, sprinklers are one of the main weapons. All of these techniques, except for the special case of sprinkler protection of high density storage, are covered in this chapter. The protection of high rack storage is dealt with in Chapter 6.2.

It is generally thought that fire protection is a simple procedure which does not require the continued application of specialist engineering expertise. This is not so.

Characteristic details

Sprinklers

Sprinkler systems are generally recognised as one of the first means of dealing with fires in buildings like warehouses. They are also widely used in other building types. The first well publicised demonstration of how sprinklers might work was given in the Theatre Royal, Drury Lane, after its refurbishment in 1794. Some impressive features were shown to an audience to allay their fears about the fire risk. After an opening address an iron fire curtain was lowered and struck heavily with a large hammer to show how solid it was. The curtain was raised and immediately water cascaded from tanks that had been fitted in the roof[18].

There are two main requirements of sprinkler systems today:
● to detect and control a fire by preventing its vertical and horizontal spread, enabling fire fighters to deal with it as a small outbreak. According to the Fire Protection Association, 40% of fires in warehouses are dealt with in this way

Sprinklers

Sprinklers are essentially heat detectors which release water when their thermally sensitive elements reach a particular temperature. Water is normally prevented from flowing by a seal held in place by a thermal element. When the preset temperature is reached, the element releases the seal and allows water to flow. A sprinkler system consists basically of:

- a supply of water, which may be contained in a tank, or taken from the mains. In any event, the system has to be under some pressure head provided by gravity or pumps
- a network of large diameter pipes and small diameter pipes for distributing water to all or parts of the building
- sprinkler heads incorporating a means of detecting fire and releasing the water in the system. The heads are fitted at appropriate intervals in the pipework, sometimes on extended arms

When one of the sprinkler heads opens in response to a fire, the fire alarm can be made to trigger by the flow of water past a sensor.

There are two main types of thermal element, the soldered link (Figure 4.68a) and the glass bulb (Figure 4.68b).

Soldered link

The soldered link type consists of a metal strut holding a seal in place to prevent water flow. The strut is held by a mechanism containing two metal components soldered together. The eutectic solder mix is so selected that at a certain temperature it melts and allows the metal components to separate. As a result, the strut is forced out of position by water pressure and the seal opens. The six sprinklers shown in Figure 4.68a are as follows (left to right):

- typical fusible link, a design first developed in the early part of the twentieth century. The link is held out from the sprinkler to enhance heat transfer (1976)
- a fast response sprinkler with lower mass link than earlier designs (early 1980s)
- another sprinkler of similar, but smaller, design (early 1980s)
- first fast response sprinkler with a very thin link (mid-1980s)
- the ESFR sprinkler which produces larger droplets than the other sprinklers (late 1980s)
- fusible link sprinkler designed for residential use; when it operates, the deflector drops 20 mm to create the water spray

Figure 4.68
Different patterns of sprinkler heads used over the years: soldered link heads (above) and glass bulb (below)

Glass bulb

The glass bulb type has a liquid filled bulb which directly holds the seal in position to prevent water flow. Heat from a fire causes the liquid in the bulb to expand until the pressure is sufficient to break the glass and water flows. To increase the pressure that fractures the glass, the liquid does not completely fill the bulb[308]. The six sprinklers shown in Figure 4.68b are as follows (left to right):

- pendant sprinkler made in 1976, but typical of the designs of the 1920s – operating temperature 68 °C
- upright sprinkler – operating temperature 93 °C

- reduced bulb size to improve response times (early 1980s)
- smaller bulb size than previous item (also early 1980s)
- fast response sprinkler (mid-1980s)
- sprinkler used in commercial premises with false ceilings – when the bulb operates, the deflector plate is forced down by the water pressure

● to cool surfaces at the perimeter of
a fire to prevent or delay their
ignition

The main function of sprinklers is to
contain fire and to reduce the
frequency of large fires. They are
usually effective, and they certainly
have played a major role in the
protection of property against fire.
However, it has increasingly been
recognised that they also may have a
role in life safety, albeit with features
additional to those required for
property protection.

The necessary condition for early
suppression is that the amount of
water that penetrates the fire plume,
and is delivered to the top surface of
burning building components,
materials and goods, has to exceed
the water density required to
suppress the fire at that stage in its
development.

If sprinklers could be relied upon
to extinguish fires rapidly and
completely, there would be little
need for vents in sprinkler fitted
buildings. More often, sprinklers
control fires but do not extinguish
them immediately, and the fire
continues to produce volumes of
smoke and steam. Venting then plays
an important role in reducing smoke
damage and helping the firefighters
to put out the fires[309].

Sprinkler systems may be wet or,
rarely, dry. That is to say, either the
systems are permanently charged,
ready at all times to discharge water
into the seat of the fire, or they are
dry. A dry system can be used where
the water might freeze if allowed to
stand in the pipes. The effectiveness
of sprinklers in this case relies on an
extremely rapid filling of the system
once the alarm is triggered. Water
leaking from sprinkler heads is
comparatively rare, though the risk
might need to be assessed for
building contents of high value.

The most recent development is
the early suppression fast response
(ESFR) sprinkler. A fast response
thermal element is combined with a
sprinkler which can penetrate the
fire plume to deliver water at higher
rates, and in bigger droplets, than
standard sprinklers. The key to this is
the early operation of the system to
suppress the fire in its developing
phase. The ability of an ESFR system
to suppress a fire in a particular
situation is determined by full scale
tests where the required water
supply and that actually delivered to
the fire are measured. ESFR
sprinklers are designed primarily for
use in high hazard situations such as
warehouses.

Prompted by a fire which in 1983
destroyed a large Ministry of
Defence (Army) storage depot, and
its contents, FRS developed a rapid
response sprinkler system for use in
high-rack storage. The system
proved very effective in full-scale
trials (see also Chapter 6.2).

Gas extinguishing systems

Gas systems are normally used
where people are not present, and
where the stored contents are very
valuable, such as computer rooms or
strongrooms. Halon has been used
extensively in the past to fight fires in
these enclosed situations, but the use
of halons is being phased out
following international agreement.

Carbon dioxide extinguishing
systems may also be found where the
principle, basically, is to starve the
fire of oxygen.

Smoke and heat exhaust ventilators (SHEVs)

Vents are primarily intended to
remove heat, smoke and toxic gases
from fires. They do not extinguish a
fire but allow people to escape and
firefighters to see to tackle the source
of the fire. While fire does the
damage to the building, it is usually
the toxic gases that kill the
occupants. When the Kentucky
Beverley Hills Supper Club suffered
a fire in 1978, an eye witness said that
the screaming stopped only after the
smoke had enveloped the panic
stricken people. In this case the fire
could only be got under control after
smoke vented itself through a hole in
the roof. In 1903 at the Iroquois
Theatre in Chicago, 571 people died
when a fire started on stage;
members of the audience in the
gallery were killed by the toxic gases
even before they had time to rise
from their seats.

Most shuttered vents are powered
to open when triggered by the alarm
system. They are often fitted into the
sloping part of roofs, but can also be

Figure 4.69
A proprietary multi-functional roof ventilator
installed in patent glazing. This old type
was designed to provide protected
ventilation for normal weather conditions,
but with the flaps opening fully in very hot
weather. In fire conditions too, smoke could
be expelled through the automatic
operation of the flaps

fitted into walls at appropriate positions. They can be found in many forms, from roof lights fitting into glazing systems to complex shapes in pressed metal. Some take the form of aluminium or steel pivoting windows powered to open by gas filled struts, and triggered by the centrally located alarm system. These windows cannot be used for normal ventilation purposes, since they pivot in a reverse mode (top out, bottom in instead of top in, bottom out) and are only weatherproof when fully closed[310,311]. They need to operate in all weathers, even under snow loads!

Natural ventilators must be 2 m clear of obstructions. Powered ventilators incorporating fans also need to be clear of obstructions for a distance of at least 1.5 times the fan diameter.

Smoke ventilators should be powered to both open and close. Those powered only to open are unlikely to be tested regularly if they need to be closed manually after each test, especially if weather conditions are poor. The closure switch should be operated only by an authorised person.

Smoke curtains are sometimes specified for large spaces (eg shopping malls) where smoke containment, or channelling is wanted. The contained smoke is then more efficiently vented away from the enclosed spaces[312].

Ducts

Ducts and their hangers for smoke and heat exhaust ventilation must be of a material which will be unaffected by the heat of a fire; surrounding combustible carcassing must also be protected. Moreover, allowance must be made for thermal expansion without causing splitting of the duct walls.

The operation of dampers fitted into ducts needs to be easily checked and maintained.

Dry risers, hosereels and handheld extinguishers

Dry risers are for use by the fire brigade after its arrival on site. The risers are connected to the pumps on fire appliances, and serve outlets positioned on each floor.

On the other hand, hosereels and hand-held extinguishers are for use by building occupants in the early stages of an fire when the outbreak is small and there is little personal risk.

Main performance requirements and defects

Outputs required

Sprinklers

Computer modelling of fires has shown that the preferred droplet size from sprinkler systems is 1 mm diameter. These fall to the floor of the compartment relatively unscathed, but those of 0.5 mm are greatly influenced by the rising thermal plume above the fire, and relatively few of them reach the fuel.

One of the main potential problems with sprinklers when they operate is that the supply system must have sufficient capacity to allow full functioning of all the heads that open. This provision is covered in the basic design of the system.

The performance of sprinkler systems is covered in BS 5306-2[313] †.

Pressurisation

Pressure differential systems are designed to protect against smoke leakage through small gaps in passive

---

† It is expected that BS 5306 will become obsolete when European Standards are published in the near future.

protection of an escape route. Air is pumped into an escape route, usually a stairwell or corridor (Figure 4.70), to maintain pressure sufficient to oppose the buoyancy of hot gases on the fire floor, the stack effect due to the building's own warmth, and wind pressure effects. BS 5588-4[314] recommends a pressure level of between 50 Pa and (approximately) 100 Pa as represented by the maximum 100 N force required to open a door.

The lower limit is large enough to overcome the combined effects of fire buoyancy in the room of origin, the stack effect where storeys are connected by shafts other than pressurised shaft and stairwells, and wind induced pressure differences. These pressure differences apply across internal doors between a pressurised shaft or stairwell and accommodation spaces. Wind pressures apply to an internal door where the building's perimeter is not breached. If an upwind window falls out when winds are strong, pressures much greater than 100 Pa can be expected even though they are not considered in design. The air supply must be sufficient to hold back cool smoke at an open door, typically giving an air speed of 0.75 m/s at the door, for systems designed to protect means of escape. The corresponding value for a firefighting shaft is 2 m/s to hold back hotter gases during firefighting.

The pressurisation technique is complicated but is the most effective for multi-compartment buildings. It is suited to the protection of core stairwells used as escape routes in tall buildings and can be useful, for example, in historic buildings.

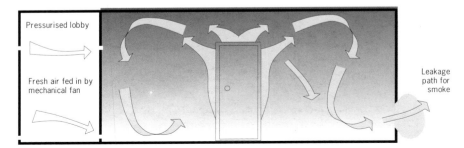

Figure 4.70
Pressurisation to force hot gases away from an escape route

Depressurisation

In the process of depressurisation, gases are removed from fire compartments to maintain the desired pressure drops across, and air speeds at, escape doors. The method is not often used in the UK, though it is included in BS 5588-4. Fire rated fans and extract ducts are usually essential. Depressurisation works best when fire compartments are relatively well sealed. If fire compartments and protected shafts are leaky, pressurisation is difficult.

A related form of depressurisation can be used for buildings containing atria which have a central common space onto which offices or shops face, provided that the atrium façade has only relatively small leakage paths. It relies on a reduced pressure in the atrium itself so that the net airflow is moving from the accommodation areas into the atrium, preventing smoke passing into the escape routes on the periphery of the building (Figure 4.71). The amount of depressurisation is governed by the ratio of the ventilation area in the atrium roof to the inlet areas at low level.

Depressurisation does not protect the atrium space itself. It protects the adjacent accommodation spaces. Depressurisation of atria can be combined with smoke exhaust ventilation to form a hybrid system.

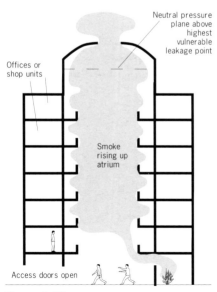

Figure 4.71
Smoke control in an atrium by depressurisation

The system can be used to protect escape routes open to the atrium's lower storeys as well as adjacent spaces on higher storeys separated from the atrium by leaky façades.

Smoke clearance

Normal practice is that smoky gases are removed from the affected parts of a building after the fire has ceased to add any significant quantities of smoke to those parts. The reason for this exists primarily to assist firefighting operations. It is not required by building regulations, but can be required by various local byelaws. Design to remove smoke from buildings is commonly ad hoc, although wall mounted openable vents with an area of 5% of the floor area are specified in some byelaws.

Dilution ventilation

Dilution ventilation is used when the fire is likely to be in the same space as the people, contents or escape routes being protected. The aim is to mix sufficient clean air into the smoke to enable people to see their escape route as well as to breathe while making their escape. Distances of 3–5 m have been suggested for familiar buildings, but up to 20 m may be more appropriate for unfamiliar buildings. For stair wells and corridors, which are usually of restricted size, 10 m is appropriate. Where smoke has been diluted to these visibility levels, toxicity should not be a problem for the duration of escape, but the method does require very large volumes of clean air[315].

The main problem is the very high rate of dilution of combustion products needed to achieve a safe visibility. With very smoky fuels, such as certain polymers, the dilution level may need to be one thousand times the initial volume of combustion gases. This may only be practicable where the fire is small and the building volume is large, and there is no compartmentation (as will be found in atria[316], shopping malls, exhibition halls and some industrial premises).

Smoke and heat exhaust ventilation

The principles are simple. Hot buoyant gases from a fire rise to form a layer at sufficient height that allows a layer of cooler, clear air to remain at lower levels for sufficient time for occupants to be safely evacuated. This clear lower layer can be maintained by exhausting smoke from the high level either by natural or mechanical means[115].

As fire or smoke plumes rise, air mixes into them, increasing the total volume of smoky gases (Figure 4.72). The gases flow outwards below the ceiling until they reach a barrier, such as a wall or a downstand. At the barrier the layer of smoky gases deepens, but has sufficient buoyancy to drive the gases through natural vents. Alternatively, the smoky gases can be removed from the layer by exhaust fans, but in any event the smoke must be above peoples' heads on the highest exposed escape route. Unless sufficient air enters the space below the layer to replace the gases being removed from the layer, the smoke ventilation system will not work.

Other fire precautions, especially the use of sprinklers, play a major role in increasing the range of building design options for containing a fire of a given size.

Smoke exhaust ventilation systems are often required to operate very early from the onset of a fire – well before the fire brigade can arrive – to protect escape routes. An effective system can also enhance the efforts of the brigade by allowing them to see what they are doing.

The time needed for safe evacuation is usually unquantifiable but may be much longer than the 2.5 minutes often used in the UK for calculating the minimum widths needed for exits, stairs etc[315].

A potential problem has been identified with air conditioned (ie cool) atria when the weather is hot. In these circumstances the operation of a naturally driven, stack effect, smoke and heat exhaust system reverses. The relatively cool air in the upper levels of an atrium is replaced by warmer ambient air drawn in through roof inlets; the cool air is pushed down, taking with it any smoke from a fire, and exhausted through open escape routes at low level. There is something to be said, therefore, for powered systems to prevent this happening[115].

Smoke exhaust ventilation cannot, by itself, control a fire. Additional measures are needed, such as a sprinkler system, the effect of which is to limit the size of a fire to that with which the exhaust ventilation system can cope.

Unwanted side effects

The advantages and disadvantages of the combined use of sprinkler and smoke ventilation systems has been intensively discussed for at least 35 years. The main subjects at issue have been, and are still:
● the safety of life
● the cost efficiency of smoke ventilation systems
● interaction between sprinkler and smoke ventilation systems which has detrimental results[317]

Sprinklers and smoke exhaust ventilation systems do interact with each other, and not always beneficially. Building designers have in the past been offered conflicting advice; it is important, then, for building surveyors, or those professionals looking after existing buildings, to establish precisely what the design originally called for before they consider any possible upgrading or refurbishment of the systems.

BRE's interim view is that the protection of life is the paramount criterion in a fire; the smoke exhaust system should operate, therefore, immediately smoke from fire is detected and before a sprinkler system can be triggered.

On the other hand, where high-hazard occupancies are identified, means of escape should not depend on exhaust ventilation, but on compartmentation and short escape routes. Any installed smoke exhaust system for these conditions should be operated by the fire service, and not automatically. It is also relevant how long the brigade takes to arrive[115].

There may be a potential conflict between a SHEV system and the security system for the building. This could happen when external doors, normally kept closed or locked, are needed to be open to operate as part of the exhaust system. The SHEV control system should over-ride the security system in these circumstances to release the doors.

Commissioning and performance testing

The design and installation of firefighting systems is normally in the hands of specialist companies. The Loss Prevention Certification Board maintains a list of approved suppliers and installers[318].

A SHEV system should be fully operational within 1 minute of its triggering. Smoke dampers, operating either open-to-closed or closed-to-open positions, should also conform with this criterion, and be leak-proof. Dampers should be very reliable. If they are to withstand testing once a week for 10 years, this will equate to a regime of at least 1000 cycles for high risk safety applications.

European Standards for equipment for use in SHEV systems are being developed (PrEN 12101 series).

Natural ventilators for use in SHEV systems should not be capable of being forced open by wind; otherwise there will be a natural tendency for building managers to lock them, making them inoperative in fire. Nor should they leak rainwater. General criteria for wind loadings and suctions and for rain penetration have been discussed

in *Roofs and roofing* and *Walls, windows and doors*.

Air inlets for SHEV systems should operate within 2–3 seconds of receiving the signal to open.

Where a SHEV system control has been incorporated into a building management system (BMS), any updating of the BMS for reasons of environmental control must take full account of the needs of the SHEV system. Even where a dedicated computer is used for fire control purposes, there will still need to be interaction with the BMS.

Mains operated systems may need emergency power backup, and switching between mains and backup needs to be automatic.

Smoke curtains, especially of the automatic retractable kind, need to be checked for closeness of fit when in the deployed position. Ideally there should be no gaps at all, but it may be more practicable to form overlaps between adjacent curtains or between curtains and carcass to minimise gaps. It has been suggested that the total area of gaps should not exceed 1% of the curtain area[115].

Commissioning tests should cover:
● reaction time for the SHEV system to open
● airflow measurements for powered ventilators
● simultaneous opening of ventilators and air inlets
● pneumatic leakages
● reaction of automatic and manual controls
● start-up time for emergency power supplies
● behaviour of automatic smoke curtains

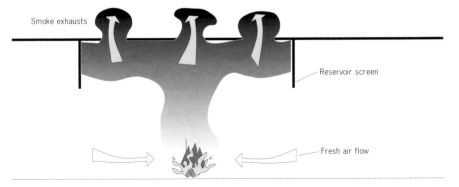

Figure 4.72
Smoke ventilation

Durability

Sprinkler heads last for a considerable time, although they need to be replaced when they have operated. Conventional wisdom is that they should last at least 25–35 years, but many have survived intact for longer.

Fans, as with all moving parts, are subject to wear, and their longevity will depend on the design and materials used.

Maintenance

As a fire may not occur for many years after protection systems have been installed, the equipment must be capable of surviving in a condition that permits proper functioning when eventually required; this has implications for regular maintenance, and even for testing. Smoke control systems should be serviced at least once a year, and alarm systems tested every week (BS 7346-1). A full-scale test by blowing smoke into a smoke detector should be carried out whenever there has been any amendment to the SHEV or BMS.

Detection and operating components should be fitted where they are accessible for maintenance and replacement purposes.

Figure 4.73
Hosereels need to be inspected for deterioration of the hose material, and extinguishers need to be appropriate to the types of fire that they would be called on to deal with

The frequency of routine testing of systems will depend on circumstances and building type. It is suggested that the operation of a SHEV system in a large shopping complex should be tested once a week, whereas that in an office building may be tested once a month. Where only property protection is involved, once a year may be sufficient[115].

Keeping a log book to record the initial design concepts, specification of components, and maintenance of the systems, for all aspects of fire safety is recommended.

Sprinkler systems are subject to rigorous maintenance and inspection regimes. Spare heads of the correct type and temperature rating must be kept on the premises to replace those which have operated (or failed to operate). Inspections need to be carried out at defined intervals: daily checking of connections and water levels in stored systems, weekly pressure gauge readings, weekly testing of alarms and pumps, quarterly reviews of changes in occupancy and therefore in the adequacy of the system, annual tests of pump flows and tanks, and regular inspections of various parts of the system and equipment for corrosion. There are also requirements for certain 3-year and 15-year inspections (BS 5306-2).

Although considered more as firefighting equipment, hosereels and extinguishers need to be checked regularly (Figure 4.73).

Work on site

Workmanship

Even the best equipment can be installed wrongly, and there have been many instances where the installation has not been satisfactory when called upon to function.

Since different parts of a SHEV or other fire protection system may be fitted by specialist contractors, it follows that there could be a division of responsibility, and lack of co-ordination. It is recommended that a single organisation has overall responsibility for the installation; that organisation might also have responsibility for the design.

Inspection

The problems to look for are:
◊ provision of sprinkler heads not adequate
◊ hoses in hosereels perishing
◊ inspection certificates out of date
◊ exhaust ventilators unable to operate in wind or snow
◊ exhaust ventilators covered with ad hoc rain shields or covers, rendering them ineffective
◊ combustible roof coverings, which are within 0.5 m of smoke ventilators, not given protection
◊ material used for sealing gaps in carcassing for cables, ducts etc not appropriate

Chapter 4.6 Gas, including storage of LPG

This chapter deals with the supply of mains and stored LPG systems. Gas central heating boilers and systems, and their flues, have already been dealt with in Chapter 2.2, and the storage and distribution of medical gases is outside the terms of reference of this book.

Gas lighting was first introduced in the UK in the early years of the nineteenth century. It is now rarely seen. One or two street lighting installations have survived – for example, at Malvern in Worcestershire.

Some 85% of dwellings in England now have mains gas supply, but, as might be expected, the percentage is much smaller in rural areas where only two-thirds of dwellings have a mains gas supply. Over the whole of England, some 1 in 8 dwellings have no gas supply of any kind[7]. Also, the perceived risk of explosion damage since Ronan Point in 1969 has been a factor in decisions not to provide gas installations in, or to discontinue gas supplies to, multi-storey blocks.

Until the 1960s, the gas first used for lighting and later for cooking, and distributed mainly in urban areas, was manufactured from coal; coke was produced as a by-product. Since the 1960s, coal gas has been supplanted by natural gas piped from the North Sea.

The main legislation relating to the safety of gas installations is the Gas Safety (Installation and Use) Regulations 1972, revised in 1984 and 1990. Many of the Approved Documents to the Building Regulations (for example B, E, F, G and J) are concerned with the wider aspects of gas safety.

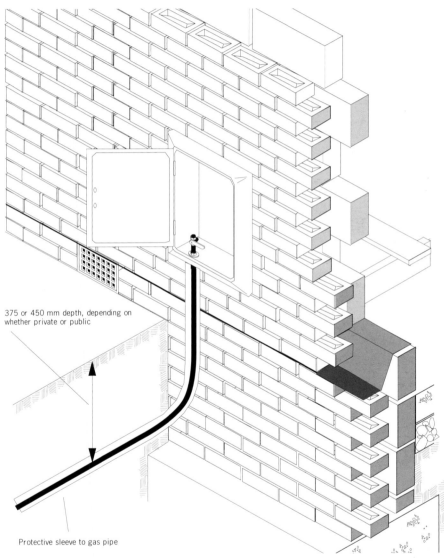

375 or 450 mm depth, depending on whether private or public

Protective sleeve to gas pipe

Figure 4.74
A built-in gas meter box

For those areas outside the range of mains distribution, there is the possibility of using propane or butane, stored either in small portable pressure cylinders, or distributed by tanker vehicles for transfer to large fixed storage tanks.

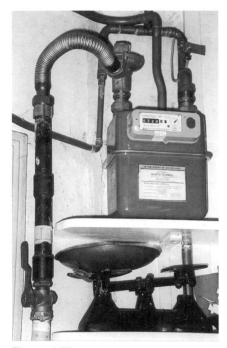

Figure 4.75
A very much adapted gas installation situated at high level in a turn of the century kitchen. The photograph was taken just before the meter was replaced by new equipment in an outside box (Photograph by permission of B T Harrison)

Characteristic details

Space requirements for major items of plant (dimensions etc)
Gas can be supplied at different pressures according to location and type of buildings served. It may be necessary to provide space for pressure reducing plant near the buildings to be served.

The gas service pipe to the meter in older installations should have been bridged where it enters the building below floor level so that it is not affected by any subsequent movement in the structure of the building or the ground beneath. In more recent years it has become standard practice for the sleeved external pipe to enter the approved meter box above floor level. Meter boxes (accessible from the outside) can be built into external walls (Figure 4.74 on the previous page), or fixed to the outside of external walls.

Old installations might have the gas meter almost anywhere in a dwelling – exposed in the kitchen was very popular at one time (Figure 4.75). Gas meters are now invariably placed in lockable weatherproof cabinets on the external walls of dwellings where they are accessible for meter reading and maintenance purposes. In older dwellings they are rarely found in accessible positions.

Carcassing a dwelling for gas normally used to include providing a supply to each of the following positions:
● a cooker
● a refrigerator
● a tap for central heating apparatus, even if not initially used
● a tap for a fire in the living room

It is not now normal practice to include the provision for a gas refrigerator.

Pipe and trunking runs
Up until the 1970s it was usual to bury gas pipes in concrete oversites or screeds preventing access, although access ought to have been provided. Until this time too, it was usual to keep distribution pipes within the dwelling, either buried in floors or in vertical runs up internal wall surfaces. Practice has now changed somewhat, and service risers to large buildings or blocks of flats, or supply pipes to individual appliances, can be found running on the outsides of buildings (Figure 4.76). In either case, inside or outside the building, the void around a sleeved pipe must be ventilated.

The main relevant Standard for service pipes not exceeding 28 mm diameter is BS 6891[319]. Also, installations must accord with the Gas Safety (Installation and Use) Regulations 1998. Of course, old installations will not have benefited from these requirements.

Flues, chimneys and terminals
Great care should be observed when installing gas fires into the hearths of former solid fuel burning appliances and open fires. In particular, the gas feed pipe should not be inserted into the masonry chimney, since flue condensate salts absorbed in the masonry will attack the gas pipe. There have been failures of copper pipes leading to gas leaks.

See also Chapter 2.2.

Fuel storage
Propane cylinders (propane is explosive and toxic) are normally stored outside the building (Figure 4.77), whereas butane (which is explosive but not toxic), is normally used in portable appliances and can be stored indoors. Large domestic heating installations are fuelled from a fixed tank to which deliveries are made by road tanker.

Figure 4.76
The gas supply pipe, run outside the house, for a new room-sealed boiler installed at high level in the kitchen. Although now common practice and undoubtedly convenient, the pipe is unprotected and potentially is vulnerable to damage

Main performance requirements and defects

Outputs required

The installation will be governed by the Gas Safety (Installation and Use) Regulations.

For buildings built before the 1970s, incoming service pipes were normally in steel or even wrought iron, protected by bitumen wrapping; since that time plastics pipes have become almost universal. Pipe diameters for small installations could be as small as 25 mm, although larger domestic installations, or those at some distance from the mains, might need to be larger with special arrangements being made for their design[320].

Internal installation pipes need to be sufficiently large to limit the pressure drop between the meter and the appliance to no more than 1 mbar; generally this means 20 mm diameter steel or 22 mm diameter copper pipe.

Unwanted side effects

The only long term trends that are clearly apparent, and not simply related to changes in consumption of different fuels, are linked to the conversion from manufactured (town) gas to natural gas. This has been associated, directly or indirectly, with a marked reduction in the risk of damage by an explosion resulting from a leak of gas within a building. It has, however, led to a small increase in the risk of damage by an explosion of gas entering a building from an external leak[321].

Commissioning and performance testing

Gas pipes must not be run in wall cavities, and where they pass through cavity walls they must be sleeved. They must have electrical equipotential bonding. (See also the Gas Safety (Installation and Use) Regulations and BS 6891).

Health and safety

Gas explosions in buildings make headline news in national newspapers and so might seem frequent. Incidents of leaks from mains resulting in explosions and fire are rare, though. Statistics from the Fire Research Station show that they made up only 0.8% of the 63,000 fires in UK homes in 1990. Mains gas has been responsible for about twice the number of explosions as stored gas. There were 541 gas explosions during the period 1972–77 resulting in death (57 fatalities) or damage of more than £100, and 654 explosions during 1985–91 caused by mains gas, of which 317 caused structural damage, and 264 from stored gas, of which 103 caused structural damage[80].

Average annual data for accidental deaths in UK homes show that less than 1.5% of them are due to fires and explosions caused by piped mains gas. In many instances fires and explosions are the consequences of faults in gas burning appliances rather than gas supply.

There are around 40–50 deaths per year in domestic buildings due to using gas installations, with inadequate air supply and inadequate flue maintenance the more specific causes. There was a significant reduction in deaths following the change from town gas to natural gas owing to the absence of carbon monoxide from natural gas. As an example, over the period of the changeover, 1969–75, the annual number of fatalities fell from 294 to 43[80].

Although external meter boxes generally are fit for the purpose, some installations may lead to pest problems unless doors are made to fit tightly, and the pipes connecting to the internal distribution pipework are tightly built in and properly sealed.

Durability

Corrosion of gas service pipes may be seen, particularly in domestic kitchens. Floor screeds are likely to lose their ability to protect steel pipes over time, and any moisture in screeds can produce corrosion of the outer surfaces of the pipes. Since the products of corrosion expand to something like seven times the volume of the source metal, they can produce cracking and lifting of the screed along the line of the pipe. Areas of raised screed, being narrow, can easily be seen in thin plastics floor coverings.

Work on site

Supervision of critical features

Adequate air supply to open flued gas burning appliances is crucial to safety (see Chapter 2.2).

Inspection

Reference should be made to the Gas Safety (Installation and Use) Regulations for items which must be checked under specific conditions.

Otherwise the problems to look for are:
◊ gas pipes in unventilated voids
◊ gas pipes in unknown locations or vulnerable to damage
◊ gas cooker supplies not terminated in an approved fitting to receive a flexible hose
◊ pipes not easily accessible
◊ air supplies to open flued appliances inadequate
◊ flue maintenance inadequate

Figure 4.77
Propane cylinders for a small domestic installation

Chapter 4.7   **Refuse disposal**

Practice in this field is evolving rapidly, and it is likely that there will be further changes in recommendations and legislation in the early years of the twenty-first century. Waste collection authorities in the UK were required to produce recycling plans under the Environmental Protection Act 1990[322]. Most of the authorities in England and Wales have submitted to the Government the recycling options they planned to develop[323]; nearly all of these have indicated they would introduce 'bring' schemes, and over one-third of them that they would introduce kerbside collection recycling schemes. A change in the means of storage of waste at the household level may be required in order to implement efficient recycling. Currently, only about 5% of household waste is recycled in the UK[324].

Figure 4.78
A large bin designed for mechanical handling

In 1982, domestic refuse amounted to around 1 tonne per household per year[18]. Current figures are complicated by the fact that the average size of households is falling, nearly halving in size between 1961 and 1991, and is expected to fall further, leading to a smaller output of waste per dwelling, even though per capita output might be increasing.

Until the late 1970s the usual receptacle for domestic waste storage for individual dwellings was a small galvanised iron dustbin of up to 0.092 m³ to BS 792[325], located outside the kitchen door. Since that time, however, the attentions of the Local Government District Audit and its successor the Audit Commission have been focused on improved value-for-money methods of collection, which has seen the gradual introduction of wheeled bins of larger capacity, suitable for mechanical handling.

In 1990, a Government White Paper on the environment[326] proposed to recycle or compost 25% by weight of household waste by the year 2000 (this represents about half the household waste that potentially could be recycled). In 1995, a second White Paper, *Making waste work*[327], set out a strategy for achieving more sustainable waste management. Five levels in a waste hierarchy were identified: reduction, reuse, recovery (recycling, composting, and energy generation), and final disposal. Targets were set for England and Wales for the waste management sector, including central Government and local authorities. Household waste recycling proposals included:

- recovering 40% of municipal waste by 2005 (municipal waste includes household, street cleaning, and some commercial and trade waste)
- all waste disposal authorities to cost and consider central composting schemes by 1997
- providing easily accessible recycling facilities for 80% of households by 2000
- composting one million tonnes of organic household waste each year by 2001

Currently, approximately 50% of local authorities use plastics wheeled bins[324]. Examples of containers used include:

- 120 and 240 litre plastics wheeled bins for both mixed waste (no segregation of recyclables) and segregated wastes (Figure 4.78)
- 40 and 50 litre plastics boxes (with or without lids) for the storage and collection of dry recyclables such as paper, cans, and plastics bottles
- subdivided 120 and 240 litre wheeled bins with compartments used to store different types of wastes, for example recyclables and residual wastes (non-recyclables)
- separate plastics bags for residual wastes (usually black bags) and recyclables (usually green, blue, or clear bags for different categories of recyclables)

Approved Document H of the Building Regulations gives guidance about the minimum volume of a waste container, 0.12 m³, which is based on the waste produced by a typical household of 0.09 m³ per dwelling and assumes collections at weekly intervals.

Households with wheeled containers do seem to generate a greater weight of waste (an average of 978 kg each year) than do households using plastic bags (an average of 645 kg each year). Other Standards which could be relevant are BS 3735[328] and BS 4998[329].

By the end of the twentieth century there was a significant growth of interest in recycling as much as possible of household refuse – certainly paper, and possibly metals, plastics and glass. Centrally organised collection of sorted refuse is still, however, in 1999, in its infancy in the UK; in the meantime people have been exhorted to take their refuse themselves to local facilities. See the BREEAM criteria given later on this page.

Refuse disposal in tall blocks of flats has always been handled differently, with hopper-fed chutes from floor landings leading to very large receptacles on ground floors or in basements. The receptacles or bins are only suitable for mechanical handling.

There have been isolated examples of waterborne refuse collection systems in blocks of flats – the Garchey system being one example – with excess water being extracted from the effluent before the solid matter is incinerated in a central plant. These plants are small compared with the municipally-owned 'refuse destructors'. The calorific value of household refuse has been estimated at around half that of other solid fuels[18].

From around the mid-1960s, some buildings may have been equipped with central vacuum cleaning plants, with duct sizes ranging from 50–150 mm bore, buried in wall and ceiling voids leading to terminals sited on walls and floors to which flexible tubes with nozzles could be connected. While no records of these systems have been found, it is evident that the noise created by the large volumes of air being transmitted by comparatively small ducts would be tremendous.

Characteristic details

Space requirements (dimensions etc)

Communal refuse systems

Access to hoppers serving refuse chutes, normally installed in buildings over four storeys high, must be unobstructed. The clear space in front of hoppers should be not less than 1.5 m deep.

The preferred height of hopper hinges is 0.6 m above floor level (Figure 4.79). A shelf on which rubbish may be placed before putting in the hopper can be useful; it should be adjacent to the hopper at the level of the hinge. Any supplementary refuse store for dry goods too bulky for disposal by chute should be accessible without having to negotiate steps.

Dustbins

Disposal arrangements for domestic refuse are extremely varied and depend on the policy of each local authority. The arrangements range from large wheeled bins to plastics bags on frames. Whereas in former years the domestic dustbin was the basis of waste storage and collection, it no longer takes a significant role.

Where old style dustbins are still used, there should be an unobstructed area in front of them which is a minimum 1.2 m (but preferably l.5 m) deep.

Where there is a policy of separation of domestic refuse, and recovery of useful items, space allocation has to be significantly increased. To encourage recycling of domestic waste on a larger scale, and to increase the re-use of non-renewable resources, the BREEAM criteria call for a set of four containers for household waste, space for them, and appropriate access for removal to a collection point.

- At least four bins should be provided which are distinctly identified as being for different purposes (eg different colours) and have a collective capacity of at least 240 litres (0.24 m³) per household – that is, an average of 60 litres per bin

- The bins should be in a suitable hardstanding area, within 3 m of an external door of a single-family house and within 10 m of an external door of other dwellings

- To accommodate the possibility of additional bins being required in the future, the bin storage area must normally be at least 2 m² without interfering with pedestrian or vehicle access

- Bins should be protected from wind on at least two sides by a wall, fence or suitable hedge at least as high as the tallest bin

- Access for a car or refuse collection vehicle to within 20 m of the location of the bins

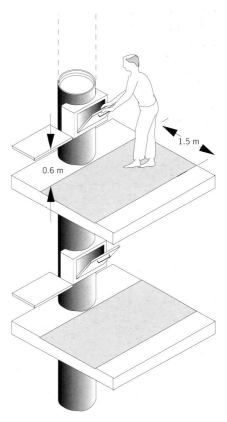

Figure 4.79

Minimum access dimensions for a hopper-fed refuse system

Main performance requirements and defects

Outputs required

All buildings, but especially those housing food preparation areas, should be provided with adequate food waste disposal facilities. Problems are created by the growing use of polyethylene bags for waste disposal; these are not sufficiently stout to keep rodents out and may also be damaged by domestic pets or foxes; they also allow insects access to the food scraps within. Where wire mesh stands are provided they often do not exclude pests. Stands should be made with 5 mm x 5 mm or smaller mesh and completely surround the bags. They must also be sufficiently large to easily accommodate the bags when fully filled, and be provided with well fitting solid lids. Modern, lightweight polyethylene or plastics lids to dustbins are not always rodent proof and they have even been known to be destroyed by grey squirrels seeking access to the contents.

Since the design of pneumatic and waterborne refuse disposal systems varies so much, it is not possible to give simple guidance on their characteristics. Manufacturers should be consulted, or independent third party advice may be available.

Case study

One city's pneumatic refuse disposal system

Measurements were made on a pneumatic refuse disposal system serving a housing estate by BRE investigators over a period of 12 months ending in May 1975. Previously refuse had been collected by manual methods. This period occurred when the numbers of new dwellings on the estate had increased from about 300 to 1000. Data on the amounts of refuse were accumulated, together with details of day-to-day use in terms of frequency of operation, types and frequency of malfunctions, pipeline blockages and the operational problems encountered, together with a breakdown of operating costs. An analysis of maintenance work was included. On the whole the study showed that the pneumatic installation provided an acceptably efficient alternative to manual collection methods, but that there was scope for improvement in the methods of handling refuse in the plant room[330].

Requirements for domestic refuse disposal are covered by the Approved Document H of the Building Regulations[331] and by BS 5906[332].

Noise and other unwanted side effects

Noise from metal dustbin lids being removed and dumped on pavements led to their replacement by rubber and plastics to BS 3735. However, local practice in refuse collection has considerably altered in character since the late 1970s and systems involving large wheeled and lidded bins made of plastics are inherently much quieter.

Chute systems can emit unpleasant odours if decaying refuse is not wrapped before disposal and this has led to problems when a single hopper serves more than one dwelling.

Health and safety

Health and safety hazards related to the handling and storage of household waste can be subdivided into physical, chemical and biological categories.

Physical hazards can occur during the handling of waste. Although the amount of materials used in packaging has been reduced, known as 'lightweighting', this causes certain materials to become more dangerous to handle, for example thinner glass breaks more readily.

Paints, batteries, garden chemicals and certain DIY products – items with chemical constituents and known collectively as household hazardous waste – present a risk to both the householder and the waste collector. Explosions have occurred at materials recovery facilities due to methods of handling, separating and processing, and particularly shredding. The separate identification and safe handling of these materials is therefore necessary.

Biological risks mainly arise from the storage of bio-waste. When biowaste is collected and stored separately, the concentration of decaying matter increases (since there is no dilution with other types of waste); combine this with the longer storage times at dwellings (from one week, the normal for the UK, to two or more) and the potential hazards to health are increased. This can be as bio-aerosols or an odour nuisance, as well as providing conditions for flies and vermin. A biowaste container with a drainage channel and air vents with filters (to prevent aerobic action), such as the 'bio-bin', might provide a solution.

Access between the storage and collection areas for refuse should avoid entry into the building. Doors to bin storage areas should ideally be of metal, tightly fitting and self-closing to deny access to pests unless they open directly to the outside in which case they should be kept locked when not being used. In blocks of flats, hoppers, chutes and containers for waste need careful design and detailing to ensure that their walls are hard and smooth, and provide no traps for waste material or footholds for climbing rodents. Polyethylene and paper sacks should only be used in specially designed holders or as liners for dustbins, otherwise they can be damaged by and harbour pests able to feed on the contents.

Hoppers contain ample supplies of food as well as harbourage for all types of pest and vermin. It seems unlikely that the chutes would provide access for rodents into the rest of the building if they comply with building regulations; a smooth inner surface, without rivets or rough joints, prevents lodgement of garbage and footholds for climbing rodents.

There is always a high risk of infestation in the room where collecting bins are housed. The door should be of a tightly fitting flush type constructed of metal or provided with metal kicking plates. It should be hung in a frame protected in the same way and be fitted with a self-closing device.

Paper for recycling presents a potential fire risk, particularly if stored loose or in bundles. A dedicated storage container with a close fitting lid may reduce this risk, as recommended in the Approved Document H of the Building Regulations.

If the waste collection authority requires the placement of the containers at the curtilage of the property on collection day, containers placed on the pavement at the side the public highway can cause a hazard. This problem becomes more acute with multiple bins for segregated waste[324].

Maintenance

Chute systems, including hoppers and terminal arrangements, require regular cleaning and disinfecting if they are to remain acceptable to users and collectors. Chutes are prone to becoming blocked and they cannot be used to convey segregated waste easily. Overflowing of receptacles has been reported to BRE. There is great potential for abuse of chutes and other waste disposal systems. Only meticulous attention to the design of disposal systems, cleansing of storage and collection areas, and maintenance arrangements can keep a check on health and safety hazards.

Work on site

Inspection

The problems to look for are:
◊ bins which are damaged or misused being accessible to rodents
◊ chute systems not regularly cleaned
◊ bins unprotected or not covered
◊ bin areas not within 20 m of vehicle access

Chapter 5 **Wired services**

This fifth chapter deals with all kinds of wired services, including mains electrical services, telephone, TV and radio, security and alarms systems, and speech reinforcement systems. Most, if not all, systems of these kinds demand the services of specialist designers and installers, particularly in non-domestic buildings. Insofar as the actual electrical installations are concerned, this chapter cannot be comprehensive. To give just one example, BS 7671[3] refers to no less than 94 other British Standards.

All that this chapter can do, therefore, is to point out the basic characteristics of installations and the nature of some of the problems that can arise. The solutions to those problems may demand specialised knowledge not normally within the province of the ordinary building professional.

The provision of lightning protection was dealt with briefly in *Roofs and roofing*[23], Chapters 1.6 and 4.1, and *Walls, windows and doors*[22], Chapters 1.8 and 3.4.

Figure 5.1
Wired services being installed in a large multi-storey building

Chapter 5.1   **Electricity**

This chapter is mainly about public supply, but there are brief mentions of aerogenerators and photovoltaics. The chapter does not deal with particular requirements for special building types such as, for example, hospitals, which are covered in many Standards (eg BS EN 60601[333]).

Since electrical interference in buildings is usually associated with malfunction in electronic building services equipment, it is dealt with in Chapter 5.2.

Homes in the UK, from the earliest days of electrical installations, were for the most part equipped with single phase 230–240 volt supplies.

Figure 5.2
This aerogenerator supplies only a small amount of power, a fraction of the total requirements for a dwelling

But there were exceptions; for example, in the late 1950s, the Watford area was supplied at 200 volts alternating current (AC). When it was altered to 240 volts, the vast majority of appliances had to be converted. A few dwellings in the late nineteenth century and early twentieth were supplied from industrial plant at 110 volts, but they were comparatively rare. Supplies at 110 volts require twice the current to supply the same power for the load and this in turn means that cables require a higher current rating; they are therefore larger and take up more space. Although it might be argued on safety grounds that an extra low voltage supply is desirable for most domestic applications, this is hardly now a practical proposition since it would presuppose a fundamental change in national practice.

Lighting circuits were protected with 5 amp fuses; power circuits, with round pin sockets, all on a radial system, were protected mainly by 15 amp fuses, but some with 10 amp. Not all sockets were earthed. The first edition of the Institution of Electrical Engineers' (IEE) Regulations for electrical installations dates from 1882, but it is profoundly to be hoped that no installations remain unaltered from that date.

The sixteenth edition of the *Regulations for electrical installations*, commonly known as the IEE regulations, was published as BS 7671[3] (*Requirements for electrical installations: IEE wiring regulations*) in 1992 and amended in 1994.

Shuttered sockets were invented in the 1920s, although it was not until 1947 that their use became mandatory. Some industrialised housing systems of the 1960s had wiring harnesses, which makes replacement not entirely straightforward.

Practically all dwellings in England now have a modern electrical installation, which for simplicity is defined as PVC insulated cables and rectangular pin sockets[7]. Nevertheless there are still a few, in both urban and rural areas, which rely on out-dated installations.

Although the criteria for Scotland were not exactly the same as for England, it would seem that the provision of a modern electrical installation is around the same level as in England, with only about 1 in 200 dwellings still having round pin sockets. However, inadequate provision of 13 amp sockets has been recorded in two-thirds of dwellings in Scotland[9].

Nominal voltages in Europe are now standardised at 230 or 400 volts, although there are still exceptions. Larger buildings may be supplied at 400–415 volt 3-phase, used directly on certain items of machinery etc; some of this higher voltage supply is transformed down to 230–240 volts for lighting and domestic areas.

Power from the wind
From ancient times, people have harnessed the not inconsiderable power of the wind to drive their industrial machinery and ships. But it was only many years after the invention of electricity that thoughts turned to using the wind directly to

Figure 5.3
Wind energy in the UK at an effective height of 10 m over an
open site in GJ/m^2 per annum (1 GJ = 278 kWh)

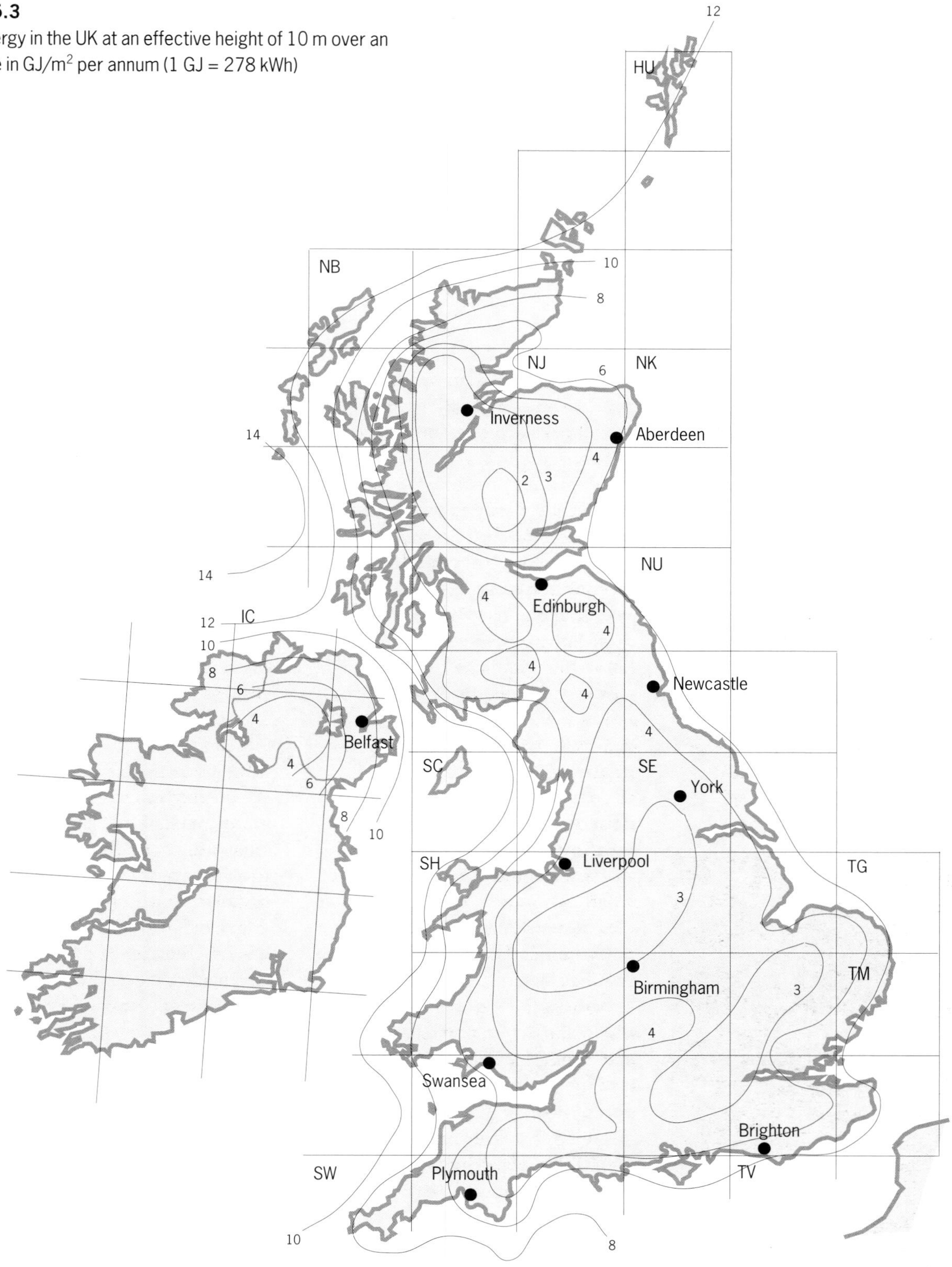

generate light and power for individual buildings.

Although in theory an attractive proposition, in practice development has barely proceeded to the stage where wind generated electricity can be considered as an alternative to mains powered, even for isolated dwellings. There may also be

environmental objections to large diameter aerogenerators. However, technology is moving quite quickly.

BRE carried out studies in the 1970s to identify the potential for power generation following pioneer work carried out by the Electrical Research Association (ERA), now ERA Technology Ltd. This was a

time of great interest in the autonomous house. It was shown that at that time only large scale installations of 1000 kW or more were likely to prove to be viable economically. The prospects for systems of around 100 kW were not good, and the same applied too for even smaller ones of around 10 kW.

Green power for The Body Shop
The Body Shop's new headquarters office and laboratory building at Littlehampton, Sussex, features natural ventilation, and energy saving lighting controls which can be overridden by the occupants. A 15 kW wind turbine and solar panels meet the bulk of the building's space heating and domestic hot water needs. It was monitored for the BRE research project 'Minimising/avoidance of air conditioning' and scores highly in the BREEAM environmental assessment [334].

Figure 5.4
This 1950s overhead transformer serves two installations

Since that time there have been significant developments with small installations, largely designed for marine applications. These are seen occasionally in isolated locations, where planning permission has been obtained. Power output is still small in relation to the average consumption of even a single dwelling, although a very highly insulated dwelling may prove more worthwhile (Figure 5.2 on page 198). Some aerogenerators have outputs as low as 1 kW, and are mainly used for battery charging, or specialised applications such as running heat pumps. At the largest scale, wind farms can now be seen, but these feed into the National Grid.

Wind does not blow consistently, even in the most exposed locations of the British Isles. Consequently either the electricity generated by this method has to be used immediately, such as for space or domestic hot water heating, or there has to be a system which stores it for supply at some future time when, say, the aerogenerator is not working. This is seen as a significant drawback, at least for the smaller scale plant. Where the electricity is being used immediately for heating water, all that is required is that the heating elements have a receptive capacity greater than the maximum output of the aerogenerator; but electricity generated for storage requires an expensive battery system.

For most of the land area of the UK, the total electricity available from each aerogenerator, if all wind energy is tapped, will be less than 1600 kWh per m^2 of rotor area per annum [335] (Figure 5.3 on the previous page). Of course, aerogenerators at exposed coasts will generate more power.

Power from the sun

Two kinds of apparatus have been extensively investigated for use in the UK: solar collectors, usually positioned on the sloping roofs of buildings, in which circulating water is heated to provide or supplement hot water requirements; and photovoltaic cells or plates, placed on roofs or walls, providing electricity for immediate consumption or storage.

Characteristic details

The single phase 230–240 volt, 60, 80 or 100 amp, 50 Hz, earthed neutral installation to BS 7671 is still the usual type provided in individual dwellings. These installations all come within the definition of low voltage installations (extra low voltage is up to 50 volts AC or 120 volts DC, low voltage up to 1000 volts AC or 1500 volts DC between conductors). Dwellings built during the 1980s and 1990s, especially those with electric space heating provision, were provided with 80 or 100 amp installations, although 60 amp installations can still be found in some older properties. A few dwellings, served in the main by overhead lines, are restricted to 30 amps (Figures 5.4 and 5.5). Dwellings with large power consumption – for example for space heating – may need 100 amp supplies.

For certain purposes of BS 7671, system types are referred to as TN, TN-C, TN-S, TN-C-S, TT, and IT, depending on the relationship of the source and of exposed conductive parts of the installation to earth; before any work on a system begins, the type must be correctly identified. (Further information is given in the Standard.)

The installation should not prejudice the supply in any way. In other words, no failure within the building curtilege should cause failure to the mains, and within the

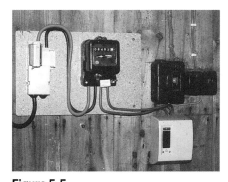

Figure 5.5
A 30 amp installation served by the overhead line shown in Figure 5.4, originally from the 1950s but since updated. The whole supply is protected by a 30 amp current operated RCD

installation it is usual for its various parts to be separately protected so that complete breakdown does not occur. Most existing domestic installations are protected by fuses, though an increasing number by residual current devices (RCDs) miniature circuit breakers (MCBs) or residual current breakers with overcurrent protection (RCBOs).

The consumer, or distribution, unit – the box that contains the terminals for the individual circuits and the fuses or circuit breakers that protect them – should provide for at least one ring circuit (in a modern system) per 100 m^2 of floor area. In an average two-storey dwelling this usually means:

- two separate power circuits, each serving half of both floors or the whole of one floor
- two separate lighting circuits, again each serving half of both floors or the whole of one floor
- electric cooker, emergency water heater, and any electric space heating normally have dedicated circuits although they may not necessarily be ring mains

Each circuit must be protected by a device rated according to the maximum load on the circuit. Each power and lighting circuit is usually limited by the number of outlets or appliances served by that circuit. Old distribution systems will sometimes be found to be radial circuits which generally are smaller than ring mains in terms of the outlets or appliances served.

Many garden sheds, garages and greenhouses are now supplied with electrical power. Freezers are often situated in garages and therefore require constant power.

Dimensional and other requirements for major items

Ideally, electricity meters should be accessible from the outsides of houses in weatherproof lockable cupboards. In many existing dwellings, especially those built before 1919, externally accessible meters buildings were hardly ever provided. Even dwellings built during the interwar period had them hidden away under staircases, in cellars or in cupboards (Figure 5.6). It may be feasible with these installations to retrofit a slave meter which is capable of being read from the outside of the house.

Where a meter is fitted outside, it seems standard practice also to place the main fuse alongside it. Sometimes the meter and the main fuse box are mounted in an outside cupboard that also houses the refuse bin. This can allow prospective intruders to tamper with or disable the house electrics, perhaps even to disable an alarm system by switching off the mains supply (although alarms should have battery backup), or vandals to disturb or distress occupants.

After a fault in a modern system protected by a circuit breaker has been corrected, current is restored by switching back on at the affected circuit breaker. An owner an old system in which circuits are protected by rewirable fuses might be strongly recommended to replace it with modern equipment and cabling.

RCDs should be provided in suitable positions for serving garden and external electrical appliances. Arguably all ground floor power circuits shopuld be protected by RCDs. An RCD should not be used on a circuit that supplies a freezer.

Prepayment meters, where installed, should be easily accessible. Coin slots should not be higher than l.4 m above floor level and not lower than 0.8 m.

The largest items of electrical equipment that need to be provided for in buildings include transformers and substations. They require special compartments sized according to the needs of the building, and are usually provided with separate and secure means of access. (See non-domestic wiring etc in the next section.)

Figure 5.6
An older installation with the electricity meter hidden away in a cupboard. Even in some refurbished properties, meters can only be read by standing on a ladder or by getting down to floor level

Wiring, lighting points and power sockets
Domestic

Radial lighting circuits will probably be restricted to 5 or 6 amps, and, for radial power serving single 13 amp sockets, to 20 amps; for directly wired fixed items such as domestic hot water immersion heaters and space heating radiators, and for ring mains, to 30 or 32 amps; and for radial circuits to cookers and instantaneous showers, some of which can be rated at 10 kW, to 40 amps or more.

It is sometimes argued that switches by doors should be horizontally aligned with door handles so that the locating of a switch relies on its direct relationship with the door handle. The recommended fixing height of door handles is 1040 mm above floor level. For housing which is accessible to disabled people, the preferred height of door handles is 900 mm.

Switches with built-in time delays are available, and are covered in BS EN 60669-2–3[336].

Toggles for ceiling switch pull cords preferably should be 1000 mm above floor level and certainly no higher than 1500 mm. Ceiling switch cords beside walls should pass

Figure 5.7
A substation being installed within a very large building

through a screw eye as a user may have difficulty catching hold of a swinging cord pull, particularly in the dark.

Rocker action switches are recommended for lighting controls. To permit easy manipulation, switches should be as wide as possible; a width of not less than 10 mm across the dolly is suggested. An illuminated switch or plate identifying the switch position can help where the switch may need to be located in darkness. Not more than two switches should be grouped together as multi-gang switchplates can be confusing to the user.

There should be an ample provision of socket outlets. Parker Morris recommended 30 years ago in his report, *Homes for today and tomorrow*[337]:

- working area of kitchen   4
- dining area   2
- living area   5
- first (or only) double bedroom   3
- other double bedrooms   2
- single bedrooms   2
- hall or landing   1
- store, workshop or garage   1

Current recommendations are still approximately the same with the exception of the living room which changes to 4; the utility room is added to the list at 2. These figures should now be treated as absolute minima, and can, bearing in mind the proliferation of electrical appliances in modern homes, with advantage be doubled. To minimise the need for adapters, twin or triple socket outlets are recommended in locations where more than one appliance might be connected.

Special sockets for electric razors may be fitted in bathrooms, but the current must be restricted to 1 amp.

Socket outlets should be carefully positioned in places where they are most needed Where there are only two sockets in a room they should preferably be placed on opposite walls. Socket outlets in low positions are difficult to reach, and may be hazardous. Some authorities argue that socket outlets should be at the same level as light switches and door handles (ie at approximately 1040 mm above floor level); they may be lower in accessible housing. At these heights they may present additional

hazards for young children. Low fixing heights may be preferred in some situations, but, although not yet standard practice in the industry, it is suggested that no socket outlet should be lower than 500 mm above floor level – other authorities recommend a 300 mm minimum. It could, however, be argued that mounting outlets much above skirting level could produce trip hazards, especially if flexibles are not long enough.

Socket outlets over kitchen worktops may be difficult to reach; a location on the fascia of the worktop may be considered, particularly where the depth of the worktop is 600 mm. Where the depth is 500 mm, outlets over worktops may be more easily reached. In kitchens to be used by elderly or ambulant handicapped people, the preferred height for socket outlets over worktops is approximately 1200 mm above floor level.

In the past, many different varieties of protection to cables have been used, ranging from screwed surface mounted steel conduit, and round, flexible corrugated, oval-shaped or top hat section PVC-U conduit suitable for burying in plaster. Metal conduit normally needs to be earthed. Installations may very occasionally be found in mineral insulated, copper sheathed cable (MICC), particularly for electrical cables for firefighting purposes and for evacuation lifts.

Non-domestic
Normally electricity is distributed to large non-domestic buildings at higher voltages than domestic single phase 230–240 volt supplies. The most common will be 3-phase 400 volts, from which 230–240 volts can be obtained between a phase and neutral in a four-wire system, but can be higher. As demand increases, so the incoming supply will also increase in voltage. Sometimes the supplying agency will provide a transformer in a substation either totally within the building (Figure 5.7), within the building but accessible from outside only (Figure 5.8), on the site of the

building, or sometimes outside the site altogether. Switchgear is generally proportionately cheaper the higher the voltage, other things being equal.

The components of a complete installation can include the following items:
● main switch gear, meters etc
● primary distribution cables and switchgear
● transformers, rectifiers and voltage regulators
● secondary distribution cables, busbars, switchgear and distribution boards
● motors and motor control gear for lifts and industrial machinery
● service equipment, fans, pumps, compressors and mechanical handling equipment
● space heating equipment, unit heaters etc
● lighting equipment
● communications and emergency services, alarms etc.

The following points may need to be considered:
● primary distribution cables need to be as long as possible in order to keep secondary cables short
● substations should be kept near the centre of the area they serve
● standardisation of as many items of equipment as possible

Secondary distribution systems can be encountered using:
● metal sheathed or armoured cable
● cable in conduit
● cable in trunking, either underfloor or overhead and suspended from the roof structure in single storey shed-type buildings
● busbar trunking, of copper or aluminium supported on insulators and kept well out of reach of unauthorised people

Cables need to be supported within the carcasses of buildings. In non-domestic buildings the usual support is by means of a tray, either enclosed galvanised sheet metal or unenclosed protected welded mesh steel wire cropped to length by the installer, used vertically or

horizontally. Alternatively, large capacity PVC-U trunking with snap-on lids may be used. For cables carrying large currents, physical separation may be needed to reduce the risk of mutual overheating when the cables are run in close proximity or touching. Slow bend trays are available for those cables which cannot be formed into tight bends. Metal supports will normally need to be earthed.

Uninterruptible power supply (UPS) systems

Certain installations, usually in non-domestic buildings, will require supplies to be maintained continuously (eg in health buildings). These installations can be provided with stand-by generators which cut in when mains supplies fail. Alternatively UPS systems may generate AC power from a battery by using an oscillator known as an inverter.

There are basically two types of UPS systems:
● off-line
● on-line

With the off-line UPS, the equipment runs directly from the mains supply and will switch to the inverter in the event of a power failure. This means that there will be a small interruption in the supply between the mains failing and the inverter starting up.

The on-line system is the preferred option since the load runs from the inverter the whole time while, at the same time, the batteries are being charged from the mains. When the batteries are fully charged they are 'floating' – that is to say they are neither being charged nor discharged. The advantage with this is that the load is completely isolated from the mains supply and therefore protected against mains glitches and spikes.

Figure 5.8
A substation for an industrial building, accessible from the outside only

Main performance requirements and defects

Outputs required

Older systems often have inadequate numbers (or rating) of power points; unsuitable positions for lighting points, switches or power points; and no cooker or immersion heater circuit. Consumer units may be badly positioned, and have insufficient fuseways and rewireable fuses. Some systems may have no earth leakage protection and earth bonding may not be to the latest standards.

There may be considerable potential for improvements in safety and economies in operation where old installations are still being used (see the case studies on the next page).

Solar collectors
See Chapter 2.6

Unwanted side-effects
Electric shock
Undoubtedly one of the main risks from electrical supplies is electric shock from faulty installations or appliances. Protective devices designed to prevent or minimise shock should function when any leakage to earth is detected; current is cut almost instantaneously, usually sufficiently quickly to avoid death, although not necessarily injury. (Provisions against electrocution are covered in some detail in BS 7671, but see also the section, Health and safety, further on in this chapter.)

Shock (or fire) risk could result from drilling and putting nails or screws through cables. This happens sometimes because electrical

contractors try to economise by running cables diagonally instead of in vertical or horizontal runs that can be traced from the positions of electrical fittings and appliances. Other contractors may not be aware of the positions or depths of cable runs. Awareness by electricians, and by other contractors in buildings, of the effects of their work methods and competence, and the work of the other skilled trades on site, is probably a contributory factor in reducing the risks of electrocution.

Fire
Another risk is from fire, again caused mainly by faulty installation practices. Not much can be done about illegal practices such as the wilful replacement of fuses by ordinary wire, which has been seen on BRE site inspections; nevertheless the major risks here are from three causes:
- careless and imprecise baring of conductors within cables when making connections, leading to short circuiting, particularly in damp conditions, and to sparks igniting combustible materials
- loss of insulation when wiring is attacked by rodents, with similar risks to those described above
- cables not being de-rated when passing through thermal insulation, leading to overheating[342]. Further information is given in the section, 'Work on site', later in this chapter.

Instances of arcing have occurred in installations with plastics-covered flat twin-and-earth cable due to the inner sheaths being cut when removing the outer sheathing. The defect responsible is probably quite common but, when arcing results, the charring of cable insulation is often blamed on an overheating connection. Every year, on average, over 2000 domestic fires are linked to fixed installations of electric cables. If the wrong technique is used for removing outer sheathing, the inner insulation may be imperceptibly cut. In the presence of even a small quantity of moisture, tracking and arcing can then be initiated between

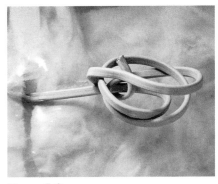

Figure 5.9
The normal heat dissipation from a cable is sometimes impeded by thermal insulation. The cable may then need to be increased in size or deliberately restricted in its loading (commonly referred to as de-rating)

the live conductor and the bare earth conductor. (Bare earth wires should be sheathed, but, unless carefully done, there may still be a bare section between the sheathing and where the wire emerges from the outer sheathing of the cable.) Once arcing has occurred a permanent path is formed, and arcing may then continue, even in dry conditions. Molten copper droplets, at high temperature, can be discharged, and the heat generated then be a fire risk[343].

All electric cables give off heat in use, depending on the load they carry. The heat emitted by any cable operating within its design loading is normally safely dissipated. Overheating due to electrical load alone should not occur if installations comply with BS 7671. Sometimes, however, the normal heat dissipation from a cable is impeded by thermal insulation or because the temperature of its environment is raised by other heat sources such as proximity to boilers. Consequently, overheating of the cable may damage its insulation. Cables in power circuits which may be loaded to full capacity are more at risk than those in lighting circuits. If these conditions are not taken into account, there can be a risk of short circuiting or fire (Figure 5.9).

Cable exposed to high temperatures should be of appropriate type (eg MICC). All cables should be de-rated where the

Table 5.1 De-rating of electric cables buried in thermal insulation	
Length buried in insulation (mm)	**De-rating factor**
50	0.89
100	0.81
200	0.68
400	0.55

ambient temperature exceeds 30 °C. De-rating means that a smaller maximum current is carried in an existing cable; another way of looking at it is that the cable size should be increased so that the required current can be carried with a smaller heat output. Where there is thermal insulation on one side of a cable the de-rating factor should be 0.75 (Figure 5.10), but where even short lengths are to be completely buried, this should be reduced to as low as 0.55 (see Table 5.1); this probably entails using the next larger size of cable. On this same basis, luminaires can overheat where they have been covered with thermal insulation (Figure 5.11).

Cables are normally de-rated on installation; that is to say their loadings will be kept within specified limits, or thicker cables should be installed, able to carry the anticipated current without heat buildup. Where cables are to be completely surrounded by insulation, the buried lengths need de-rating according to Table 5.1.

Electromagnetic compatibility
The Electromagnetic Compatibility Directive[344] which came into force during 1992 applies to all electrical and electronic products in the European Community. Its two aims are to remove barriers to trade by harmonising electrical interference standards and to reduce electrical interference by controlling not only 'electromagnetic pollution' but also equipment susceptibility to interference. The Directive has two 'essential requirements', namely that equipment:

- should not generate excessive electromagnetic disturbances that could interfere with radio and telecommunications equipment or any other electronic equipment
- should have adequate immunity to electromagnetic disturbances to enable it to operate properly in its normal environment

Products that have been designed to comply with relevant harmonised European Standards for interference emissions and immunity are deemed to meet the requirements of the Directive. The harmonised interference standards do not, however, cover installation of equipment and the onus is on installers to follow manufacturers' instructions and to adopt good installation practices (see also BRE Digest 424[345]).

Commissioning and performance testing
If an existing electrical system is to be reused when full scale refurbishment of the building is carried out, it should be thoroughly surveyed. No work should be carried out without a basic safety check including, at least, verification of correct phase and neutral connections, earthing effectiveness (Figure 5.12), insulation resistance and ring main continuity.

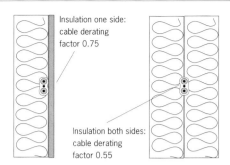

Figure 5.10
Where there is thermal insulation on one side of a cable, the de-rating factor may need to be 0.75, but where it is completely buried, this may need to be reduced to as low as 0.55 – in other words only half the carrying capacity of cable not affected by thermal insulation

Figure 5.11
Luminaires can overheat if covered with thermal insulation

Figure 5.12
Equipotential bonding needs to be checked for its effectiveness. Here, a bath panel and duct cover have been removed prior to carrying out a check

It is the maximum demand for electrical energy, the connected load, that determines the minimum overall capacity of the system, and the necessary sizes of all distribution systems and other appliances. Whether or not the system should take into account the non-simultaneous use of appliances (the diversity factor) depends on the importance of having a non-interrupted supply, clearly crucial in some building types such as hospitals.

It should be possible to estimate the average likely demand for power when considering either replacing or upgrading installations. The load densities for lighting will be comparatively straightforward, but, at the other extreme, the load density for manufacturing processes and workshops will not be straightforward. The load density depends on building types and processes involved; it could be 3–10 kVA/100 m² for light industry and double that for certain types of heavier industry.

A topic of considerable interest in large installations is the power factor: the ratio of power actually provided to a system to that which would be provided if the same current was in phase with the voltage. The dictionary definition is *'ratio of the total power (in watts) dissipated in an electric circuit to the total equivalent volt-amperes applied to that circuit'* (*Chambers Science and Technology Dictionary*, 1988). It is expressed as a decimal less than unity, and it is important that the power factor should be as near to unity as possible; correction of a low power factor will sometimes be possible by using capacitors. Power factors much less than unity call for heavier cables than would otherwise be the case.

Operating temperatures for electrical cables buried in screed for floor heating purposes should not exceed 70 °C for normal PVC and 85 °C for heat resisting PVC. These temperatures easily will exceed the maxima for several kinds of floor finish, and further guidance is available in *Floors and flooring*[23].

Health and safety

Although electrical installations are designed to be as safe as is reasonably practical, absolute safety is not possible, and death and injury still occur. To give a few examples, during the years 1989–96, on average there were 22 deaths and 2,400 injuries each year from electrical installations, including fixed wiring, and portable and non-portable appliances. The largest category of accidents relates to portable appliances. Although the number of fatal accidents in homes is falling – for example, in the 1980s they averaged around 40 each year, whereas in 1994 there were 12 – the number of serious injuries seems to be increasing. In work situations, deaths from electrical accidents recorded by the Health and Safety Executive in the late 1990s averaged just over 24 each year, with serious injuries averaging nearly 340. The majority of the deaths were attributable to fixed wiring, with about half the injuries from fixed wiring and half from other causes, including appliances.

The maximum voltage that can be transmitted by touching faulty electrical equipment must not exceed 50 volts AC or 120 volts DC, and the duration should not exceed 5 seconds. For mains voltages of 240 volts AC, the duration must not exceed 0.4 seconds. There are lower figures for highly conductive situations such as swimming pools[80] (Figure 5.13). It must be emphasised that these voltages relate to fault conditions on equipment. Where conductors are exposed and are within touching distance, voltages will need to be much lower: 24 volts or less for safety extra low voltage (SELV) supplies.

These requirements are enshrined in regulations, of which BS 7671 is the most comprehensive. There are also requirements in the Building Standards (Scotland) Regulations 1990, the Electricity Supply Regulations 1988, and various items of legislation covering building types such as cinemas, and agricultural and horticultural buildings. From the safety point of view, the regulations

Figure 5.13
Special precautions are needed for electrical safety in situations such as swimming pools, because of the high humidity

are designed to offer protection against electric shock, fire, burns and injury from moving parts driven by electricity.

All buildings should comply with BS 7671 and also with the Electricity at Work Regulations 1989, though installations complying with former editions of the IEE regulations do not necessarily fail to achieve conformity with the relevant parts of the Electricity at Work Regulations. Special requirements apply to rooms such as bathrooms, and, again, to buildings such as swimming pools, and agricultural and horticultural buildings. The Regulations do not necessarily apply to those aspects of lift installations which are covered by BS 5655[346].

Shocks from the discharge of static electricity in environments with RH less than around 60% are frequently experienced and can cause stress problems.

Fires can also be caused by faulty fixed electrical installations such as mains cabling, switches and meters. In 1990, defects in such installations in dwellings caused 2,100 fires; around 6% of these fires caused personal injury.

The Building Regulations 1991 make no direct reference to the safety of electrical installations or equipment. Existing levels of safety are established by three means:
- Electricity at Work Regulations 1989
- Electricity Supply Regulations 1988 which control the electricity supply companies
- Fitness Standard for Human Habitation of Dwellings, in Section 604 of the Housing Act 1985, under which the electrical installation is included in the assessment of whether a dwelling is free from serious disrepair

In practice, installations are tested to the current edition of the IEE Wiring Regulations (ie BS 7671).

Under the Health and Safety at Work Act 1974, the onus is placed on employers to ensure that electrical installations are safe. Completion and inspection certificates to BS 7671 are required for all installations to certify that the installations comply with legal requirements. There are a number of types of certificate specified in the Standard; for example, certificates applying to emergency lighting, alarms and petrol filling stations.

The fifteenth edition of the IEE regulations (published in 1981) introduced requirements for equipotential bonding of 'extraneous metal'. This is an onerous requirement in housing because of the large number of metal components that may require bonding, such as metal baths, shower fittings and all metal pipework, though it is not usually necessary to bond individual taps. There are two different cases, where bonding, or protective multiple earthing (PME), does and does not apply. Procedures to meet the requirementsof the Standard are complicated by the lack of simple guidance on how to comply, but will often require a separate earth conductor instead of bonding to earth via the metal pipework. Different views are held on the need to bond metal window frames or individual steel radiators. However, there is a site test which can be applied involving the measurement of resistance between the item and true earth, which should be 0.25 megohms or greater for it not to be considered to be an extraneous conductive part.

It is no longer permissible to use the incoming water main as a primary earth and, increasingly, the electrical supply authorities no longer provide a separate rod type earth (Figure 5.14).

PME poses a considerable problem for the installation of short wave antenna systems where there is a requirement for a good earth system for the antenna to work against. The chassis of the radio equipment will be at the potential of the radio earth. This means that there could be a large potential difference between that and the earth provided on the PME system, and they must not be connected together. Many amateur radio enthusiasts are unaware of this and will either connect them together or isolate them. Under certain fault conditions the former could be a fire hazard and the latter could be a shock risk.

An important area from a safety point of view is the operation of controls for building services. These controls are usually part of a network, and an electrical fault in one control unit can affect the whole network. Maximum temperatures for all services installations for protection against burns range from hand controls of metal, which are limited to 55 °C, to parts which do not need to be touched which can go as high as 80 °C if made of metal and 90 °C if made of plastics.

External meter boxes may give rise to pest problems unless doors are made to fit tightly and the wires and conduits connecting the internal supplies are tightly built in and adequately sealed.

Figure 5.14
An earthing arrangement to a rural property. It was provided by the electricity supply authority in the late 1960s and is still in use some 35 years later. The rainwater downpipe helps to ensure a damp subsoil into which the metre long rod has been driven. Electricity suppliers no longer provide earthing arrangements of this kind

Durability

Rubber insulated (tough rubber sheathed or TRS) wiring has not proved to be durable. Even if it has been sheathed externally in lead, it will need replacing. The most durable material seems to be mineral insulated copper covered cable (MICC).

PVC insulated wiring in domestic installations should have a lifespan of 30-plus years, but adaptation over the years can lead to unsatisfactory results (Figure 5.15). Some fittings (eg pendant and batten lampholders in plastics) will embrittle and discolour well before 30 years.

In some work carried out in the late 1950s, BRE identified around 20 years as the life expectancy of electrical installations in industrial buildings. With installation costs running at up to 20% of total building costs, this represented a significant part of the total investment in any building[347]. Since then, things have probably got worse rather than better, with acceleration in the speed of change of requirements (from a safety angle, and by occupiers acquiring more electrically driven equipment and demanding more sophisticated building services).

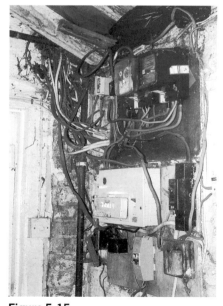

Figure 5.15
Domestic installations in older dwellings have frequently been found during BRE site investigations to be potentially dangerous and inadequate for their uses, with loose wiring and out of date equipment not conforming to current requirements

Maintenance

Even if manufactured with modern materials and maintained in good condition, existing electrical systems may require adaptation; for instance, by installing extra socket outlets or fused spurs for fixed appliances. If any changes are carried out then these and, arguably, the whole of the installation, will have to comply with the latest edition of BS 7671.

In these circumstances, bringing the whole of an existing installation up to present standards may be appropriate; protective devices such as MCBs and RCDs can sometimes be fitted to existing consumer units. The advice of the supply authority should be sought on changing supply characteristics and on overhauling an installation.

Where a fuse has blown, and replacement fuses also blow, the installation and appliances connected to it should be checked for the fault causing the problem. Fuses sometimes blow for no apparent reason, but it is reasonable to assume that they will not continue to blow unless there is a fault in the system. The other point to consider is that fuses do not always blow at the rating stated. In some tests carried out at BRE, particularly where current buildup was slow, some cartridge fuses blew at considerably more than their nominal ratings.

RCDs must be trip tested at least every three months to ensure that they continue to protect the installation.

Where equipment has battery backup, it is advisable to maintain the ambient temperature at or below 20 °C. Battery life can be shortened dramatically by higher temperatures; but by the same token they do not perform well at temperatures near to freezing. Discharging lead acid and nickel cadmium batteries occasionally (approximately every six months) helps to prevent them losing capacity. Newer systems tend to be controlled by computer and do this automatically. Some installations also lose capacity due to a 'memory' effect.

Work on site

Workmanship

It is most important that the correct size of wire or cable is used for the maximum potential load on a circuit. The following specifications may be used for guidance:
- 1 or 1.5 mm^2 for lighting, depending how many fittings are connected
- 2.5 mm^2 for power circuits and ring mains
- 4, 6 or 10 mm^2 for cookers and showers, depending on current demand and output temperatures

It is important that wires and cables are passed though timber floor joists in the correct positions. Where new or replacement electrical services are to be run within existing timber floor depths, and the joists need to be drilled to accommodate them, care should be taken that the recommended sizes and positions of holes are not exceeded (Figure 5.16).

Where cable trays in plastics are used, sufficient space must be allowed between trays for thermal movement; the trays should be fixed at centres recommended by manufacturers to avoid sagging.

Supervision of critical features

No power points should be retained unless they are 13 amp rectangular pin on ring circuits or permitted spurs. Existing rectangular pin sockets may be weak mechanically as a result of extensive use and require replacement. Stationary appliances should have their own fused spur connections; and immersion heaters, fixed electric fires and cookers their own dedicated circuits.

In rehabilitation work, pendant flex and lamp holders, unless fairly recent, should be replaced. Earth bonding should be replaced regardless of condition. Functional and safety considerations, damage and wear, will probably indicate whether consumer units should be replaced or not. Cables covered (or liable to be covered) in insulation should be particularly checked, as heat dissipation will be reduced.

TRS (rubber insulated) wiring or even older types, obsolete power points or fittings, worn or damaged modern fittings, pendant flexes and shrouds damaged by heat may all be dangerous and present serious defects. Often, heat resisting flex was not used for heater connections. There may be excessive numbers of junction boxes in power or lighting circuits, or unprotected low level wiring. PVC insulated and sheathed cable can be (potentially) damaged by heat or loss of plasticiser.

Installations having an large number of junction boxes, particularly where these will become inaccessible, should be identified; the more there are, the greater the risk of faults or failures. Cables less than 1.2 m above ground level that are unprotected or in positions vulnerable to contamination or damage (including contact with expanded polystyrene or hot water pipes) should also be replaced. Any wiring to be retained should be tested for electrical continuity, and conductor and insulation resistance. Fuses in consumer units should be examined to see whether they have been tampered with, or is there evidence of scorch marks. The replacement of fuses by inappropriate wiring has already been mentioned.

Electricians undertaking work should be appropriately qualified, either being members of the Electrical Contractors Association (ECA) or registered with the National Inspection Council for Electrical Installation Contracting (NICEIC). Alternatively, qualified electricians certify that installations carried out by unqualified people comply with regulations and Standards.

The qualifications of electrical installation contractors are covered in EN 59004[348]. The ECA runs an assessment scheme which provides for regular updating of experience. Categories of work include:
- installations up to 1 kv, including most buildings (except industrial), and alarm, surveillance and control systems
- installations up to 36 kv, including industrial and public lighting
- installations above 36 kv, mainly very large industrial installations

The ECA also have a 6-year warranty scheme backed by a bond.

Regular inspections of installations should be carried out, and guidance on the frequency of these for various building types is available from the IEE[349]. On completion of an inspection, a form should be completed by the person carrying out the work, giving measurements, alterations required, recommendations, and advice on when the next inspection should take place.

Inspection

The problems to look for are:
◊ ducting sagging through inadequate support
◊ round pin sockets
◊ rubber insulated cables
◊ old paper insulated, lead sheathed cables
◊ pendant flexes and batten holders that have deteriorated
◊ cables laid in thermal insulation not de-rated
◊ fuses of the wrong capacity
◊ floor areas of more than 100 m² being served by one 2.5 mm² ring main
◊ floor areas of more than 20 m² being served by one 2.5 mm² radial circuit
◊ floor areas of more than 50 m² being served by one 4 mm² radial circuit
◊ number of spurs or sockets on a ring main are excessive
◊ spurs feeding more than one single socket or one double socket or fused unit
◊ more than 12 lighting outlets connected to a lighting circuit
◊ metal light fittings not earthed
◊ wiring to boilers not protected with heat resisting sleeves
◊ no heat resistant cables to central heating boilers
◊ wet or damp conditions, especially in loft spaces
◊ main equipotential earth bonding cables less than 10 mm² to mains gas and electricity conduits at entries to buildings
◊ supplementary equipotential earth bonding cables less than 4 mm² where conductors are not mechanically protected
◊ equipotential earth bonding clamps not labelled as such
◊ ends of MICC cables not sealed against water vapour
◊ PVC insulated cables in contact with expanded polystyrene insulation
◊ signs of arcing from badly made connections
◊ low oil levels in transformers
◊ overheated pendant fittings
◊ broken fittings
◊ PVC-U cable trays and ducting distorting (no allowance for thermal movements)

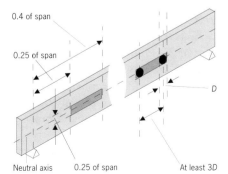

Figure 5.16
Permitted zones for drilling holes for cables

Chapter 5.2

Telephone, radio, TV and computer

Figure 5.17
The keypad seems destined to become an indispensable part of modern life – until the next development in control systems arrives

The second half of the twentieth century has seen a massive explosion in the use of communications media of all kinds. The rate of technological change, and therefore the rate of obsolescence in installations, is an accelerating one.

Many Standards have been published, and more are being developed in the European Telecommunications Standards Institute (ETSI).

This chapter includes reference to electrical interference in buildings, since it is usually those installations dealt with in this chapter that are affected.

Characteristic details

Types of internal wiring

This chapter deals with both voice and data transmission.

The wiring scheme most commonly used for data is structured cabling known as category five (CAT-5) unshielded twisted pair (UTP) cabling. This is terminated at RJ45 modular outlets and is designed to support a data transmission rate of 100 Mb per second to each outlet.

In many cases the same wiring is also used for voice to avoid the need for a separate telephone wiring system. If this method is adopted then it will be necessary to use either a telephone with a plug-in adapter or one that is specially wired for CAT-5 systems. Simply changing the plug will not work.

Some people may prefer conventional (CW-1308) telephone wiring to eliminate any confusion by users plugging equipment into the wrong outlet. It also prevents the voice and data networks from being inadvertently connected together. This is particularly important on a campus site with copper telephone cables running between buildings. There could be quite a high potential difference between earthing points, especially during nearby lightning strikes.

Where there may be interference or where eaves-dropping on radiated signals may be a problem, shielded twisted pair (STP) could be used instead. This obviously increases the installation costs, but the foil screened (FTP) type may be a cheaper and lighter alternative.

The earlier 10BASE2 or thin-wire Ethernet computer wiring used a coaxial cable with devices connected in daisy-chain fashion. Although this is a much simpler wiring scheme in terms of cabling costs, the segments are limited to 185 m and can support only a maximum of 30 connected devices. A fault anywhere on a segment would cause all devices on that segment to be affected. The equipment hubs are more expensive and are harder to obtain due to the increasing popularity of structured cabling. This type of wiring presents problems when adding devices as they require to be connected with two cables, unless they are added at the ends of the segment. For example, if the equipment to be connected is 5 m from an outlet, this will increase the length of the segment by 10 m. The maximum data rate for this wiring is 10 Mb per second.

Space requirements and cable containment
Domestic
Telephone lines in rural areas, and in some urban areas too, will often be found strung overhead from distribution poles to terminals at eaves level on dwellings. To many people, these lines are unsightly. It is only since the 1960s that underground cabling has come into general use. Domestic installations, which were formerly hard wired to appliances owned by telephone companies, now have sockets for the customer to plug in privately owned devices.

Television rod and dish aerials now abound, mostly individual installations, but some blocks of flats have communal aerials. Their design and installation is the responsibility of specialist firms.

The rapid increase in the number of different systems and wiring needs leads to an increase in the space which needs to be allocated to house them (Figure 5.18)

Non-domestic

Conventional telephone wiring consists of a multi-pair cable run from the building distribution frame (BDF), where the cables enter the building, to a distribution point mounted in a convenient location closer to where telephone outlets are required. This has the advantage that a large multi-pair cable requires less space than a bundle of individual cables but the possible disadvantage that wiring is dedicated and cannot be shared with data without using a dial-up service.

Some telephone systems now have a central processor connected by fibre-optic cables to peripheral cabinets situated in various locations. These house the cards to which the telephones connect. They are cheaper to cable than centralised telephone systems with copper cabling to distribution frames, and optical cables eliminate problems of segregation and induced surges such as lightning. However it could mean that costly uninterruptible power supply (UPS) systems would need to be installed to supply each of the cabinets. Also in the event of a system fault there would be no way of connecting directly to an exchange line via the optic cable.

Communications equipment must be installed in rooms which allow sufficient space for cooling and ventilation. The location of this equipment is constrained by the maximum length of cabling that is permitted to each outlet. It is standard practice when installing CAT-5 cabling to 'flood' the area with outlets to avoid the need for costly additions later.

CAT-5 cabling is a star-wired system from the communications equipment. There is a separate four-pair cable from each outlet, terminated directly on a patch panel located by the communications equipment. A patch cord (or cable) with two modular plugs is used to connect this to the appropriate communications port.

The panel is normally incorporated within the equipment cabinet or wiring closet with cable management bars to keep patch cords neat and tidy. Communications equipment is also normally mounted within the cabinet which should be, ideally, 800 mm wide to accommodate the communications hubs and to give space down each side of the cabinet for routeing the patch cords.

Many large office buildings built in the 1980s and 1990s have platform or false floors designed to provide sufficient space for the multitudinous wires and cables required to serve modern communications installations (see Figure 5.19 on the next page and *Floors and flooring*, Chapter 2.6).

A platform floor, albeit complete with its supporting deck, is required to provide the full range of performance and loading characteristics in addition to special requirements relating to demountability. These floors normally provide a system of loadbearing fixed or removable floor panels, supported by adjustable pedestals or jacks placed at the corners of the panels, to provide an underfloor void for the housing and distribution of services.

Access to the underside of the floor may be provided intermittently instead of making every panel removable. Some designs employ traps cut into the deck, supported by angled shoulders. In full access systems, most or all of the panels are removable. Partial access systems have runs of removable panels or individual traps, or both. Platform floors have become widely used in recent years in offices, telephone exchanges, data processing rooms, electronic control rooms and conference rooms where easy access to underfloor services is required.

Figure 5.18
Wiring being installed in the control room of a highly serviced dwelling

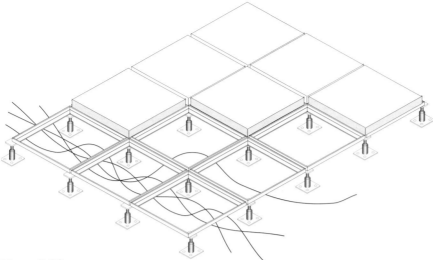

Figure 5.19
A platform floor

Since the suspension system for platform floors is usually in metal, there will normally be requirements for earth bonding to meet the requirements of the regulations for electrical installations[3].

Earth continuity bonding is often provided from a terminal on the bases of pedestals, but installations have been seen with the clips clamped to the threaded shanks of pedestals. It has not been found necessary to bond each and every panel since continuity is provided by the sheath of the panel; apparently missing bonding may in fact still prove to be satisfactory. Earth bonding of concrete supported panels is not easy to achieve, and these installations may need to be checked carefully.

Main performance requirements and defects

Outputs required
Domestic
For television and radio reception, permanently installed screened cable (to at least three points in the living areas of a dwelling) will normally be required for connecting to an aerial or cable supply. It is much better if these cables can be concealed.

For telephones, consideration should be given to installing permanently wired standard line jack sockets in at least the entrance hall, living room and main bedroom.

Non-domestic
Many so-called electronic offices now have very large numbers of personal computers and ancillary devices served from overhead wiring. Failures have occurred. The Building Services Research and Information Association has observed that the probable reason for these failures is that the harmonic currents generated, which affect power supplies, were not foreseen when the electrical systems were designed. Fires have resulted, in addition to loss of power supplies.

Unwanted side effects
Electrical interference in buildings has been identifed as one of the main causes of malfunction in electronic building services equipment.

The EC have issued the Electromagnetic Compatibility Directive, which requires installations not to cause interference with other installations[344].

Electrical disturbances on power cables can couple into signal and data cables to cause interference to computer networks, BMSs, and other systems used in building services. To avoid interference, power and signal cables must be segregated, with a minimum separation distance maintained between them; if they cannot be separated, ideally they should be screened. (See BS 6701-1[350], BRE Digest 424[345] and *Building Services (The CIBSE Journal)*[351].

Techniques to reduce the effects of electromagnetic fields include screening, filtering and earthing. Radio frequency screening is normally provided by high conductivity metal enclosures, although some modern enclosures are made from conducting plastics. In most practical cases, screening performance is limited by discontinuities (eg 'windows' in

Figure 5.20
Testing a controller for immunity to electrical interference

Electrical interference in buildings

Interference occurs when electrical disturbances are coupled into equipment that is not immune to those disturbances. The environment determines the nature of the disturbances and the installation introduces the coupling mechanisms. The immunity of the equipment is dependent on its design.

Low-frequency electric and magnetic fields are present throughout buildings, near power distribution cables, lighting cables, transformers and motors. Radio frequency (RF) electromagnetic fields are generated by transmitters, switching circuits and RF heating equipment. Mains power lines are capable of large voltage transients produced by switching heavy inductive loads, by lightning, and by sags and surges when heavy loads are connected or shed.

Disturbances can enter a system or piece of equipment by:
● resistive, or direct, coupling (eg along wires and through components)
● capacitive coupling (eg of transients between primary and secondary windings of a transformer, or between signal wires and power cables)
● inductive coupling (eg between signal wires and power cables carrying heavy, mains frequency current)
● electromagnetic coupling (eg between wires acting as transmitting and receiving aerials for RF electromagnetic waves)

Capacitive or inductive coupling between two circuits will cause the signal in one to be superimposed at a lower level on the other. In communication circuits this is referred to as cross-talk.

Electrical disturbances can be divided into three types: conducted, electrostatic discharge and radiated.
● conducted disturbances are those on power cables, leads to sensors and relays, and leads used for communication
● electrostatic discharge is caused by the buildup of static electricity due to friction
● radiated disturbances include the magnetic and electric fields close to conductors carrying currents and voltages; and, at greater distances, radio frequency fields associated with electromagnetic waves that travel through space

Devices which give limited protection against power line surges are available.

A lightning strike on a building can produce 4 MV, 200 kA pulses along the lightning conductor to earth (Figure 5.21).

Figure 5.21
Lightning can be a source of interference

A typical severity level cited in equipment immunity standards is 10 volts per metre. The distances at which various transmitters will produce a field strength of 10 volts per metre are given below:
● handheld transmitter (1W) 0.5 mm
● mobile transmitter (10 W) 1.7 m
● base station (100 W) 5 m
● radio broadcast transmitter (100 kW) 170 m
● television broadcast transmitter (1 MW) 550 m

In addition to deliberately generated electromagnetic signals, there are spurious signals produced by devices such as light dimmers, fluorescent lights, thyristor controllers and electric motors. Arcing or sparking contacts will also produce interference.

Electric or electrostatic charge produced by friction on certain insulating materials, in a dry atmosphere, can result in potentials of several thousand volts. A discharge of this potential can cause interference if the discharge path is close to sensitive electronic circuits. A typical example of electrostatic charge occurs when a person wearing insulated shoes walks on a carpet made of a different material; discharge occurs when the charged person touches a conducting object at earth potential, such as a filing cabinet. The discharge causes an intense arc which will produce high electromagnetic fields capable of causing malfunctions and even destruction of some circuits in electronic equipment [352].

enclosures for data read-outs or inspection) which reduce the screening effect at high frequencies. Enclosures used to screen sensitive equipment in severe electrical environments should have electromagnetically sealed doors; this is sometimes achieved by using conductive gaskets. The circuits within the enclosure that are likely to produce emissions (such as power supplies and thyristor controllers), or circuits which are likely to introduce emissions, should be screened from microprocessor circuits susceptible to interference.

Screening can be an expensive solution; sometimes physical separation is an effective and economical alternative.

The trend in modern electronics is towards semiconductor technologies with greater electrostatic discharge (ESD) sensitivities. Precautions against the effects of ESD may be included in installation design by effective screening, earthing and bonding. This should ensure that discharges are conducted to earth by the shortest electrical path and not via sensitive circuits. Furthermore, the environment can be improved by

humidifying the air or increasing its conductivity through ionisation, and by avoiding carpets based on synthetic materials which develop greater levels of charge than do natural fibres.

Paying attention to such details as grounding, earthing, routeing of leads and layout of components on printed circuit boards will go a long way to ensuring that high levels of immunity are achieved.
Components, such as suppressors or filters, can be added if greater immunity is required [352].

Commissioning and performance testing

Many types of platform floors are used in computer rooms where it is important that any electrostatic buildup which can affect equipment is avoided. This particular aspect will need to be checked for individual applications, though normally the requirement will be in the range 2×10^{10} to 5×10^5 ohms resistance[353].

There is likely to be a considerable increase in fibre-optic cable runs to individual computers or workstations; therefore a lot of thought needs to be given to allowing enough capacity in cable containment systems.

Figure 5.22
Wiring in the space under an access floor

Fire

The space beneath a platform floor is sometimes used as a plenum for ventilation systems, and this will need to be borne in mind in relation to the spread of fire and smoke (Figure 5.22).

During the 1980s and 1990s, wiring under access floors had to be specified to be fire retardant and of potential low smoke emission. However, these may not behave in fire situations as expected.

Recent years have seen a rapid growth in cabling and wiring for communications, but with recognition of the risk of overheating and fire, particularly where the underfloor void acts as a plenum for heating and air handling systems. When upgrading information technology systems it has been common practice to leave old cables in place, simply inserting new ones into the available spaces which leads to overcrowding. This situation is bound to get worse rather than better in future, with local area network (LAN) cabling being replaced, on average, in three-year cycles. Only rarely will cables be contained within metal conduit providing some protection in case of fire.

Although PVC is the most widely used electric cable insulation material, its use for several applications has been questioned. It is known that when PVC is involved in fire, hydrogen chloride – an acidic, toxic gas – together with relatively large quantities of carbonaceous smoke is produced. The reliance upon microelectronic control circuits which are liable to severe damage if exposed to the acidic or smoky atmosphere generated by burning PVC, and the possibility of higher toxic hazard in certain applications, has lead to the consideration of rigorous alternative specifications for cable insulation in sensitive areas.

BRE has developed a full scale test rig to investigate the fire risks[354]. A series of tests were carried out using the rig in a variety of configurations, concentrating on low smoke zero halogen (LSZH), low fire

performance cables, with surprising results. The cables readily ignited, developing a fire ball peaking at 800 °C, which not only affected horizontal runs, but also vertical runs, implying transmission between storeys. On the other hand there were other cables which had superior performance in the test (eg communications metallic plenum (CMP) cables[†], which showed very low optical smoke densities[355]).

Health and safety

People using display screen equipment have the protection of the Health and Safety (Display Screen Equipment) Regulations[356]. Employers have duties to assess workstations where display screens are used:

- to identify and reduce risks
- to observe minimum requirements for the correct ergonomic use of the equipment
- to offer suitable breaks in using the equipment
- to provide suitable training and spectacles if required

As already noted, since the suspension system for platform floors is usually metal, for earth bonding in accordance with BS 7671 will normally be required.

Maintenance

The strength of electric, magnetic and radiated electromagnetic fields decreases as the distance from the source increases. Sensitive equipment can be protected simply by placing it well away from disturbance sources.

† Copper communications cable for use in plenum spaces as defined by the US National Electrical Code.

Work on site

Supervision of critical features

The following basic precautions should be taken when installing electronic equipment

● Electronic equipment should not be connected to mains power supply lines used for loads that are sources of electrical disturbance and emission, including all inductive and heavy current loads, transformers, motors and thyristor controllers. Where possible a separate supply line should be connected directly to the main distribution point

● Electronic equipment should not be located close to loads, or the supply lines to loads, that are sources of electrical disturbance and emission

● Power cables and signal cables should be routed separately, and all crossover points should be made at right angles. Where parallel runs are unavoidable, it is advisable to ensure that the cable spacing is more than 10 diameters of the power cable

● Low frequency signal cables should be screened twisted pairs, with the screen earthed at one end

● High frequency signal cables should be coaxial, with the screen earthed at both ends

● All earth connections within an enclosure should be made at a single point or ground plane

● If separate signal and safety earths to a power supply input are specified, care should be taken to avoid the creation of an earth loop

● The mains supply earth should be of high electrical quality

Workmanship
Internal cabling

Communications cabling should be installed to BS 6701-1 and 2[350]; it must be segregated from power cables either by a distance of at least 50 mm or by a physical barrier, even where crossing at right angles.

Cabling should be installed in a containment system where possible, or laid on cable trays, or attached directly to the building fabric. It should not rely on support from other building services such as electrical conduits, pipework or suspended ceiling hangers. These may be removed at a future date and leave the cables with no means of support. Cables in vertical runs should also be supported.

Installation of CAT-5 cabling

CAT-5 cabling should be installed and tested to Telecommunications Industry Association TSB-67[357]. Tight bends and kinks can adversely affect performance (especially at the higher data rates) so this must be allowed for in the planning of cable routes. It requires specialist contractors with equipment to install and test for TSB-67 compliance.

The maximum length of cable runs from the patch panel to the outlet point (known as horizontal wiring) must not exceed 90 m and patch cabling should not exceed 10 m.

Whereas it is normally quite a simple matter to add a power outlet to an existing circuit, this not the case with communications wiring and therefore consideration should be given to allow access to cabling routes and leave adequate room for future expansion (eg an increase in capacity of 25%).

External cabling

Cable ducts and cables should be installed with reference to BS 6701. Cables for external use must be waterproof, especially if they are to be routed underground. Many telecommunications copper cables have polyethylene sheathing (since PVC is not waterproof) and are 'jelly filled' (ie grease filled). Polyethylene cables cannot be used internally as in a fire they give off toxic fumes. They

must be jointed to a suitable internal cable normally within 3 m of the entry point. Where this is not possible then the cable must be enclosed in metal trunking with intumescent sealant.

Fibre-optic cables

Installation of optic cables requires specialist tools and skills but it is important that system designers and specifiers know something about them.

To avoid jointing fibre-optic cables at the entry point they should be of internal/external grade. If cables are to be run in ducts then they should be at least steel wire armoured (SWA) to resist attack by rodents which are attracted by certain types of insulation material, especially PVC. They must also contain a water barrier, usually an aluminium foil wrapped around, overlapped and sealed.

It is very important to observe the bending radius of optic cables to prevent damage and degradation. The optic cable currently used is 62.5/125 μm and terminated on patch panels using ST-II or SC connectors.

Inspection
The problems to look for are:
◊ LSZH cables in underfloor plenums
◊ unscreened sensitive electronic equipment
◊ wiring situated near sources of interference
◊ cables not armoured against rodents
◊ cables not waterproofed
◊ cables not segregated
◊ cabling with tight bends, or which is kinked or pinched
◊ cables of excessive length
◊ cables supported by other services – which might be altered later (eg plumbers can leave cables unsupported)
◊ cables lying on ceiling hangers

Chapter 5.3

Security and fire detection systems

This chapter includes intruder (burglar) alarms, fire detection systems and alarms, and passive protection from industrial espionage.

The main standard for fire detection and alarm systems for buildings is BS 5839. Part 1[358] deals with system design, installation and servicing. BS 4737-1[359] defines an intruder alarm system as an electrical installation to detect and indicate the presence, entry or attempted entry of an intruder into protected premises.

Firefighting systems have been dealt with in Chapter 4.5.

Characteristic details

Closed circuit television (CCTV)

CCTV has been installed in a great many buildings, particularly department stores and other buildings where security is an identifiable risk. Quite inconspicuous and inexpensive cameras are available. Specialist designers and installers need to be consulted.

Where the security system for a block of flats which has common access areas includes a CCTV camera link, the following must be considered:

- the type and position of the camera, and the lighting of the entrance, to give occupants an adequate view of callers
- signal transmission to the flats via a separate video link or via the common television aerial

The choice of system will depend on the degree of privacy or security required. Systems using the common aerial normally allow any resident to monitor all comings and goings. This can increase security but cases have been reported of residents taking advantage of the system to break into neighbours' flats when they have gone out, or using it as an early warning system of the presence of police in the building.

Naturally, installations are not found solely in flats. Detached dwellings in high risk areas may need provision for viewing callers or for monitoring children's activities. Images may be called up on television screens within the dwelling (Figure 5.23).

Intruder alarm systems

An alternative, though less preferred, name for these systems is burglar alarms. The relevant British Standards include BS 4737[359], BS 5979[360], BS 6707[361], BS 6799[362], BS 6800[363], BS 7042[364] and BS 7150[365].

The separate parts of BS 4737 cover the specification of components for systems, including foil for application to glazing, protective switches, microwave Doppler detectors, ultrasonic movement detectors, acoustic detectors, passive infrared detectors,

Figure 5.23
A television screen displaying an image from a CCTV camera mounted outside the dwelling

Figure 5.24
Dismantled battery operated smoke detectors

Smoke and fire detectors

The first fire detection device, based on a bimetallic strip activated under fire conditions, and completing an electrical circuit, was patented in 1884[18].

Two-thirds of dwellings in England now have one or more smoke detectors, as opposed to one-third in 1991. Of these, only about 7% are mains powered. Most dwellings have only one detector, but 10% have three or more[7] (Figure 5.24).

The proportion of dwellings in Northern Ireland having smoke detectors is almost exactly the same as that for England[10] †.

Most readers will be familiar with the wired systems with manually operated switches often protected by glass shields, which of course need to be broken when fire is detected by the building's occupants (Figure 5.25) They are covered in BS 5839-2[366]. These manual systems can be augmented by automatic sensors of various kinds,

† Data for Scotland and Wales was not available at the time of writing.

volumetric capacitive detectors, pressure mats, beam interruption detectors, capacitive proximity detectors, and Codes of practice for installation and maintenance. Part 1 includes requirements for the system comprising detectors, control equipment and power supply, and additionally covers construction of containers, flexible connectors and anti-tampering provisions. Mains supply is transformed down to 12 volts DC.

The other Standards cover the more specialised applications such as high security and wire-free systems, but also DIY installations.

One of the more common forms of detection is the passive infrared (PIR) sensor which detects the body heat of the intruder. When two or more detectors are housed within the same unit, the incidence of false alarms can be reduced. False alarms may also be reduced by using dual technologies, again incorporated within the same housing. A PIR detector may be combined, then, with an ultrasonic or microwave movement detector.

Another common form of detector is the breaking glass detector. There are three basic kinds:
● vibration detectors, used mainly on large panes
● metallic foil tape, which breaks when the glass breaks – a technology that has been used for many years
● acoustic detectors, which listen for the sound of breaking glass

Intruder alarm systems may be an effective deterrent to an intruder, but they require careful design to be effective in flats, especially at entrance doors, lobbies and easily-accessible windows. Unless there is a resident caretaker, a system relying on a local audible alarm (ie a sounder-only system) is unlikely to be effective; connection to a central monitoring system is therefore necessary. Systems for each individual flat may be more cost effective if combined with a fire alarm system. To be effective the alarm system should be able to indicate which flat is the source of the alarm. For single dwellings, the systems can be much simpler.

Figure 5.25
A manual switch to operate fire alarm systems

of which one of the most common is the fusible sprinkler head, which is described below.

Local switches are commonly used for manual operation, and take little space. They should be sited at a height of around 1.5 m to allow operation by disabled people.

Smoke detection systems are normally based on ionisation or optical principles. In the latter case it is the opacity of the air containing the smoke particles which triggers the alarm. Devices of this kind can be very sensitive indeed, monitoring the reflected light scattered by smoke particles, and detecting levels of smoke (say around 0.2%) invisible to the naked eye. Other devices operate at around 5% obstruction. Optical beam detectors with ranges from 1 m to 100 m are covered in BS 5839-5[367].

Other devices operate on heat sensors, either of fixed temperature or rising temperature. There are also infrared flame detectors.

Domestic, self-contained battery operated alarms rely on batteries being replaced when exhausted. There is evidence from inspections that the majority of battery operated alarms are exhausted and useless; therefore mains operated alarms with battery backup are preferred. Alarms can be linked so that when one is triggered, they all sound.

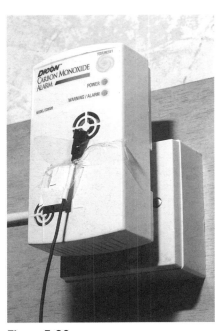

Figure 5.26
A carbon monoxide detector under test

Carbon monoxide

Using combustion appliances in the home can produce levels of carbon monoxide that affect the health of the occupants, often fatally. Most incidents occur with open flued appliances rather than room sealed appliances. If an open flued appliance or its flue is damaged, poorly installed or poorly maintained, carbon monoxide may be released into the living space.

Electrical detectors for carbon monoxide are readily available for domestic use in the UK, and are covered by BS 7860[368] (Figure 5.26).

The number of dwellings having carbon monoxide detectors is not recorded in the various House Condition Surveys, since they are comparatively newly introduced.

Detectors should be sited in any room where there is a combustion appliance. The best position for a detector is on the ceiling at least 300 mm from any wall, and between 1 m and 3 m from the potential source. If it must be located on a wall, then it should not be sited within 150 mm of the ceiling, and it must be higher than the tops of doors and windows[369].

Fire alarms

In domestic buildings, alarms may be integrated within the same unit as the detector, and powered by batteries. Alternatively, and preferably, they can be powered from the mains. They need to be sited within effective range of occupants, particularly sleeping occupants, and this may involve installing repeater alarms triggered remotely.

Fire alarms in non-domestic buildings are normally connected to a central board which is placed in a prominent position in the building (Figure 5.27), and accessible by those responding to the call. They can be as small as an electricity consumer control unit. Alarms are obtainable in a variety of types, the most common being bell or siren, mostly operating on low voltage circuits.

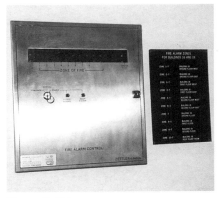

Figure 5.27
A fire alarm control point sited in the entrance lobby of an office building

It has frequently been found that a simple bell or sounder system is insufficient to persuade untrained building occupants to begin evacuating a building when fire breaks out. A voice alarm seems to be much better in initiating escape, and an adapted public address system, in which a pre-recorded message can be broadcast, can offer the opportunity of delivering clear instructions and motivating people quickly to leave the building. They need to be sufficiently loud to be heard over the noise of powered exhaust fans.

Release mechanisms

Magnetic door retainers, which operate to retain fire doors in the open position until released automatically when triggered by an alarm, may also be obtained in both low and 240 volt varieties. They are covered in BS 5839-3[370].

Luminaires used in lighting means of escape routes are covered in BS 4533-102.22[371].

Firefighting systems were dealt with in Chapter 4.5. For further information on human behaviour in fire, see BRE Digest 388[372].

Figure 5.28
The entryphone linked to a method of releasing the lock on a door has become a popular method of limiting access to buildings. Some models can be vulnerable to tampering to gain unauthorised entry

Door entry controls

The entryphone linked to a method of releasing the lock on a door has become a popular method of limiting access to buildings (Figure 5.28). Self-evidently it should be positioned close to the entrance door to which it refers. It should be:

- clearly visible
- of substantial construction
- firmly fixed in position using concealed or tamperproof fixings
- positioned at a height suitable for use of the key pad and the audio link, and, if possible by disabled people
- able to provide direct audio link between each flat and the entrance door

All cabling junction boxes should be concealed and protected from attack or interference. The element at the occupant's end should be positioned close to the entrance door of the flat.

Depending on the type of access control chosen, consideration must be given to providing:

- automatic door release for each flat
- an override on the locking system for use by the emergency services
- an override, with or without a time window, for deliveries etc
- a duplicate system on the inner foyer door

Measures to prevent industrial espionage

Electromagnetic screening can be applied to rooms and offices in commercial and industrial buildings, using techniques that are unobtrusive and do not impair the quality of the working environment, while providing an acceptable level of performance at reasonable cost. The effectiveness of the shielding depends on the characteristics of such details as the quality of seams and joints around doors and windows; in materials; it depends on corrosion prevention, bonding of cable screens and service pipes, and filtering of signal and power cables. A screened office shielded with a combination of a metallised glass window and copper mesh lining to the walls was constructed at BRE to evaluate screening methods[373].

A brief description of electromagnetic screening was given in Chapter 1.10 of *Walls, windows and doors*, together with an illustration of measurement in progress.

Main performance requirements and defects

Outputs required

Where smoke control systems are provided for the primary purpose of protecting life, all parts of the system must come into operation immediately the fire is detected. This implies automatic operation, since having a human being in the decision-making process introduces the risk of confusion and delay. Smoke detection with an automatic initiation of smoke exhaust is a particularly effective combination in most circumstances (Figure 5.29). There should also be some provision in the design for override facilities which can be operated by the fire services to assist them during firefighting operations.

The level of sound emitted from warning devices is important. BS 4737-1, for example, specifies a sound level of not less than 70 dB(A)[359].

Unwanted side effects
Magnetic fields

Magnetic fields from power cables and electrical equipment can cause interference to computers and other electronic equipment. Even low-level magnetic fields can cause interference to computer display screens[374].

Magnetic fields produced by equipment such as busbars, transformers and lift drives can extend up to 5 m or more from their source before falling to background levels; even fields from pumps and motors can extend up to 1 m. Fields from overhead power lines can extend 50–100 m, and from underground power cables up to 20 m.

Actions to alleviate the problems include:

- locating sensitive equipment well away from sources
- appropriate configuration of low-voltage power cables
- avoidance of neutral earth faults
- good balancing of 3-phase supplies
- shielding of sensitive equipment
- use of liquid crystal display screens

Figure 5.29
Scale model studies of venting smoke in shopping malls

Figure 5.30
A test in progress at BRE Cardington Laboratory to measure the effects of a fire in furniture

False alarms

The high incidence of false alarms, from both intruder and fire alarms, are a common cause of complaint, probably second only to false car alarms. Approximately 98% of all alarm calls are false, and police authorities may refuse to respond to calls from centrally monitored systems which are persistent offenders.

There are limitations inherent with all intruder detection devices. For example, PIR detectors will not penetrate glass, and they do not work above 30° C; and slight rises in the surface temperature of building materials caused by sunlight interrupted by leaves blowing in the wind, by unit heaters switching on, may cause problems. This is where the use of dual technologies can offer certain advantages, although manufacturers may have developed single technologies which reduce the risks.

Acoustic detectors are prone to false alarms where a high ambient noise level contains particular frequencies to which the detectors are sensitive. Vibration detectors may also generate false alarms when used on glass in inappropriate conditions, and become detached from the glass over time.

Following the deaths of ten people in a fire in the furniture department of a Manchester store, experiments were begun in 1981 at the Fire Research Station to evaluate the effectiveness of sprinkler systems in detecting and controlling fires in furniture (Figure 5.30). It was thought at the time to be essential to increase greatly the confidence that could be placed in automatic fire detection and to minimise the risk of false alarms; considerable progress has been made since then.

If the building environment is particularly 'dirty' then point smoke detectors will be prone to false alarms, and it may be necessary to reduce their operating sensitivity. It might be possible to use a system based on the responses of a group of detectors, or on a sampling system.

Commissioning and performance testing

Apart from the simplest devices for both intruder alarms and smoke alarms designed for amateur installation, the installation of these systems is usually placed in the hands of specialist companies. The Loss Prevention Certification Board (LPCB) maintains a list of approved products and installers[375]. Also, the National Approval Council for Security Systems (NACOSS) maintains a list of approved installers for intruder alarms, access control systems and CCTV systems[376].

Systems which are required to be monitored by the police should conform to the requirements of the policy on intruder alarms set out by the Association of Chief Police Officers. In practice these requirements can be expected to be met by NACOSS approved installers.

Smoke detection devices may not be satisfactory if mounted in the roof of very tall air conditioned atria, and would be better installed in the various rooms leading off the atrium where a fire is more likely to start[115].

Security

In many cases door entry systems are insecure since the control box is mounted outside; by removing two screws, the door release solenoid wires can be found. Connecting a battery directly to these wires will release the door. The switched side of the door release mechanism should never be fitted inside the control box.

Durability

The durability of systems seems to depend on design as well as the effectiveness of maintenance. The availability of spare parts depends on manufacturers' policy and continuity of the business.

One small domestic burglar alarm installation seen by a BRE investigator had been supplied and fixed by a one-man firm. After functioning without any problems for 15 years, the system failed and could not be repaired because the installer had died and his business closed down; the system had to be replaced.

Maintenance

Entryphones have often been vandalised, perhaps because of insufficient interest by building owners and occupants in safeguarding and maintaining the systems. In some cases it has been necessary to introduce a concierge system. The costs of the concierge can be partly offset by a saving on the costs of repairs.

Work on site

Inspection
The problems to look for are:
◊ systems inadequately maintained
◊ exhausted batteries in battery powered systems not replaced
◊ equipment vandalised
◊ sensitive equipment within range of magnetic fields
◊ fixings not tamperproof
◊ detectors not in recommended positions
◊ alarms not in recommended positions

Chapter 5.4

Rooms for speech and sound reinforcement systems

Many spaces or rooms are used both for speech and for music, although the requirements are different. The acoustics of auditoria for music are complicated and are not dealt with in this book. The text, therefore, deals only with spaces for speech, with speech intelligibility and reinforcement, and with public address systems.

Characteristic details

Rooms for speech

Rooms where natural speech is important should be provided with surfaces that ensure good acoustic conditions for audiences. Rooms should not be too dead or absorbing that makes speaking tiring or listening difficult, nor should they be too lively or reverberant which can degrade intelligibility and clarity. Background noise levels from mechanical services such as air conditioning and ventilation systems, and from external sources such as local road traffic, must be controlled so that they do not mask or degrade the clarity of the received speech. The main factors influencing speech clarity are:
● reverberation time (absorption)
● room size (volume)
● background noise level
● distance between the person speaking and the listener

For medium-to-large sized rooms such as lecture theatres, auditoria and multi-purpose halls, a reverberation time of 0.9–1.1 seconds is best for speech. In smaller rooms such as classrooms, meeting rooms or conference rooms, lower values ranging from 0.5–0.7 seconds are best. Many school classrooms are too reverberant and often have extensive areas of sound reflecting surfaces (eg glazing, and hard floor and wall surfaces) which allow sound (noise) levels to build up and intelligibility to be impaired. Many larger rooms – for instance multi-functional halls, school halls, and sports halls – are often not acoustically treated. They are frequently too reverberant to provide good speech listening conditions.

The ceiling in many rooms is the only available surface that can readily be acoustically treated: in many cases it is the most effective surface for aiding speech by providing useful early reflections to most parts of rooms. Conversely, the side and rear walls generally offer little useful contribution to intelligibility but could help with reverberation control. However, these surfaces are often vulnerable to damage, and they may contain windows which cannot be treated, leaving the ceiling as the only suitable surface for treatment. A practical solution can be to leave the central area of the ceiling sound reflecting, and add absorption to the side margins and, where possible, to the upper walls.

For the best speech conditions, it is essential that there are good sightlines between the speaker and listener – particularly in larger rooms, theatres and halls. Either raising the speaker onto a dais or stage or raking the audience seating can be considered (Figure 5.31). Curved (concave) surfaces should be avoided if possible as these can focus late reflections and echoes back to a

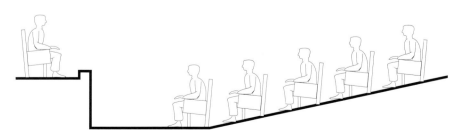

Figure 5.31
Raising the speaker or raking the audience seating may be necessary in large auditoria

speaker or to listeners. This tends to happen when the acoustic path difference is greater than about 10–15 m. Shorter focused path differences can lead to strong sound colorations when certain frequencies are emphasised. Sound reflecting surfaces close to the speaker (1 m or less) should also be avoided if at all possible as, again, this can lead to coloration of the sound due to interference effects from local surface reflections. Interference effects can be minimised if the flat profile of surfaces is broken up by diffusing elements. Physical modelling and testing of surfaces can also help to determine the best speech conditions for particular circumstances.

In large rooms, where the distance between the speaker and listener becomes too great, it may help to install a sound reinforcement (SR) system. This enables the loss in sound level projected to the rear of the room or space to be compensated for. If carefully designed, it can also overcome other acoustic problems such as overly long reverberation times or high levels of background noise. There are, though, distinct limitations as to what improvement can be achieved with a sound system.

Sound reinforcement and public address systems

The object of an SR system is, as its name suggests, to reinforce or assist speech (or music) within a space by making it louder, or by compensating for the distance speech travels between source and listener. This means that the pick-up microphone (or microphones) are in the same room as the loudspeakers, so producing the risk of acoustic feedback or 'howl round'. The aim of a reinforcement system is to sound as natural as possible which places extra requirements on the system (eg sound quality and perceived direction of sound origin).

In practice, public address (PA) systems tend to be of slightly lower sound quality than SR systems. The main difference is that, with the PA system, there is no direct reinforcement of the natural sound – the microphone (and announcer) being located away from the listeners. This means that the likelihood of acoustic feedback is greatly reduced, though in some systems where high levels of sound transmission are required (eg at football matches and other similar sporting events where high crowd noise levels need to be overcome), feedback can still occur due to poor isolation between the microphone and loudspeakers. (In most PA paging or announcement systems, it helps that the person speaking is able to hear the sound being broadcast as this assures them that the announcement is actually being transmitted). Although usually having lower sound fidelity to SR systems, PA systems nevertheless must provide a high degree of voice clarity if speech is to be heard and understood (Figure 5.32).

Where high quality background or foreground music reproduction is required, with or without live voice announcements (eg in clubs, pubs, bars etc), good sound quality is an important aspect of the system's performance.

Figure 5.32
Flat panel loudspeakers installed at North Greenwich bus terminal (Photograph by permission of P Mapp)

Design principles

Sound reinforcement and public address systems follow the same basic design principles and are described here together, with additional information relevant to each system being added as required.

All SR and PA systems are based on the following basic equipment chain (Figure 5.33) comprising:
● microphone
● preamplifier or mixer
● amplifier
● loudspeaker

Microphones and microphone control

In public address systems, there may be several microphone stations at various locations within a building or complex. These will normally operate on a priority basis, whereby certain units are given a higher priority than others; for example, the security control room microphone may have a higher priority than the front desk, and will automatically cut in and interrupt messages if required. Alternatively paging and announcement systems can operate on an equal priority 'first come, first served' basis.

In SR systems, multiple microphones may be required; for example, one for each person taking part in a discussion or performance or, in churches, at different locations within the building – say lectern, pulpit, altar, and choir. Roving microphones or wireless microphones are also frequently used. All microphone signals are fed into a mixer which enables the level of each signal to be adjusted and to be switched on or off as required. Other sound sources – tape or CD players, or background music machines – may be connected to the mixer, depending upon the system and user requirements. Mixers can be simple mixer-amplifiers, controlling perhaps four or six inputs, or be complex consoles as found in theatres and similar venues. Mixer-amplifiers provide a cost effective and simple means of control but may not provide the flexibility of operation required by

some users. Separate mixing consoles can generally control anything from around 8 to 32 or more channels or input sources. While they offer far more flexibility of control, they need to be operated by trained personnel and generally be constantly attended.

To reduce the likelihood of feedback and to obtain maximum clarity, it is important to have as few microphones as possible operating or live at any given time. There are a number of ways in which this can be achieved. Either an operator can individually control each microphone or the person using a microphone can switch it on and off – either by means of a switch on the microphone itself or through a local control switch. Alternatively an automatic microphone mixer can be employed. This automatically detects when a microphone is being used and switches it on, and then off when the person has finished speaking.

There are many types of microphone, ranging from a tie-clip wireless or handheld units, and their cabled equivalents, through to permanently fitted gooseneck microphones such as are found on lecterns or paging stations. Microphones may be:
● omni-directional: sound is picked up equally from all directions
● uni-directional: sound is picked up over a restricted angle. This directional response helps to reject unwanted sound and can reduce susceptibility to feedback

Amplifiers and loudspeakers

As already mentioned, in small systems the power amplifier may also incorporate the mixer section. In large systems however, a separate input control (and routeing) system will normally be employed. There are two primary forms of loudspeaker signal distribution and corresponding amplifier formats. These are:
● low impedance (similar in concept to domestic systems with typical loudspeaker impedances of 4–8 ohms)
● high impedance or 100 volt line systems

Low impedance systems are usually used for high quality and high power systems found in theatres, concert halls and other entertainment venues. Up to two loudspeakers are connected to an amplifier channel, with a dedicated cable connecting each loudspeaker or set of two units to the amplifier (Figure 5.34). Cable lengths must be kept as short as possible to minimise power losses within the cables. Amplifiers should therefore be located as near to the loudspeakers as possible. Cable conductors sized at 2.5–4 mm^2 are common in professional entertainment systems and installations.

High impedance systems, commonly called 100 volt line systems (70 volts is used in the USA and Japan – sometimes referred to in descriptions of American or Japanese equipment available within the UK), operates in an entirely different way and is ideally suited to large, distributed public address and sound reinforcement systems. With a 100 volt line system, each loudspeaker is fitted with a power matching transformer and connected to a single loudspeaker line in a parallel circuit, daisy-chain

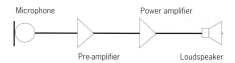

Figure 5.33

Basic sound system components

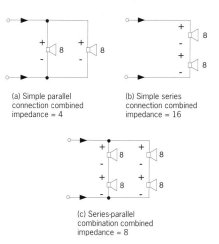

(a) Simple parallel connection combined impedance = 4

(b) Simple series connection combined impedance = 16

(c) Series-parallel combination combined impedance = 8

Figure 5.34

Low impedance loudspeaker connections

fashion, up to the available power rating of the amplifier (Figure 5.35).

In emergency sound systems or voice alarm systems (VAS), dual circuits are normally employed whereby adjacent loudspeakers are connected to different amplifiers and loudspeaker lines, minimising loss of announcement coverage in the event of an amplifier or cable fault. The 100 volt line distribution system is well suited to large sites or buildings as cable losses are significantly reduced in comparison with low impedance systems.

Loudspeakers intended for 100 volt line operation usually use matching transformers with a range of power tappings, enabling local individual adjustment of sound outputs to be made. Historically, the quality of 100 volt line loudspeakers has not been as good as low impedance systems – often due to the use of low cost matching transformers with limited frequency response and power handling characteristics (that is, how much power the system can handle without distortion). Because loudspeakers can be easily added to an existing circuit, it is very easy to load an amplifier circuit beyond its rated capacity, particularly in facilities that have undergone expansion or changes of layout. This can lead to distortion of the sound and eventually to amplifier unreliability and failure. Systems used for emergencies normally employ loudspeaker line monitoring or surveillance. This continuously monitors the loudspeaker lines and automatically detects major system faults. Where loudspeaker systems are used as part of a fire alarm system, the installation and equipment installed must meet the relevant and specific fire alarm Standards and Codes of practice[358].

Loudspeaker selection and placement

There are two main approaches to the design of loudspeaker systems. These are sometimes referred to as:
- centralised systems: point source or high level systems
- de-centralised systems: and distributed systems or low level systems

In a centralised system, either a single directional loudspeaker or cluster of directional loudspeakers is used. This approach works well in large spaces with relatively high ceilings such as theatres, concert halls and similar auditoria, and some churches. Adequate and even coverage of the space is required. A disadvantage with the cluster approach is that a relatively high sound level can build up (particularly at bass frequencies) leading to premature feedback and reduced clarity. Rear wall reflections should be well controlled to prevent late, disturbing reflections (echoes) from arriving back at the front of the room or stage. Either full range loudspeaker cabinets are arranged around a central position to cover the space, or an array of individual directional high frequency horns and bass loudspeaker enclosures may be used. In the past, arrays of directional column loudspeakers were sometimes used. For high quality sound reinforcement this latter practice is now little used, although it can still be applied to public address systems.

De-centralised or distributed systems employ a number of loudspeakers distributed around the space. These may be directional devices, such as column loudspeakers, often found in a traditional church mounted to the structural columns or walls (Figure 5.36), or less-directional devices such as conventional ceiling loudspeakers. Distributed ceiling systems have the advantage that they do not necessarily affect the orientation of speech in the room (that is, in which direction in the room the speaker and audience are facing). As a rule of thumb, ceiling heights of up to 5–6 m may be suitable for a conventional distributed ceiling system. At greater heights, either directional units or a centralised approach will need to be adopted. The acoustics of the space determine the working limits of the system.

It is also common in large spaces with a main loudspeaker cluster also to have supplementary or infill loudspeakers. These may be

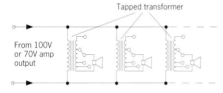

Figure 5.35
High impedance constant voltage distribution

Figure 5.36
A column loudspeaker installed in Hereford Cathedral in the 1960s

directional and used to cover the rear of an auditorium or space, or be of low directivity to infill below balconies or similar shadowed overhangs. High power, distributed directional arrays also now tend to be the common format in modern football stadia and similar venues. Enclosed stadia and arenas may use either format or a combination of the two.

Centralised systems generally tend to be of low impedance format, whereas distributed systems more commonly adopt a 100 volt line approach.

When selecting a loudspeaker for a particular application, many factors have to be taken into account. These include acoustic performance characteristics and environmental and aesthetic, and mechanical and electrical considerations.

In reverberant spaces, directional units would normally be used. In large systems, the efficiency and sensitivity of the unit are important as this can have a major impact on amplifier power requirements. In high noise areas, the units must be capable of producing a sufficiently high sound pressure level to adequately and reliably overcome the background noise level (Figure 5.37). Loudspeakers being used externally or in equivalent locations need to be of weatherproof construction, capable of withstanding wide ranges of temperature and humidity as well as atmospheric corrosion.

Loudspeakers need to be mounted so that they have a clear, unobstructed 'view' of the area they are intended to cover. In reverberant spaces, they need to be located so as to be as close as possible to the audience and directed to minimise overspill onto reflective surfaces (Figure 5.38). When mounted overhead, they must be fixed so that they cannot fall and injure people below. (Secondary fixings and safety chains may be required.) Equally, if wall mounted, loudspeakers must be located so as not to cause an obstruction or a potential injury hazard.

Systems for the hard of hearing

It is important when designing loudspeaker installations to consider the needs of the hard of hearing, particularly in buildings used by the general public. In some buildings it may be appropriate to cover only specific or limited areas (eg at information desks). In auditoria, practical limitations can restrict coverage. Two main forms of system are available.

● The most widely used is the audio frequency induction loop system (AFILS). This is relatively simple to install and provides an audio frequency transmission signal that can be picked up directly by many hearing aids without the need for any additional equipment being required by the listener. The system comprises a loop cable, located around the area to be covered and a special induction loop drive amplifier. Single turn loops should be used in preference to multi-turn loops which almost always suffer from a degraded frequency response characteristic

● An infrared transmission system is often preferred in concert halls, theatres and cinemas; not only does it usually provide a higher quality signal, but it is far less likely to cause interference in other sound or video installations. However, a special headset, earpiece or hearing aid coupler may be required

Figure 5.37
A cluster of loudspeakers installed in an ice rink (Photograph by permission of P Mapp)

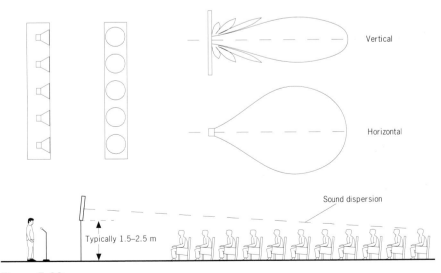

Figure 5.38
Column or line source loudspeakers

Table 5.2 Typical buildings and spaces, and speech intelligibility

Space	Reverberation time	Noise level	Adverse factors which may affect speech intelligibility
Domestic living room	–	*	TV
Office	–	*	Machinery and computers
Cafeteria	*	* *	>75 dB(A)
Theatre, court, council chamber, cinema	–	*	Poor design
Concert hall	* *	*	Orchestra
Museum	* *	*	Children noise
Church (modern)	*	*	–
Church (traditional)	* *	–	Reverberant
Cathedral	* * *	–	Very reverberant
Town hall	* *	*	–
Sports hall	* * *	*	>75 dB(A)
Football stadium	* *	* * *	>100 dB(A)
Swimming pool	* * *	* * *	>90 dB(A)
Ice rink	* * *	* *	–
Railway station	* * *	* * *	RT + noise
Shopping mall	* *	*	>75 dB(A)
Atrium, lobby	* * *	*	>75 dB(A)
Airport concourse	* *	*	>75 dB(A)
Aircraft hangar	* * *	* *	>75 dB(A)
Warehouse	* *	*	>75 dB(A)
Factory, workshop	* *	* *	NB 90 dB(A)

Grading in the above table:

–	Not a problem	
*	May be a problem	Take care
* *	Will be a problem	Consult expert - take great care
* * *	Significant problem	Expert advice required

Table 5.3 Acoustics: reverberation times

Reverberation time (seconds)	Comments
<1	Excellent intelligibility likely
1.0–1.2	Good intelligibility likely
>1.5	Careful design required (loudspeaker selection and spacing)
1.7[†]	Good intelligibility in large spaces may be limited with distributed loudspeaker systems (eg shopping malls and airport terminals)
>1.7[†]	Directional loudspeakers required in churches, multi-purpose auditoria and highly reflective spaces
> 2[†]	Very careful design and high quality directional loudspeakers required. Intelligibility may be limited in concert halls, churches, treated sports halls and arenas
> 2.5[†]	Intelligibility may be limited. Highly directional loudspeakers required in large (stone built) churches, sports halls, arenas, atria, enclosed railway stations and transportation terminals.
> 4[†]	Highly directional loudspeakers are required in very large churches, cathedrals, mosques, large and untreated atria, aircraft hangars, untreated enclosed ice sports arenas and stadia. The loudspeakers should be located as close to audiences as possible

† Expert acoustic advice should be sought.

Table 5.4 Background noise levels

Space	Recommended values($L_{Aeq}T$[†])
Concert hall	25–30
Theatre	25–30
Commercial multi-purpose hall	25–35
Music teaching room	25–35
Audio visual room, cinema	30–35
Cathedral	30–35
Churches	30–35
Council chamber, courtroom	30–35
Lecture theatre	30–35
Banqueting hall	35–40
Classroom	35–40
Conference room, meeting room	35–40
Executive office	35–40
School hall, multi-functional hall	35–40
Club, public house	40–50
Exhibition hall	40–50
Museum, library, gallery	40–50
Office (cellular)	40–50
Sports hall	40–50
Restaurant	40–55
Office (open plan)	45–50
Corridor	45–55
Washroom, toilet	45–55
Canteen, kitchen	50–55
Shopping mall, department store	50–55
Swimming pool	50–55

† See BS 8233[(377)].

Main performance requirements and defects

Speech intelligibility

Table 5.2 indicates where reverberation times and noise levels are likely to interfere with speech intelligibility, and whether expert assistance will probably be required to cure problems.

Reverberation times

Table 5.3 lists typical reverberation times which are encountered in internal spaces, and provides an indication of intelligibility of speech within those spaces.

Curved surfaces and long path echoes should be avoided.

Where an SR system is used in reverberant spaces, making the system too loud is likely to reduce rather than improve speech clarity and intelligibility.

Background noise levels

In high noise areas, a sound reinforcement system should be capable of producing a sufficiently high sound pressure level to adequately and reliably overcome the background noise level. Table 5.4 lists typical background noise design levels found in various types of spaces. (Occupational noise levels may be considerably higher.)

Work on site

Inspection

The problems to look for are:

Loudspeakers
◊ inadequate coverage (spacing between distributed loudspeaker systems too wide)
◊ blurring of speech (distance between loudspeakers, >35–50 milliseconds (m s) or 10–17 m, too great)
◊ distinct echoes (distance between loudspeakers, >60 m s or 20 m, too great)
◊ poorly directed (eg column loudspeakers directed over listeners' heads)
◊ poorly sited (too far away from listeners, or obscured by structure or fixtures and fittings)
◊ undesirable or late reflections or echoes (loudspeakers too close to microphones or aimed at microphones)
◊ undesirable or late reflections or echoes (loudspeakers aimed at reflective surfaces or distant (eg rear) walls)
◊ causing injuries, potential safety hazards or obstructions
◊ not sufficiently directional in reverberant spaces or sufficiently powerful in noisy areas (inappropriate types of loudspeakers)
◊ poor frequency responses and range for good intelligibility and quality
◊ physically damaged (particularly damaged grilles)
◊ cable connections damaged

Microphones
◊ poorly sited to pick up voices of speakers (fixed microphones too far from speakers)
◊ muddying of sound and premature feedback (too many live microphones in use)
◊ distortion and feedback (microphone (gain) controls poorly set)
◊ intermittent sounds, hums or crackling (damage or wear of microphone cables and connectors)
◊ hums, instabilities or oscillations (microphone cables routed too close to mains cables or loudspeaker cables)
◊ microphones too close to reflective surface
◊ microphone cables not screened

Amplifier control systems
◊ distortions or feedback (poorly set mixer (gain) controls)
◊ inadequate ventilation or airflow around amplifiers
◊ unsafe connections of power and 100 volt line cables (exposed or non-shrouded cables)
◊ equipment inappropriately earthed

System wiring
◊ electrically unsafe
◊ physically unsafe (eg trip hazards)
◊ microphone and signal cables not routed well away from power, lighting, data and speaker cables
◊ wiring types inappropriate (low smoke or fire proof cables not provided)
◊ microphone cables unscreened
◊ conductor sizes for loudspeaker circuits inadequate (particularly for low impedance systems)

Chapter 6

Mechanical handling devices

This sixth chapter deals with mechanical handling devices installed within buildings, but not plant for constructing buildings. Included, therefore, are passenger and goods lifts, escalators and travelators; and, to a limited extent, goods handling devices such as powered platforms and overhead conveyors. Travelling cranes installed within industrial workshops and factories are subject to special health and safety considerations, and might have required special structural features. They are not dealt with in detail here, though some relevant criteria are cited.

BRE has been concerned only with particular aspects of mechanical handling devices installed in buildings, as opposed to mechanical handling devices used during the construction of buildings (eg cranes, fork lifts, pallet lifts etc) which are outside the terms of reference of this book. The reduction of BRE involvement in mechanical handling equipment coincided with the setting up of the Heath and Safety Commission's test facilities at Buxton, where they took over the former Safety in Mines Research Establishment site at Harpur Hill.

Aspects such as reliability, performance and safety, particularly fire safety, have, however, figured significantly in the BRE workload over the years. Research into the design and installation of mechanical handling devices has been mainly restricted to two aspects:

● the overall quality of service provided to the user
● the impingement of that service on the layout of the spaces within the building, the provision of adequate strength and stability, and the accuracy of construction for the carcass of the building to which the services are to be fixed

As with other aspects of construction, the legal framework for mechanical handling devices is changing from a purely UK context to a European one. For example, as a result of an EEC Directive, all lift installations completed after 30 June 1999, together with their components, have had to carry a CE mark to indicate that they conform to the relevant harmonised standard. This will not necessarily bring in its wake an overall increase in quality. Where installations are carried out by contractors who are quality assured to a recognised standard (eg to IS 9001) their conformity with EC standards can be certified by the contractor. Where the installer is not quality assured, the design and installation must be certified by an independent third party body.

One of the more important requirements which has not yet been harmonised in the European context is performance in fire conditions. These requirements are governed in the UK by the Building Regulations and by BS 5588[378].

Although there is no attempt in the following two chapters to give a complete guide to the use of mechanical handling devices, some of the more important issues facing building owners, and their professional advisers, when they consider whether to upgrade or replace installations are discussed. However, inspection and maintenance of installations by untrained personnel are both illegal and undesirable.

Figure 6.1
A lift shaft awaiting the installation of the lift cars and doors

Chapter 6.1

Passenger and goods lifts, escalators and dumb waiters

Passenger lifts serving multi-storey buildings were at first powered hydraulically by water under high pressure. Hydraulic lifts were said to be able to work satisfactorily up to heights of around 80 m. The hydraulic power could be either produced by local pumps, or by pressurised mains. In late Victorian times London was provided with a system of mains covering quite a large area.

Electric passenger lifts were introduced into the UK from the USA in 1910, but their numbers did not increase significantly until a decade later[18]. It was the electric passenger lift that made office buildings above four or five storeys viable, and the escalator that made city centre multi-storey shops popular. However, in housing and

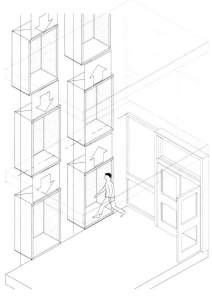

Figure 6.2
The method of operation of a paternoster installation. The lift shaft is not shown

other residential buildings, since the end of the 1939–45 war, there has been increasing use of domestic sized inclined ramp stair lifts by physically disabled people, enabling them to remain in their homes for more years than might otherwise be the case. Stair lifts can be installed in commercial and public buildings where there is no reasonably practical alternative, but the advice of the building control and fire authorities should be sought.

There have been few installations of paternosters in the UK. A paternoster is a continuously but relatively slowly moving set of passenger cars with open doors on an endless chain, arranged so that adjacent openings at each floor level served reveal one set of cars rising, and another set falling (Figure 6.2). Passengers board and leave the cars as they move. They are unsuitable for the nervous and the infirm, and, for that reason, in the past have tended to be restricted to multi-storey educational and similar establishments. Where paternosters are installed there should be an alternative enclosed lift for passengers unable or unwilling to use the paternosters.

Escalators have been used in the UK since the years between the wars. They operate continuously, moving large numbers of people one storey at a time, either upwards or downwards, and are used mainly in building types such as department stores, airports and underground stations. Indeed, they are said to be able to carry many more passengers than lifts in a shorter time and for a similar amount of energy used.

Where used in the horizontal mode only they are known as travelators.

Dumb waiters are small service lifts for moving goods short vertical distances, and are normally installed in restaurants where kitchens and dining rooms are on different floors. They can be powered by hand by pulling up or down on suspension ropes, or, in more sophisticated form, by electricity.

BRE, or rather its predecessor BRS, first began to investigate lift provision and performance in buildings in the early 1960s, and it was clear at that time that lift specification was a very empirical process. Unfortunately, once built, buildings are not very easy to alter with respect to lift shafts. There are, however, cases where new lifts have been inserted into existing buildings which were not previously provided with lifts. Feasibility depends on a particular building's structure and layout, and the amount of spare space available. One successful retrofitted installation was that at the Natural History Museum where a new 30 m two-car shaft was installed in the 1881 building; even the finishes to the lobby matched the original terracotta.

The ways in which buildings are organised and used – for example, timetabling in educational establishments and staggered hours in offices – significantly affects the perception users have of the lift services provided.

Figure 6.3
Escalators installed in a department store

Lifts research at BRE and a number of other institutions was increased significantly in the mid-1970s; as a result, models were developed to simulate and predict lift performance, and to improve the basis of their operation and the perception of them by users. The first use of computer lift design programs for these purposes dates from around this time.

Some parts of BS 2655[379] and BS 5655[346] are still current, though being superseded by BS EN 81[380].

Firms of consulting mechanical engineers specialising in lift and escalator specification, installation and maintenance, independent of manufacturers are able to offer advice to building owners in these specialised fields.

Characteristic details

Space requirements (dimensions etc)

Lifts

Passenger lifts powered by electricity can be used for any height of building. The lifts are normally divided into categories according to their carrying capacity:

- service lifts not exceeding 1 m² floor area
- residential and domestic lifts, including balanced manually driven personal lifts, up to 115 kg, and vertical or inclined ramp powered lifts for single person (with or without a wheelchair), up to 150 kg (Figure 6.4). Some types may go up to 400 kg
- occasional passenger lifts, up to 830 kg
- light passenger lifts, up to 1000 kg
- general purpose passenger lifts, up to 1600 kg
- intensive passenger lifts, up to 1800 kg
- bed/passenger lifts, up to 2500 kg
- general purpose goods lifts, up to 3800 kg
- heavy duty goods lifts, up to 5000 kg

It is usually taken that an average passenger weighs around 80 kg, so a light passenger lift will take 13 people, and an intensive passenger lift 21 people.

Full enclosure of either the lift platform or the aperture in the carcass of the building is not always needed. There are several categories; for example, partial enclosure of platform or aperture or both, or full enclosure of either or both.

Car sizes will vary according to need, from around 350 mm × 350 mm for single person standing lifts, to wheel chair lifts of around 1 m × 1 m, to those suitable for multi-passenger use which are much larger; light passenger lifts are usually around 2 m². Bed/passenger lifts are usually found in hospital and health care buildings where overall car size is the main criterion, with at least 1.8 m × 2.7 m being needed. Circulation and access spaces associated with lifts – the dimensions of the landing for entry and exit – depend on the use of the lift, but a landing depth in front of the double-leaf sliding doors should be at least 1.5 m is usually a minimum requirement. In theory doors can open either in the horizontal or the vertical direction, but the latter case

Figure 6.4
Stair lifts have been a considerable help to people who cannot use, or have difficulty using, stairs. These lifts enable them to stay in their homes for more years than would otherwise be the case

is extremely unusual. Where there are two or more lifts in a bank, particularly where they have a common landing, the overall width of each set of doors will need to be greater: about 2 m or more.

A clear width of door of at least 800 mm is normally required for small-capacity cars, though clearly this will be insufficient for larger lifts involving the movement of beds or goods. Multiple leaves may slide telescopically to either hand, or have a central opening.

Where a lift car travels in a fully enclosed shaft, the shaft size is directly related to required car size. No size restriction applies where the car projects from the building carcass, but there may be other restrictions. A shaft or well may accommodate more than one lift car.

The size of machine rooms is normally determined by the rated capacity of the lift cars served. A light passenger lift will need a motor room of around 15 m², rising to around 50 m² for a heavy duty goods lift. A single motor room may serve more than one lift car, with a proportionate saving in some floor area. Cars can be supported in a variety of ways, though wire ropes or chains are the most common.

Variable frequency control systems for lifts are generally considered to be the most suitable. They give accurate levelling at floors, and offer improved life expectancy.

There used to be two principal arrangements for hydraulic lifts. For low-rise buildings, the passenger car tended to be supported on top of the ram, which either needed to be as long as the height of travel, or could be telescopic. In both cases this demanded a deep sump to accommodate the ram when the car was at its lowest level. For higher buildings, the ram was positioned within the shaft alongside the car. The car was suspended from a multi-fall pulley system driven by the ram. The ram travel distance was therefore reduced by the pulley systems.

Both electric and suspended hydraulic lifts are counterweighted to reduce power requirements.

In high-rise buildings in which lifts are used intensively and where space is limited, two-storey or double-deck lift cars may be installed. These provide a service only for every other floor, and intending passengers need to realise that a walk of one flight will be required, either at the beginning or end of travel.

Escalators and travelators

Escalators and travelators are normally described by their width, with the height of the escalator of course governed by the height between the floors served.

Widths of 600 mm are described as one person escalators, 800 mm as one and a half person, and 1 m widths as two person. Escalators are not normally wider than this because of the need for the passenger to be within easy reach of a handrail. As with lifts, a minimum length of landing must be provided, not less than 0.85 m.

The angle of rise or fall of an escalator is usually around 30°, although smaller installations travelling at lower speeds and between limited storey heights can be steeper – up to 35°. Rises between fully extended treads should not exceed 240 mm, and for those designated as means of escape, 210 mm. Of course, the steeper the angle, the shorter the 'footprint', or length in plan, of the installation.

With a slope of 30°, an escalator rising through a floor-to-floor height of 4 m will occupy a plan length of around 8 m, with landings extra (Figure 6.5).

Travelators are sometimes used over slight inclines, up to around 10°.

Some escalators and travelators are designed to be stationary when no passengers are being carried, but starting when a passenger passes a sensing device. Clearly, in these cases, the direction of travel needs to be prominently displayed.

Main performance requirements and defects

Outputs required

Lifts

Single person lifts, whether AC or DC or battery powered, are restricted to rated speeds not exceeding 0.15 m/s, with braking distances of up to 20 mm.

Other categories of lift can be found with rated speeds far in excess of this figure, particularly those serving tall buildings. Early installations of the 1920s and 1930s were usually rated at around 1 m/s, although some installations achieved over 2.5 m/s. In more recent years speeds have tended to increase, and 3 m/s is now around the average. For tall buildings, the speeds may be increased still further, but even these pale in comparison with speeds reached by winding mechanisms in mine shafts. One typical colliery in Warwickshire was winding with

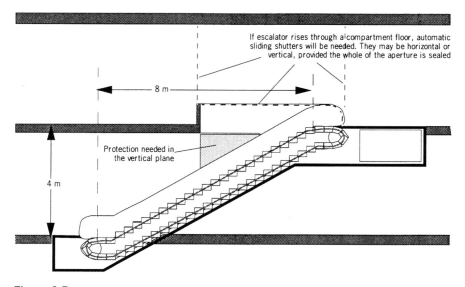

If escalator rises through a compartment floor, automatic sliding shutters will be needed. They may be horizontal or vertical, provided the whole of the aperture is sealed

8 m

Protection needed in the vertical plane

4 m

Figure 6.5
A typical escalator installation

steam power at 8 m/s in the early years of the twentieth century, but the accident rates were higher than would be acceptable for lifts in buildings. The criterion for passenger comfort is not so much the maximum speed as the rates of acceleration and deceleration; variable speed motors are essential in this respect. In contrast, paternosters operate at 0.3 m/s.

Waiting time rather than speed is crucial to the overall perception of service by users, and, indeed, to overall journey time on other than non-stop journeys over several floors. Reduction of waiting times is often arranged for tall buildings by stopping only at alternate floors, or by non-stop services at higher speeds to intermediate lift lobbies every few storeys; passengers then transfer to slower, stopping cars for onward travel to intermediate floors.

BRE was heavily involved in the 1970s in recording lift system performance by means of automatic data loggers . The concept of lift response time – the interval between the registration of a call from a particular landing and the arrival of a lift at that floor, having free capacity and travelling in the correct direction – enabled direct comparison to be made between different types of lift control systems using their mean response times. As a result of improved design methods, buildings built during the 1980s and later should have a much better service from their lift systems than those built earlier.

Where there are several lifts serving a high-rise office building, it is common to provide a central progamming system. The program alters the pattern of service to correspond with major traffic flows: from, say, peak upwards transit at the start of the working day, then equal use based on demand during the day both upwards and downwards, to peak downwards transit at the end of the working day. The lifts 'home' onto the appropriate floors in anticipation of the major traffic flow, and many lift cars have sensing devices so that when a full load is being carried, calls from intermediate floors are over-ridden. It is obviously essential for adequate service that such programs and devices work consistently, and are appropriate for the patterns (or can be tailored to the patterns) of use.

There is some controversy about the accuracy with which lifts and powered lifting platforms ought to align with finished floor levels, BS 6440 quotes 12 mm either way, which some find too large a step, up or down. There is no doubt that such small differences can be successfully negotiated by wheelchair users, but the hazard of tripping by able-bodied people remains; a smaller difference in height across gaps is usually preferred. For gaps of 6–10 mm width, which is around the clearance usually found between car floor and landing, a variation in height of 2–3 mm can usually prove acceptable. There is further discussion of this problem in *Floors and flooring*[23], Chapter 1.5.

Escalators and travelators
Escalators are normally designed to travel at rated speeds of 0.5 m/s, 0.65 m/s or 0.75 m/s. A 1 m wide escalator can carry over 150 people per minute at 0.5 m/s, nearly 200 people per minute at 0.65 m/s and 225 people per minute at 0.75 m/s. The 1 m width permits overtaking.

Travelators are usually rated at 0.75 m/s, but faster speeds are sometimes encountered where special provision is made for stepping on and stepping off belts; for example, short lengths at each end travelling at slower speeds, and a faster belt in the middle (ie between the two short belts).

Noise and other unwanted side effects
Lifts
In surveys of user experience in flats the noise from lifts has often been mentioned as disturbing. Similar problems have also been reported in hotels and other residential buildings. There are two sources of noise: self-closing gates and door mechanisms, and lift motors and control gear. For small lifts, swing doors with suitable closing mechanisms are usually quieter than concertina or scissor gates. The motor and switch gear are generally housed at the top of the shaft and the necessary measures to minimize noise from them should have been taken at the design stage. Unfortunately, where the noise from lifts is found to be disturbing, there is little that the individual occupiers of flats can do to moderate it.

Escalators and travelators
By the very nature of their design, escalators and travelators cannot be completely silent. However, the background rumbling of equipment is unlikely to give rise to complaint, although squeaking from treads on the rails of elderly installations, due to inadequate maintenance, has occasionally occurred.

Case study

Replacement of lift control system
At the Hull Royal Infirmary, Humberside, electricity consumption has been reduced by over 7%, against a rising trend, by replacing a lift control system[381].

The old installation was ageing, and breakdowns were frequent. The electro-mechanical control system was inefficient, and excessively long waiting times were being experienced; maintenance costs were escalating. It was decided, therefore, that complete replacement was justified.

Four high capacity, static converter drive, variable speed lifts (two 200 kg and two 1400 kg) now serve the 13 floors of the tower block. They give improved ride comfort, especially in relation to acceleration and deceleration, quicker response times, more accurate levelling, reduced maintenance costs, and, of course, reduced electricity consumption. All the measures undertaken have a payback of under three years, and many measures are easily replicated at other sites. The variable speed control of lift mechanisms was particularly successful.

Weathertightness

Special requirements apply to lifts, escalators and travelators used on the exterior of buildings and exposed to the weather. These installations are comparatively rare and any problems of weathertightness should be referred to manufacturers and building contractors.

Lifts designated for firefighting puposes need special provisions for keeping water away from the shafts.

Commissioning and performance testing

Lifts

The doors of lift cars and landings are required not to bow by more than 15 mm under test, and should recover to operate satisfactorily afterwards.

Powered lift car doors should be equipped with automatic reopening mechanisms which operate when the doors encounter resistance to closing greater than 150 N. This value is quite different from the forces normally associated with ordinary internal doors. (See Chapter 1.12 of *Walls, windows and doors*[22].)

The Royal National Institute for the Blind recommends that artificial lighting levels on landings and lift car floors should be greater than 50 lux, and for the partially sighted, 75 lux might be better[382]. For the cars, 100 lux is recommended[383].

Figure 6.6
A large general purpose lift being installed. These doors are manually controlled

Speed governors should operate at 115% of the rated speed[346].

There are stringent requirements for the electrical safety of the drive motors and cabling of passenger and goods lifts. These are dealt with by the relevant British Standards – for example, BS 5655 – and increasingly in European Standards produced by CENELEC.

Lift controls should be sited at 1–1.3 m above floor level, and close to the door frame. Buttons should require no more than 7 N pressure to operate. For disabled people, each installation serving more than three floors should have audible indicators. In addition, lift control buttons should have the number 5 key embossed to aid those who are visually impaired[382].

Escalators and travelators

Artificial lighting levels on escalator landings should be greater than 50 lux, while on the escalator itself 150 lux is recommended.

Again, stringent requirements apply for the electrical safety of the drive motors and cabling of escalators and travelators. They can be found in the relevant British Standards; for example, BS EN 115.

Speed of travel of handrails should not vary by more than 2% from the speed of travel of the treads or belts. The breaking load of the handrail should not exceed 25 kN.

Escalators and travelators should be provided with a braking system which will operate, for example, if the electricity supply suddenly fails. The braking system provided should be capable of bringing the escalator gently to rest with a uniform rate of deceleration within a short distance. The maximum stopping distance for an escalator with a rated speed of 0.75 m/s, for example, is 1.5 m; for a travelator travelling at 0.9 m/s, it is 1.7 m.

Travelators should have warning signs at their ends, so that passengers, particularly visually impaired people, have adequate warning.

Health and safety

There are a number of risks attached to the operation and use of most mechanical handling devices in buildings. These may include many types of physical hazard for passengers; for example, slipping, tripping and falling caused by unsuitable or irregular surface finishes; trapping of clothing, or even arms and legs, by moving parts against the carcass of the building; electrical hazards, either by actual contact causing shock or burns; fire and smoke and harmful fluids or gases emitted by malfunctioning machinery, and so on. General provisions for assessing the importance of these risks, and dealing with them, is contained in health and safety legislation. Relevant legislation includes:

- Health and Safety at Work Act 1974
- Factories Act 1961
- Office, Shops and Railway Premises Act 1963
- Electricity at Work Regulations 1989
- Workplace (Health, Safety and Welfare) Regulations 1992
- national building regulations

However, the following section describes some of the more important considerations.

Lifts

Generally speaking, the safety record of passenger lifts is very good, and they are not a major source of accidents. No fatalities were recorded in the most recent official figures. However, reported accidents in housing and non-domestic buildings involving moderate or severe injury number more than 100 each year on average. For instance, during 1985–89 there were eight recorded accidents involving children, of which five involved trapped fingers. Building owners and users must be constantly vigilant to ensure that accidents are kept to a minimum and that a good safety record is maintained[80].

The risks extend not only to passengers riding in lift cars but also to other users elsewhere in the building, and to maintenance and inspection engineers. (Safety rules for the construction and use of electric lifts in EN 81-1 are embodied in BS 5655-1). Among the main hazards which need to be recognised and provided for are:

- shearing, when a lift car starts at the same time as the doors are open and being used, or where the roof or floor of a car, or a counterweight, passes close to the carcass of the building in a position accessible to maintenance staff
- impacts resulting in passengers being crushed or falling
- trapping of passengers in a car when it breaks down between floors
- fire, in a lift car, its drive mechanism or the shaft
- electric shock from the power or lighting systems
- failure of materials and mechanisms through wear or corrosion
- lift doors, with defective sensors, that close so fast that passengers are trapped, perhaps injured

Of course, it is difficult to provide adequate protection against wilful abuse where constant supervision is impractical.

In older designs of lift installations, some form of mechanism was normally provided at the bottom of lift shafts to reduce the rate of impact in the unlikely event of failure of the suspension or drive system. This mechanism would have been an hydraulic buffer, or a gradual constriction of the guide rail distances which squeezed the car to a gradual stop. More recent designs may use other safety mechanisms.

Most lifts installed in the UK since the 1930s have had double doors or gates to protect passengers on landings and in cars. The situation in other countries did not always demand this provision, and many installations abroad were provided only with doors on the landings, with carefully controlled clearances between cars and shafts; smooth surfaces on the backs of the doors provided some protection for

passengers. Some old unrefurbished lifts, particularly goods lifts, may still have perforated doors or scissor gates which have finger traps for the unwary passenger (Figure 6.7), but the requirement for new installations is for imperforate doors. Landing doors should not be wider than the car width, in order that vertical gaps are not revealed when the doors are in the open position.

Some of the newer installations employing extensively glazed lift cars can induce feelings of vertigo in sensitive passengers. There is no simple solution other than changing the design of the car.

The paternoster has obvious dangers for elderly and infirm people. Each car should have some form of movable protection to the gap between it and the next car (eg overlapping sheets) to prevent the the cars being used improperly, or the tops or bottoms being vandalised.

Some permanently open ventilation of closed lift cars is required to cover emergencies when a car and its occupants are trapped between floors. This is normally provided by upper and lower vents in one of the walls of the car, the sizes of the vents being calculated from the floor area of the car (usually 1%). It should not be possible to insert fingers into the ventilator; grilles should have apertures not exceeding 10 mm diameter. An alarm system should also be sited in the car; this can take a variety of forms such as a telephone or bell push, but the form

and method of response should be clearly indicated in the car. This is particularly important in cases of offices or other building types where personnel work outside normal business hours. Alarm systems must be checked regularly to ensure they operate effectively.

So far as the suspension mechanisms are concerned, the factors of safety employed will depend on the numbers of ropes or chains, typically 10 in the case of chains and 16 in two-rope systems.

It is usual for consideration to be given to the slip resistance of both car and landing floorings; Table 6.1 below illustrates typical values.

Figure 6.7
Old lift gates are a potential hazard for young fingers

Table 6.1 Guide to coefficient of friction between flooring and smooth rubber soles (broad range)		
Material	**Dry**[†]	**Wet**[‡]
Carpet	> 0.7	0.4–0.6
Cork	> 0.7	0.3–0.5
Linoleum, smooth	> 0.7	0.2–0.35
PVC, smooth	> 0.7	0.2–0.3
PVC, carborundum or aluminium oxide	> 0.7	0.3–0.45
Resin, smooth	> 0.7	0.2–0.3
Resin with carborundum in finish	> 0.7	0.3–0.5
Rubber, smooth	> 0.7	0.2–0.3
Thermoplastic or vinyl asbestos tiles	> 0.7	0.2–0.35

† Clean, no polish (contamination can greatly affect values)

‡ On very smooth floorings of all types, the coefficient of friction can be as low as 0.1. Coefficient of friction in the wet is greatly dependent on surface texture of flooring

Although values for wet floorings are given, as lifts may be situated near to external doors, every effort should be made to keep water, and perhaps even urine in some installations, away from lifts and escalators not designed to cope with such conditions.

Rodents must be prevented from entering lift shafts, not only to restrict their access to all levels of the building but to prevent damage to electrical and winding cables associated with lift equipment. They have been known to survive by feeding on lubricating grease.

There are statutory requirements for health and safety inspections of lift installations to be carried out twice a year.

Escalators

Escalators having ribbed treads parallel to the direction of travel are normally supplied with combs at each landing, the teeth of which nest, with adequate clearance, into the spaces between the ribs. Since the teeth of the comb are inclined, the forward movement of the horizontal treads is sufficient, in principle, to deposit the passenger gently onto the landing. Although simple in essence, this operation is fraught with danger for anything caught in the clearances between successive treads or between the comb and the ribs. This is why, for example, dogs need to be carried on escalators so that their claws are not at risk. A similar risk of trapping occurs at the sides of the treads, and here deflectors or brushes are normally installed to keep footwear clear of the skirt or side panels.

Escalators should have at least two level treads at top and bottom where the combs engage, to provide a reasonable space for pedestrians to adjust the position of their feet after stepping on and before stepping off.

At upper and lower landings, where the treads of escalators align themselves before and after the change of slope, especially good lighting conditions are called for. Certain ribbed treads, especially those in shiny stainless steel, exhibit strong or dazzling patterns which can camouflage (to the unwary user) the positions of nosings; one tread tends to blend visually with that beneath or above, potentially leading to falls, particularly when users walk down an escalator, whether it is moving or not.

The great danger with escalators carrying a full load of passengers is that of a falling person initiating what might be termed an avalanche of bodies on the longest flights. One of the worst accidents of this kind occurred in the London Underground system on 3 March 1943 during an air raid alert when over 170 people were killed. This type of risk is the reason why double flight escalators with intermediate landings are preferred for the larger storey heights. To comply with Standards, passenger operated emergency stop switches must be provided at regular intervals not exceeding 15 m on escalators and 40 m on travelators.

In addition to the risks associated with the moving parts themselves, there is also a need to place limits on the clearances between moving and fixed parts of the installation, to provide imperforate and unclimbable balustrades, and to interrupt inclined planes between escalators with obstructions to prevent unruly behaviour. Where balustrades are made of glass, it must be toughened and at least 6 mm thick.

Where escalators intersect, either with floors or with each other, a triangular shape results (Figure 6.5). This space needs to be filled with suitable shielding.

Fire

Kiosks and display areas, and openings to shop units and other occupancies on malls and walkways, should be situated at least 6 m from discharge points from escalators and lift doors (see BS 5588-10).

Lifts

Misuse and malfunctioning of lifts has caused a number of deaths (BS 5588-10). Causes included failure of power supplies and lifts being called to, or held at, the fire floor. If car and landing doors open in a fire area they will usually remain open, exposing the occupants to the fire. Lifts, therefore, should never be considered as suitable means of escape. Special evacuation lifts for disabled people are designed to operate during a fire.

The Building Regulations Approved Document B makes provision for firefighting lifts within firefighting shafts for all buildings more than 20 m above ground level, and more than 10 m below ground level, including the shafts being equipped with sprinkler systems to

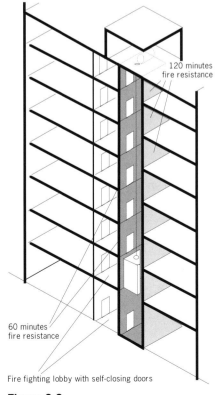

120 minutes fire resistance

60 minutes fire resistance

Fire fighting lobby with self-closing doors

Figure 6.8
Firefighting lift shaft

BS 5306-2. The number of firefighting lifts required depends on the gross floor area of the building, with one lift required for buildings up to 900 m², two for 900–2000 m², and one extra for every 1500 m² (or part) above 2000 m². For buildings without sprinkler systems there needs to be one for every 900 m². In all cases the lift should be within 60 m of every part of the floor which is served. Every firefighting lift needs to be approached through a firefighting lobby equipped with self-closing fire doors.

The walls of an ordinary lift shaft should have a fire resistance of at least 30 minutes loadbearing capacity, integrity and insulation, but firefighting lift shaft walls should have at least 60 minutes for the side facing into the shaft, and 120 minutes for the side facing outwards from the shaft (Figure 6.8).

Lift shafts should not contain any cabling which is not part of the lift installation.

Ordinary lifts must not be used for evacuation in a fire. However, a firefighting lift may be used for the evacuation of disabled people before the arrival of the fire service which then will take charge of the operation. There is also the possibility of evacuating disabled people using an ordinary passenger lift which has been equipped with a similar standard of structural protection against fire as a given by protected stairway – a so-called evacuation lift which needs to be clearly labelled as such. (See Chapter 6.3 of *Walls, windows and doors* and BS 5588-8.)

An evacuation lift has to be provided with a power supply exclusively for the operation of that lift; also with an alternative standby power supply in case the main one fails. This does not apply to hydraulically powered lifts serving two storeys where the final exit is from the lower one; the controls for these lifts, which may need to be protected by key operation, should ensure that the lifts return automatically to the final exit floor.

All substations, distribution boards and cabling should be protected for a period not less than that applicable to the structure enclosing the shaft.

There are special requirements in British Standards for landing and lift car doors in relation to fire risk.

Powered stair lifts cannot be used as designated means of escape in the event of fire.

Since paternoster lifts have no doors at either landings or cars, and therefore smoke and possibly fire can be transmitted between floors, special precautions are required with the lobbies at each landing level; the complete installation then becomes an integral part of a protected shaft.

Escalators

Escalators cannot be used as escape stairs in the case of a fire. They should be halted on discovery of a fire, although provision needs to be made for people using them when a fire is discovered.

Escalators which pass through compartment floors need to be protected during a fire by automatic shutters. At risk of stating the obvious, it is absolutely essential that they work; designing all the necessary fire protection before installation, and regular checks of operating systems during the life of the installation, are therefore vital (Figures 6.9 and 6.10).

Figure 6.9
The automatic shutter on this escalator failed to work, and the effect can be seen on the suspended ceiling

Figure 6.10
Another case of an escalator well burning out when the automatic fire shutter failed

Durability

With all kinds of powered installations using drum wound cables, wear is directly related to the lead-on and lead-off angle of transfer of the cable from pulley to drum as well as to the load carried; this angle must be closely controlled. BS 5776 restricts it to not more than 2°.

Lifts

There are no standard minimum requirements for the working lives of lift installations.

BRE unpublished investigations, dating from the early 1980s, put the expected average lives of lift mechanisms at around 40 years, with a minimum of 25, if regularly maintained. Since then, service lives seem to have reduced. The main difficulty is that parts of systems wear out, or develop problems, at different time intervals. It appears, for example, in the most basic of installations, that suspension ropes and display panels tend to give the shortest lives, around 5 years; doors, and their motors and controls, and cars generally around 10–15 years; winding gear, motors and electronic control systems around 15–20 years; and guides 35–60 years. Lifespans depend heavily on the quality of the original specification, and the amount of usage of the installation.

Shafts of masonry or concrete generally do not deteriorate but isolated instances have occurred where it has been necessary to carry out resin injection to cracks. Pits at the bottom of shafts generally seem to be trouble-free, but flooding, and health and safely risks occur from rubbish, in particular the syringes of drug users, thrown down shafts, pose problems in some areas and in poorly maintained buildings. Flooding can cause problems with electrical links. Control rooms need to be protected from the weather and extremes of temperature. Sky-lights at the top of shafts or in lift motor rooms present a high risk of vandalism and weather ingress, and so must be secure. Guide rails are expected to last the life of the building, though should be checked for plumbness and corrosion when carrying out Health and Safety PMM7 inspections[384].

There are marked differences in the durability of installations in unsupervised blocks of flats when compared with other buildings such as offices, where vandalism tends to be less of a problem or even non-existent. The service life of lift components in the building stock of at least one large owner is largely governed by maintenance arrangements, the types of buildings served by lifts, vandalism, theft, and the availability of spare parts.

In fact, vandalism seems to be the main reason for lift failure for this particular owner, and it has come to be accepted as almost un-preventable. The main problems include door failure, control panel faults especially push buttons, digital indicators, and theft of security cameras and actual lift parts. There is also a lack of user understanding of how lifts work, especially hydraulic lifts because of their relative slowness. A suitable remedial strategy might include the use of security access doors, entry telephones, roving security patrols on particular estates, CCTV, mirrors in lift cars, high level LED indicators set behind screened holes, and scrolling indicators to show where the lift is (and perthaps reduce user frustration).

Escalators and travelators

There are no standard minimum requirements for the lifespans of escalators and travelators.

Case study

Lift installation in a block of flats

A bank of lifts in a 1960s block of flats had been refurbished some 10 years before the need for further refurbishment became apparent. At the earlier refurbishment, almost all the working parts, except parts of the control systems and the guide rails, had been renewed.

At the later refurbishment, which coincided with a change of responsibility for servicing and maintenance, a full handover test including load, voltage, mechanical and electrical testing was carried out by competent lift engineers. They noted that not all lift cars originated with the same manufacturer. There had clearly been problems of vandalism. Problems were also identified with parts of the original control system which had not been renewed on the earlier refurbishment; for example, the system was giving false floor indication, and the controls were not to current British Standards requirements. Although replacement door motors had been fitted, other critical parts of door operating systems were retained and linkage problems had resulted. A lifting beam was supported at one end by an adjustable steel prop! There were also potential difficulties in obtaining replacement parts from abroad. Control rooms were well ventilated.

All the lifts were identified as health and safety risks, and recommendations were made for their upgrading or replacement. Concern in the main related to lift motors and control systems. Piecemeal replacement of lift door motors, but connecting them to existing mechanisms, would cause linkage problems. The main controls too would benefit from replacement with variable frequency controls. The preferred solution to the range of problems would be to renew the whole installation, but flat occupants would have to suffer a reduced service while works were in progress.

Maintenance

As with all mechanical devices carrying people, regular maintenance is vital for the maintenance of safety. In particular, wire ropes must be renewed at statutorily prescribed intervals, and the working of emergency telephones, stop switches and control systems checked and tested at regular intervals. The regular monthly examination of the lift stock of a large owner of buildings, for example, includes no less than 42 individual items relating to inspection, adjustment and lubrication. Continuity of maintenance teams, and their familiarity with the systems being inspected, is an important factor in maintaining quality of service.

Particularly, in mind of the disastrous fire at the Kings Cross Railway underground station, the voids under machinery should be kept clear of detritus.

Wear on moving parts needs to be regularly monitored, especially the teeth of combs on escalators, which should be easily renewable.

Lift systems in public sector multi-storey housing at one time suffered considerably from vandalism and poor performance: so much so that lifts were out of action for long periods, leading to real hardship for the occupants and dissatisfaction with their housing. A concierge system, together with more durable and easily cleanable surfaces in cars and doors, largely overcomes this problem. (Graffiti removal is discussed in *Walls, windows and doors*, Chapter 9.4.)

The maintenance costs of lift systems can be considerable. Research in the early 1970s showed that they directly related to passenger capacity multiplied by the number of floors served[385]. Things have not changed much since the 1970s. Maintenance logs might be helpful in monitoring life expectancy if levels of maintenance and repair, and the incidence of vandalism, were fully reported The logs would also provide a check on the effectiveness of maintenance actions.

Work on site

Access, safety etc

Motor rooms should have floors of non-slip materials. (See Table 6.1 earlier in this chapter, and the discussion in *Floors and flooring*, Chapter 1.5.)

A lift car roof needs to be able to support a load of 2000 N – that is to say, two people with tools.

Safe working practices for the installation, examination, inspection, testing, maintenance, repair and dismantling of lifts is contained in BS 7255[386].

Guidance for safe working practices concerning escalators and travelators are given in BS 7801[387]. This covers examination, inspection, testing, servicing, maintenance and repair, and the adequacy of health and safety training of all persons carrying out these duties. Lone working is not allowed. A site safety assessment must be carried out before any work begins, and then periodically through the progress of the works. Adequate barriers need to be provided to keep unauthorised people a safe distance from the works. Adequate temporary lighting of at least 200 lux must be provided.

Inspection

Inspections must be carried out by 'competent persons' at regular intervals defined by the manufacturers, by Standards, and by law. The list of items to be inspected by competent lift engineers is not included here.

In the case of powered domestic stair lifts, the interval between inspections should not exceed 12 months. For powered domestic lifts the interval should not exceed 6 months[388]. Aspects covered at these regular inspections include:
◊ safety devices and brakes
◊ wear in moving parts
◊ electrical safety
◊ completion of certificates

When examining a building for refurbishment potential, items relating to the lift installation to which particular attention should be paid include:
◊ components of the installations not clearly marked with manufacturers' data
◊ degree of non-conformity with current Standards (related to age of installation)
◊ test certificates out of date
◊ rubbish and detritus around motor rooms and in lift wells
◊ door operating mechanisms not functioning correctly
◊ alarms not functioning correctly
◊ waterproofing in firefighting lifts not effective
◊ means of escape in case of fire blocked
◊ instruction handbooks not available
◊ clearances around doors exceeding 6 mm
◊ positions (floor levels) of lifts in shafts not identifiable

Chapter 6.2

Goods handling devices, conveyors and warehouse storage

Lifts for handling passengers and goods were dealt with in the previous chapter. This chapter deals with other forms of goods handling devices and conveyors, including pneumatic distribution systems which are permanently installed in the building. Temporarily and permanently installed cradle systems for the cleaning and maintenance of external walls of tall buildings are mentioned. Also included, primarily because of BRE concerns with fire risks, is the storage of goods in warehouses served by tower fork lift trucks or mechanical goods storage and recovery systems. Brief mention is made also of moveable racked storage for books and other small items. Portable materials handling devices and automatic car parking by mechanical handling systems (being a very specialised application) are excluded though.

Characteristic details

Fixed lifts for road vehicles are dealt with in BS AU 161-1a[389] and platform hoists in BS 5655-14[390].

Overhead cranes

With the rundown of the engineering industries, permanently installed overhead cranes in factories and workshops are not seen as frequently as in former years. These travelling cranes were probably installed at the times the buildings were designed, since it would have been appropriate in most cases to have used the loadbearing structures of the buildings to provide the support for beams and cradles (Figure 6.11). Overhead cranes normally operate at 415 volt 3-phase supply.

The design requirements for overhead cranes are contained in BS 466[391].

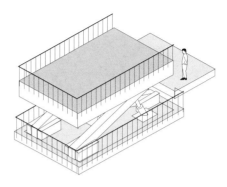

Figure 6.12
The principle of operation of a scissor hydraulic lift

Lifting platforms

Scissor lift platforms are sometimes installed as permanent fixtures in buildings and usually operate by hydraulic jacks directly connected to the arms of the scissors (Figure 6.12). They are normally used for raising heavy loads and have the advantage that they do not need deep pits. One of the largest installations is at the Imperial War Museum. The capacity of the installation is around 74,000 kg, with a platform size of 5 m × 10 m, and is capable of raising the largest military tanks through 5 m.

Permanently installed scissor lifts are also used in loading bays, and may be capable of accommodating a wider range of vehicle heights than do docklevellers (described below). They can carry up to around 8 000 kg: in other words, around half as much again as the largest heavy duty goods lift, though of course they operate over a very limited height range. They also normally operate at 415 volt 3-phase supply.

Figure 6.11
Overhead travelling crane in the BRE Heavy Test Laboratory

High-rack automated warehouse

The Central Ordnance Depot at Donnington is one of the largest automated warehouses in the UK, with 34 000 m² floor area, and the main storage areas some 20 m high. The foundations had to be specially reinforced to take the loads from the storage racks.

Control of the automated storage and retrieval system is by central main frame computer linked to smaller computers in each storage spur. Spurs are separated by 25 m spaces, and compartments within the spurs are formed by 4-hour fire resistant walls. The complex has 50,000 sprinkler heads situated at alternate levels within the racking system. Staff do not normally need to gain access to the storage areas except for servicing and maintenance purposes.

Goods are stored on pallets, retrieved by automatic stacker crane. The pallets are moved to the spurs by a combination of automatic guided vehicles and conveyors. Manual selection is involved when part pallet loads are required. Operating costs were expected to be significantly lower than alternative storage systems due to fewer staff, and low heating and lighting loads[392].

Docklevellers

Docklevellers are used in loading bays to accommodate different vehicle heights, and therefore enable loads to be loaded or unloaded efficiently. They can be portable or fixed to building carcasses, but only the later are dealt with here.

Clearly, docklevellers cannot operate over a very wide range of vehicle heights without increasing the angle of inclination of the ramp; consequently they present difficulty in moving some loads. They also need to be provided with automatically operating toe guards which physically prevent the feet of operatives being trapped under moving platforms, and with stop buttons to halt moving platforms in emergencies.

Roller and belt conveyors

Only brief reference is necessary to the two main types of conveyors:
● those operating under gravity where goods are moved on steel frameworks carrying closely spaced rollers with each operating independently on its own axle
● powered belt conveyors

Both types of conveyor tend to be independent of the structure of the building, but may need to pass through compartment walls and floors; in the event of fire, shutters linked to the fire detection system operate to contain the fire. One of the largest installations is at the British Library at St Pancras, London, which is designed to retrieve and return up to 25,000 books each day from the 25 million stored in the basements, all controlled through a central computer. The books are carried in containers which each carry the codes for routeing automatically to destinations.

Overhead chain conveyors

Usually these systems are suspended from the roof or walls of a factory or workshop; the loads imparted need, of course, to be taken into account in the original design of the structure. Any change in the loads to be carried should generate a review of the loadbearing capacity of the structure.

Designs vary so widely that it is difficult to provide general comments. However, by using simple arithmetic, a double channel supported overhead conveyor carrying trolleys at 2 m centres, with each trolley carrying a load of 1 tonne, will impart a dead load of around 500 kg/m run, without any allowance being made for the weight of the trolley, and for vibration, impact and other dynamic loadings. Any assessment of total overhead loadings will need to take account of loads from, for example, unit heaters, electric substations and distribution, and ventilation ducting, in addition to those imparted by overhead conveyors.

Library movable rack storage

Where storage space in libraries is in short supply, and books are not in continuous demand, it has often been possible to install a moveable double or single sided rack system. The racks run on gears and rails set into the floor surface, and are normally powered manually (Figure 6.13). Very large storage systems of this kind are installed, for example, in the vaults of the British Library which has over 180 miles of shelving, in the National Library of Scotland, and in the Public Record Office at Kew. Smaller installations, of as few as six racks, are quite feasible. Where the storage racks involve heavy loads, measures are needed for ensuring that floor deflections are kept within limits that will not impair operation of the racks.

High-rack automated warehouses

The high-rack system is a common form of high density storage in which a vertical framework is used to store large quantities of materials, essentially in vertical columns, on a minimum floor area.

High-rack storage in warehouses has grown rapidly since the 1970s. Many of these warehouses are serviced by mobile fork lift stackers, and some are fully automated under sophisticated computer control. The requirements for level floors were referred to in *Floors and flooring*, Chapter 3.1. Since the racks can be up to 13 m high, it is important that the truck masts are kept vertical to avoid collisions with the racks.

In some installations, the retrieval systems are supported from the racks themselves.

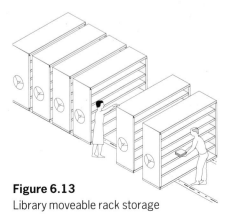

Figure 6.13
Library moveable rack storage

Air tube systems

When first introduced and operated in the first half of the twentieth century, these systems comprised small diameter tubes suspended on wires strung between sales points and central cash points in department stores. They have been replaced by 50–150 mm diameter air tube systems, which have the advantage over wire-suspended systems that they can be routed round corners. The carriers in the tubes travel at speeds of around 6 m/s, though they are still limited to payloads of around 5 kg. Power is supplied by an electrically driven fan system which creates either negative or positive pressure in the tubes. These systems are used in a variety of building types, for example laboratories and offices as well as stores of various kinds.

Cradle systems

Manually operated or power operated cradle systems are used on medium and high rise buildings to facilitate external maintenance and cleaning (Figure 6.14). The permanent tracks at roof level sometimes run round the whole periphery of the verge or parapet, with some systems providing turntables at corners. Permanent installations are covered in BS 6037[393] which covers requirements for construction, installation, testing and use. On very tall buildings provision must be made for tethering the cradle to avoid sideways movement in windy conditions, although using the equipment in high winds should be avoided whenever possible. Some installations incorporate vertical rails attached to the facades for this purpose; it may not be possible, though, to retrofit these rails unless the whole of the cladding is being replaced.

Figure 6.14
A permanent installation for access equipment on a 1970s building

Main performance requirements and defects

Strength and stability

The design of all types of racked storage systems must depend on the kind of goods to be stored, as well as the means and frequency of retrieval. The moveable types will clearly be subjected more to wear and tear than fixed ones. A major moveable rack storage system initially gave problems because the mass of stored material was underestimated, and the racks deformed when fully loaded.

Outputs required

Gravity roller conveyors usually operate on the basis of (approximately) a 5° slope. Corners may be turned by using tapered rollers arranged radially, or segmented rollers where those on the outside of the curve rotate faster than those on the inside.

Belt conveyors usually operate on the level or at low angles of slope with corners requiring separate conveyors, and vertical (downwards) transfer of goods by gravity between conveyors. Speeds and widths can vary greatly, from a fraction of a metre per second to, perhaps, 1.5 m/s. Goods may need to be constrained by side walls. The belts of inclined conveyors may have protruding transverse ribs or bars to prevent slippage of the goods carried.

Fire

The development of quick response sprinkler systems for use in high-rack storage was described in Chapter 4.5.

The nature of stored goods and their storage arrangements are important to the design of a sprinkler system. There is potential for rapid fire spread through the 'flues' formed by goods in high-rack systems, so optimum protection may be provided by an in-rack, quick response sprinkler system.

Full-scale fire tests at the Fire Research Station (FRS) have demonstrated the speed with which fire can spread within a high-rack, both vertically and horizontally. Flames within the flues formed by the racks can reach a height of 10 m within about one minute following ignition at ground level (Figure 6.15). Standard response sprinklers do not generally operate until at least 75 seconds after ignition (ie 15 seconds after the flames have reached 10 m). The most effective fast response sprinklers tested by FRS operated 56 seconds after ignition, before the flames reached the top of the racking. ESFR sprinkler heads have a significant role to play in preventing fire spread in these types of storage, especially where the goods are piled rather than racked.

Very rapid spread of flame is a key factor in determining the specification of any protection system. If standard sprinklers are located only at ceiling level, a significant fire may develop before they activate. Also, the water may not be able to penetrate into the goods in the rack. Recognising this, regulatory bodies such as the Loss Prevention Council (LPC) in the UK and Factory Mutual in the US both recommend that sprinklers are fitted within the racking for certain categories of stored goods.

However, FRS tests have shown that, even with standard sprinklers appropriately fitted in-rack, fire spreads too quickly for the system to control it. The flame front can pass sprinkler heads before they respond, causing ignition of material further up the rack. The fire then continues to spread, despite activating further heads until, in the worst case, the whole system is overwhelmed and the fire is effectively burning out of control. This could happen very quickly and, by the time the fire brigade arrives, losses might be extensive. Warehouses should be configured so that firefighting is not restricted by limited accessibility to the aisles.

Of course, where the goods are simply piled rather than racked, the fast response sprinklers have to be mounted on the ceiling.

Detector system

With racking in a single-storey building, there is no effective ceiling until the roof level is reached, and any sprinklers not fitted to the shelving units must be fixed to the underside of the roof; even with fast response sprinklers, large fires could develop. An effective system needs 'area-beam' (smoke) detectors to protect each storage aisle, and linear heat detector cables on the top of each run of storage compartments; also a projecting lip to the top of each storage will direct any smoke towards the beam detectors[394].

An important method of achieving maximum protection is to insert a layer of non-combustible material on the top of the racking, and capping the ends of the racks with vertical sheets of non-combustible material at sprinkler locations. This enhances the sensitivity of the sprinklers at these locations and ensures that they are not by-passed by an advancing flame front.

Health and safety

Clearly, all mechanical equipment which is accessible by people will involve particular heath and safety implications; handrails, effective forms of guarding, and high-visibility warning colours and notices may be required. It is not possible to provide general guidance that applies to all items of equipment, though it is worth noting that, in other contexts, handrails are provided where changes of level exceed 600 mm.

Docklevellers with smooth decks, or decks which become coated with oil or grease, or wetted by rain, can become slippery; these conditions are made worse when levellers are used at inclined angles. The combination of certain footwear materials (eg rubber), oil and water is particularly hazardous. (See Chapter 1.5 of *Floors and flooring*.) PVC or resin coatings that incorporate carborundum or aluminium oxide will give coefficients of friction between a deck and smooth rubber soles better than 0.7 in dry conditions, and 0.45–0.5 in wet conditions.

Figure 6.15
A test fire in a high racked storage facility

Durability

Docklevellers, being at least partially exposed to the weather, need to be protected from corrosion damage. Steel components need to be galvanised or given other suitable protection.

The life of a zinc coating is dependent on the environment to which it is exposed and is proportional to the thickness of the zinc; generally steel sheeting should have a zinc coating weight of not less than 275 g/m² . With over-thick coatings, the zinc may crack and spall under heavy impacts.

The cantilevered suspension rails for cradles and cranes carrying access systems to building façades impose very heavy loads on their fixings at roof level. The fixings require regular inspection to ensure that the weatherproof collars to the supports have not fractured, and that there are no signs of corrosion.

Maintenance

As with all mechanical devices, wear and tear will inevitably occur. Regular inspection and regular maintenance to manufacturers' instructions will be necessary to ensure both continuing performance and continuing safety.

Work on site

Inspection

Inspections must be carried out by 'competent persons' at regular intervals defined by manufacturers and, in certain cases, by law. Aspects covered at these regular inspections include:

◊ safety devices and brakes
◊ wear in moving parts
◊ electrical safety

When examining buildings for refurbishment potential, particular items to which attention should be paid include:

◊ installations not clearly marked with manufacturers' data
◊ degrees of non-conformity with current Standards (related to age of installation)
◊ test certificates out of date
◊ operating mechanisms not functioning correctly
◊ alarms not functioning correctly
◊ instruction handbooks not available

References and further reading

Each numbered reference below is shown only under the chapter in which it first appears in the text. An item may appear in a *Further reading* list before being cited as a reference.

Preface
[1] Egan J. *Rethinking construction.* London, Department of the Environment, Transport and the Regions, 1998
[2] BRE. Protecting buildings against lightning. *BRE Digest* 428. Garston, Construction Research Communications Ltd, 1998
[3] British Standards Institution. Requirements for electrical installations. IEE Wiring regulations. Sixteenth edition. *British Standard* BS 7671:1992. London, BSI, 1992
[4] Department of the Environment, Scottish Development Department and Welsh Office. UK Model Water Byelaws 1986. London, The Stationery Office, 1986
[5] Water Supply (Water Fittings) Regulations 1999. Statutory Instrument 1999 No 1148. London, The Stationery Office, 1999
[6] Health and Safety Commission. Managing construction for health and safety. Construction (Design and Management) Regulations 1994. Approved code of practice. (L)54. Sudbury, HSE Books, 1995

Chapter 0
[7] Department of the Environment, Transport and the Regions. *English house condition survey 1996.* London, The Stationery Office, 1998
[8] Welsh Office. *Welsh house condition survey 1993.* Cardiff, Welsh Office, 1993
[9] Scottish Homes. *Scottish house condition survey 1996.* Survey report. Edinburgh, Scottish Homes, 1997
[10] Northern Ireland Housing Executive. *Northern Ireland house condition survey 1996.* Belfast, Northern Ireland Housing Executive, 1997
[11] Harrison H W. Quality in new-build housing. *BRE Information Paper* IP 3/93. Garston, Construction Research Communications Ltd, 1993

[12] White M K, Kolokotroni M and Perera M D A E S. Trickle ventilators in offices. *BRE Information Paper* 12/98, Garston, Construction Research Communications Ltd, 1998
[13] Building Research Station. *Principles of modern building,* Volume 1: Walls, partitions and chimneys. London, The Stationery Office, 1938
[14] BRE. *BRE housing design handbook.* BRE Report. Garston, Construction Research Communications Ltd, 1993
[15] BRE. Assessing traditional housing for rehabilitation. BRE Report. Garston, Construction Research Communications Ltd, 1990
[16] Energy Efficiency Best Practice programme. Energy efficiency in the work place: a guide for managers and staff. *BRECSU Good Practice Guide* 133. London, DETR, 1994
[17] BRE. *Annual Report 1971.* The Stationery Office, London, 1971
[18] Billington N S and Roberts B M. *Building services engineering.* Oxford, Pergamon, 1982
[19] Thompson B (Count Rumford). *Chimney fireplaces.* London, 1796
[20] Ministry of Housing and Local Government. *Housing Manual 1949.* Technical Appendices. London, The Stationery Office, 1951
[21] Harrison H W. *BRE Building elements. Roofs and roofing. Performance, diagnosis, maintenance, repair and the avoidance of defects.* Garston, Construction Research Communications Ltd, 1996 (2000)
[22] Harrison H W and de Vekey R C. *BRE Building elements. Walls, windows and doors. Performance, diagnosis, maintenance, repair and the avoidance of defects.* Garston, Construction Research Communications Ltd, 1998
[23] Pye P W and Harrison H W. *BRE building elements. Floors and flooring. Performance, diagnosis, maintenance, repair and the avoidance of defects.* Garston, Construction Research Communications Ltd, 1997
[24] Grant I F. *Highland folk ways.* London, Routledge and Kegan Paul, 1961

[25] Department of the Environment. *English house condition survey 1986.* London, The Stationery Office, 1987
[26] Department of the Environment. *English house condition survey 1991.* London, The Stationery Office, 1993
[27] Henderson G. Energy efficiency means warmer homes at no extra cost. *BRECSU Information Leaflet* 11. London, DETR, 1989
[28] White R B. *Prefabrication: a history of its development in Great Britain.* London, The Stationery Office, 1965
[29] Prior J J, Raw G J, Charlesworth J L, Burke D J and Mitchell J C. BREEAM/New homes, Version 3/91. An environmental assessment for new homes. BRE Report. Garston, Construction Research Communications Ltd, 1991

Chapter 1.1
[30] Henderson G. The implications of climate change for buildings. *Building Services (The CIBSE Journal)* 14 (1) p41
[31] BRE. *Designing for production. Lecturers' notes.* Volumes 1 and 2 (1985), and 3, 4 and 5 (1994). BRE Report. Garston, Construction Research Communications Ltd, 1985 and 1994
[32] Burberry P. Space for services: 5. Distribution and sizing. *Architects Journal* (1986) (10) 63–66

Further reading
Chartered Institution of Building Services Engineers. *Explaining building services to architects.* London, CIBSE, 1999

Chapter 1.2
[33] Britten J R. What is a satisfactory house? A report of some householders' views. *Housing Review* (1977) **26** (5)
[34] Hunt D R G and Gidman M I. A national field survey of house temperatures. *Building and Environment* (1982) **17** (2) 107–24
[35] Bordass W T, Bromley A K R and Leaman A J. Comfort, control and energy efficiency in offices. *BRE Information Paper* IP 3/95. Garston, Construction Research Communications Ltd, 1995

[36] **Energy Efficiency Best Practice programme.** Energy efficiency in refurbishment of industrial buildings. *BRECSU Good Practice Case Study* 194. London, DETR, 1994

[37] **Energy Efficiency Best Practice programme.** Energy efficiency in the work place: a guide for managers and staff. *BRECSU Good Practice Guide* 133. London, DETR, 1994

[38] **Hart J M.** An introduction to infra-red thermography for building surveys. *BRE Information Paper* IP 7/90. Garston, Construction Research Communications Ltd, 1990

[39] **Rayment R.** Energy savings from sealed double and heat reflecting glazing units. *Building Services Engineering Research and Technology*, 1989

[40] **Perera M D A E S, Walker R R, Hathaway M B, Oglesby O D and Warren P R.** *Natural ventilation in large and multi-celled buildings. Theory, measurement and prediction. Final report for the period 1st October 1981 to 30th June 1984.* Garston, BRE, 1984

[41] **Henderson G.** Estimating energy costs and potential savings in housing. *BRECSU Information Leaflet* 6. London, DETR, 1991

[42] **Bromley K, McKay M and Wiech C.** Incorporating phase change materials within the building fabric. *CIBSE National Conference 1994, Brighton, 2–4 October* 1, 101–4

[43] **Energy Efficiency Best Practice programme.** *The Government's Standard Assessment Procedure for energy rating of dwellings.* London, DETR, 1993

[44] **Bromley A K R, Bordass W T and Leaman A.** Are you in control? *Building Services (The CIBSE Journal)* (1993) **15** (4) 30–32

Further reading
Hart J M. *The use of thermography in the thermal performance testing of buildings. Applications of thermal imaging* (ed Burway S, Williams T and Jones C). IOP Publishing, 1988

Chapter 1.3
[45] **Garratt J and Nowak F.** *Tackling condensation.* BRE Report. Garston, Construction Research Communications Ltd, 1991

[46] **Raw G J and Fox T A.** The environment in small homes. *CIB 89, 11th Intl Congress, Quality for building users throughout the world, 19–23 June 1989*

[47] **British Standards Institution.** Code of practice for control of condensation in buildings. *British Standard* BS 5250:1989. London, BSI, 1989

Chapter 1.4
[48] **Slater A I.** Energy efficiency in lighting. *Energy Management* (1991) **29** (2) 24–6

[49] **Littlefair P J.** The energy optimisation of window size: an area of concern? *Procs of World Renewable Energy Congress, Reading, 23–8 September 1990,* **4** 2614–21

[50] **British Standards Institution.** Luminaires. *British Standard* BS 4533: 1986–90 (various parts). London, BSI, 1986–90

[51] **British Standards Institution.** Luminaires. General requirements and test. *British Standard* BS EN 60598-1:1993. London, BSI, 1993

[52] **Aizlewood M E.** Interior lighting calculations: a guide to computer programs. *BRE Information Paper* IP 16/98. Garston, Construction Research Communications Ltd, 1998

[53] **BRE.** Lighting controls and daylight use. *BRE Digest* 272. Garston, Construction Research Communications Ltd, 1983

[54] **Department of the Environment and the Welsh Office.** *The Building Regulations 1991. Approved Document L: Conservation of fuel and power* (1995 edition). London, The Stationery Office, 1994

[55] **Slater A I, Bordass W T and Heasman T A.** People and lighting controls. *BRE Information Paper* IP 6/96. Garston, Construction Research Communications Ltd, 1996

[56] **Slater A I.** Lighting controls: an essential element of energy-efficient lighting. *BRE Information Paper* IP 5/87. Garston, Construction Research Communications Ltd, 1987

[57] **Simpson J and Tarrant W S.** A study of lighting in the home. *Lighting Research and Technology* (1983) **15** (1)

[58] **Lowry S.** Noise, space and light. *British Medical Journal* (1989) (299)

[59] **BRE.** Office lighting for good visual task conditions. *BRE Digest* 256. Garston, Construction Research Communications Ltd, 1981

[60] **Chartered Institution of Building Services Engineers.** *Code for interior lighting.* London, CIBSE, 1994

[61] **Slater A I, Perry M J and Carter D J.** Illuminance differences between desks: limits of acceptability. *Lighting Research and Technology* (1993) **25** (3) 91–103

[62] **British Standards Institution.** Luminaires. Specification for luminaires for emergency lighting. *British Standard* BS 4533-102.22:1990. London, BSI, 1990

[63] **British Standards Institution.** Emergency lighting. Code of practice for the emergency lighting of premises other than cinemas and certain other specified premises used for entertainment. *British Standard* BS 5266:1988. London, BSI, 1988

[64] **Webber G M B and Hallman P J.** Movement under various escape route lighting conditions. *Safety in the built environment* (ed Sime J D). London, Spon, 1988

[65] **Webber G M B.** Emergency wayfinding systems: their effectiveness in smoke. *BRE Information Paper* IP 10/97. Garston, Construction Research Communications Ltd, 1997

Further reading
Energy Efficiency Best Practice programme. Low energy domestic lighting. GIL 20. London, DETR, 1994

Littlefair P J. Photoelectric control of lighting: design, setup and installation issues. *BRE Information Paper* IP 2/99. Garston, Construction Research Communications Ltd, 1999

Littlefair P. Developments in innovative daylighting. *BRE Information Paper* IP 9/00. Garston, Construction Research Communications Ltd, 2000

Chapter 2
[66] **Birtles A B and O'Sullivan E F.** Reliability of underground heat mains in the UK. *BRE Information Paper* IP 1/86. Garston, Construction Research Communications Ltd, 1986

[67] **Courtney R G and Hobson P J.** The performance of 15 district heating schemes. *BRE Current Paper* CP 34/78. Garston, Construction Research Communications Ltd, 1978. (Reprinted from *Heating and Ventilating Engineer* (1977) **51** (603) 4–11, and (1978) **52** (604) p24)

[68] **Energy Efficiency Best Practice programme.** The use of combined heat and power in community heating schemes. *BRECSU Good Practice Case Study* 370. London, DETR, 1999

[69] **Energy Efficiency Best Practice programme.** Community heating in Sheffield. *BRECSU Good Practice Case Study* 81. London, DETR, 1994

Further reading
Energy Efficiency Best Practice programme. Opportunities for electricity sales to tenants from residential CHP schemes. *BRECSU New Practice Profile* 112. London, DETR, 1994

Energy Efficiency Best Practice programme. Guide to community heating and CHP: commercial, public and domestic applications. *BRECSU Good Practice Guide* 234. London, DETR, 1996

Energy Efficiency Best Practice programme. Community heating in Nottingham: an overview of a rejuvenated scheme. *BRECSU Good Practice Case Study* 312. London, DETR 1996

Energy Efficiency Best Practice programme. Community heating in Nottingham: domestic refurbishment. *BRECSU Good Practice Case Study* 313. London, DETR, 1996

Energy Efficiency Best Practice programme. Community heating in Nottingham: pipework refurbishment. *BRECSU Good Practice Case Study* 314. London, DETR, 1996

Energy Efficiency Best Practice programme. Community heating – a guide for building professionals. *BRECSU Good Practice Guide* 240. London, DETR, 1999

Chapter 2.1

[70] Clean Air Act 1956. London, The Stationery Office, 1956

[71] **Allaby M and Lovelock J.** Wood stoves: the trendy pollutant. *New Scientist*, November 1980

[72] **BRE.** Domestic chimneys: Solid fuel – flue design. *BRE Defect Action Sheet (Design)* DAS 126. Garston, Construction Research Communications Ltd, 1989

[73] **British Standards Institution.** Specification for open-fireplace components. *British Standard* BS 1251:1987. London, BSI, 1987

[74] **British Standards Institution.** Installation of domestic heating and cooking appliances burning solid mineral fuels. Specification for the design of installations. *British Standard* BS 8303:1994. London, BSI 1994

[75] **John o'London.** *London stories old and new.* London, Newnes, 1926

[76] **Daws L F.** Heat transfer and condensation in domestic boiler chimneys. *Building Research Station National Building Studies Research Paper* 40. London, The Stationery Office, 1966

[77] **British Standards Institution.** Installation of chimneys and flues for domestic appliances burning solid fuel (including wood and peat). Code of practice for masonry chimneys and flue pipes. *British Standard* BS 6461-1:1984. London, BSI, 1984

[78] **European Union of Agrément (UEAtc).** Prefabricated chimneys for dwellings. *Method of Assessment and Test* No 4. Garston, British Board of Agrément, 1980

[79] **BRE.** Repairing chimneys and parapets. *BRE Good Repair Guide* GRG 15. Garston, Construction Research Communications Ltd, 1998

[80] **Cox S J and O'Sullivan E F O.** *Building regulation and safety.* BRE Report. Garston, Construction Research Communications Ltd, 1995

[81] **British Board of Agrément.** *Directory of installers.* Garston, BBA, 2000

Further reading

British Standards Institution. Chimneys. General requirements. *British Standard* BS EN 1443:1999. London, BSI, 1999

Chapter 2.2

[82] **Department of the Environment and The Welsh Office.** *The Building Regulations 1991 Approved Document L: Conservation of fuel and power* (1995 edition). London, The Stationery Office, 1994

[83] **BRE. Energy Efficiency Best Practice programme.** *The Government's Standard Assessment Procedure for energy rating of dwellings.* London, DETR, 1993

[84] **BRE.** Improving energy efficiency: boilers and heating systems, draughtstripping. *BRE Good Repair Guide* GRG 26 Part 2. Garston, Construction Research Communications Ltd, 1999

[85] **British Standards Institution.** General guidance on the marking of gas appliances. *British Standard* BS DD 221:1997. London, BSI, 1997

[86] **BRE.** Domestic gas appliances: air requirements. *BRE Defect Action Sheet (Design)* DAS 91. Garston, Construction Research Communications Ltd, 1986

[87] **British Standards Institution.** Installation of flues and ventilation for gas appliances of rated input not exceeding 60 kW. Specification for installation of ventilation for gas appliances. *British Standard* BS 5440-2: 1989. London, BSI, 1989

[88] **Department of the Environment and The Welsh Office.** *The Building Regulations 1991 Approved Document J: Heat producing appliances.* London, The Stationery Office, 1995

[89] **British Standards Institution.** Installation of flues and ventilation for gas appliances of rated input not exceeding 60 kW. Specification for installation of flues. *British Standard* BS 5440-1:1990. London, BSI, 1990

[90] **Trim M J B.** Condensing boilers prove their case. *Building Services and Environmental Engineer* (1990) **12** (7) p17

[91] **Energy Efficiency Best Practice programme.** Energy efficiency in large residential buildings: condensing gas boilers. *BRECSU Good Practice Case Study* 79. London, DETR, 1994

[92] **British Standards Institution.** Specification for design and manufacture of electric boilers of welded construction. *British Standard* BS 1894:1992. London, BSI, 1992

[93] **British Standards Institution.** Flue blocks and masonry terminals for gas appliances. Specification for precast concrete flue blocks and terminals. *British Standard* BS 1289-1:1986. London, BSI, 1986

[94] **Rayment R and Morgan K.** Comparing conventional and electronic room-stat performance in experimental houses. *CIB Symposium S17, Efficiency of Heating Plants, Delft, September 1981* (CIE Publication No 62, 241–62)

[95] **Rayment R and Morgan K.** The performance of heating control systems in experimental houses. *Procs of CEC Intl Seminar, Brussels, 23–25 Oct 1979*

[96] **BRE.** Building management systems. *BRE Digest* 289. Garston, Construction Research Communications Ltd, 1984

[97] **Jaunzens D.** Heating controls and how to make the most of them. *Housing and Planning Review*, June/July 1992, **47** (3) 20–1

[98] **Bromley A K R and Smith G P.** User experience of BMS performance and benefits. *Heating and Air Conditioning* (1993) **6** 14–15

[99] **Energy Efficiency Best Practice programme.** Energy efficiency in hotels: energy efficient space heating and hot water. Ritz Hotel, Piccadilly, London. *BRECSU Good Practice Case Study* 245. London, DETR, 1994

[100] **Energy Efficiency Best Practice programme.** Modern domestic coal-fired boilers. *BRECSU Good Practice Case Study* 87. London, DETR, 1994

[101] **Energy Efficiency Best Practice programme.** Upgrading controls in domestic wet central heating systems: a guide for installers. *BRECSU Good Practice Guide* 143. London, DETR, 1994

[102] **Energy Efficiency Best Practice programme.** Energy efficiency in sports and recreation buildings: condensing gas boilers. *BRECSU Good Practice Case Study* 43. London, DETR, 1994

[103] **Birtles T and Shaw M.** Specifying and selecting a building management system. *BRECSU Information Leaflet* 20. London. DETR, 1990

[104] **Bromley A K R.** Installing building management systems to meet electromagnetic compatibility requirements. *Procs of Electrical Contractors Association EMC Awareness Seminar, 26 June 1990*, 31–50

[105] **British Standards Institution.** Specification for installation of gas-fired hot water boilers of rated input not exceeding 60 kW. *British Standard* BS 6798:1987. London, BSI, 1987

[106] **BRE.** Balanced flue terminals: location and guarding. *BRE Defect Action Sheet (Design)* DAS 92. Garston, Construction Research Communications Ltd, 1986

[107] **Shepherd T A.** Spillage from open flued combustion appliances. *Building Services (The CIBSE Journal)* (1994) **16** (9) 43–4

[108] **Shepherd T A.** Extract fan flow rates resulting in spillage. *Building Services Engineering Research and Technology* (1993) **14** (4) 143–9

Further reading

British Standards Institution. Heating boilers. Heating boilers with forced draught burners. Terminology, general requirements, testing and marking. *British Standard* BS EN 303-1:1992. London, BSI, 1992

British Standards Institution. Heating boilers. Heating boilers with forced draught burners. Special requirements for boilers with atomizing oil burners. *British Standard* BS EN 303-2:1992. London, BSI, 1992

British Standards Institution. Heating boilers. Test code for heating boilers for atomizing oil burners. *British Standard* BS EN 304:1992. London, BSI, 1992.

British Standards Institution. Heat exchangers. Test procedures for establishing performance of air to air and flue gases heat recovery devices. *British Standard* BS ENV 308:1991. London, BSI, 1991

British Standards Institution. Code of practice for oil firing. Installations up to 44 kW output capacity for space heating and hot water supply purposes. *British Standard* BS 5410-1:1977. London, BSI, 1977

British Standards Institution. Code of practice for oil firing. Installations of 44 kW and above output capacity for space heating, hot water and steam supply purposes. *British Standard* BS 5410-2:1978. London, BSI, 1978

British Standards Institution. Safety valves. Specification for safety valves for steam and hot water. *British Standard* BS 6759-1:1984. London, BSI, 1984

British Standards Institution. Specification for vessels for use in heating systems. Calorifiers and storage vessels for central heating and hot water supply. *British Standard* BS 853:1996. London, BSI, 1996

British Standards Institution. Specification for forced circulation hot water central heating systems for domestic premises. *British Standard* BS 5449:1990. London, BSI, 1990

British Standards Institution. Stationary circulation pumps for heating and hot water service systems. Specification for physical and performance requirements. *British Standard* BS 1394-2:1987. London, BSI, 1987

British Standards Institution. Specification for convection type space heaters operating with steam or hot water. *British Standard* BS 3528:1977. London, BSI, 1977

British Standards Institution. Specification for welded steel boilers for central heating and indirect hot water supply (rated output 44kW to 3 MW). *British Standard* BS 855: 1990. London, BSI, 1990

British Standards Institution. Safety of domestic gas appliances. Specification for central heating boilers and circulators. *British Standard* BS 5258-1:1986. London, BSI, 1986

British Standards Institution. Safety of domestic gas appliances. Combined appliances: gas fire/back boiler. *British Standard* BS 5258-8:1980. London, BSI, 1980

British Standards Institution. Safety of domestic gas appliances. Specification for combination boilers. *British Standard* BS 5258-15:1990. London, BSI, 1990

British Standards Institution. Thermal performance of domestic gas appliances. Specification for thermal performance of central heating boilers and circulators. *British Standard* BS 6332-1:1988. London, BSI, 1988

British Standards Institution. Thermal performance of domestic gas appliances. Specification for thermal performance of combined appliances: gas fire/back boiler. *British Standard* BS 6332-3:1984. London, BSI, 1984

Chapter 2.3

[109] **Hampton D.** Glastonbury Thorn First School: energy comment. *Architects Journal*, 10 November 1993 **198** (18) p51

Chapter 2.4

[110] **Lang A, Duffy F, Jaunzens D and Willis S.** *New environments for working.* BRE Report. Garston, Construction Research Communications Ltd, 1998

[111] **BRE.** Natural ventilation in non-domestic buildings. *BRE Digest* 399. Garston, Construction Research Communications Ltd, 1994

[112] **British Standards Institution.** Code of practice for accommodation of building services in ducts. *British Standard* BS 8313: 1989. London, BSI, 1989

[113] **Rayment R and Whittle G E.** Domestic warm-air heating systems using low-grade heat sources. *BRE Information Paper* IP 1/89. Garston, Construction Research Communications Ltd, 1989

[114] **Smith J T and Webb B C.** Absorption chiller efficiency. *Building Services (The CIBSE Journal)* (1993) **15** (9) 26–7

[115] **Morgan H P, Ghosh B K, Garrad G, Pamlitschka R, De Smedt J-C and Schoonbaert L R.** *Design methodologies for smoke and heat exhaust ventilation.* BRE Report. Garston, Construction Research Communications Ltd, 1999

[116] **Klote J H and Milke J A.** *Design of smoke management systems.* USA, ASHRAE and SFPE, 1992

[117] **BRE.** CFCs and buildings. *BRE Digest* 358. Garston, Construction Research Communications Ltd, 1992

[118] **Butler D J G.** Performance of air-conditioning systems with alternative refrigerants. *BRE Information Paper* IP 6/98. Garston, Construction Research Communications Ltd, 1998

[119] **Bordass W T, Bromley A K R and Leaman A J.** User and occupant controls in office buildings. *ASHRAE Conference, Building design, technology, and occupant well-being in temperate climates, Brussels, 17–19 February 1993*

[120] **Health and Safety Commission.** *The prevention or control of legionellosis (including legionnaires' disease). Approved Code of Practice.* Sudbury, HSE Books, 1995

[121] **Brundrett G W.** Building services: healthy buildings. *DoE Construction* (1989) Issue No 72

[122] **Heath and Safety Executive.** The control of legionellosis including legionnaires' disease. *HSE Guidance Note* HS(G)70. Sudbury, HSE Books, 1993

[123] **Heath and Safety Executive.** The control of legionellosis in hot and cold water systems. MISC150. Sudbury, HSE Books, 1998

[124] *The Office Environment Survey.* London, Building Use Studies, 1987

[125] **Leinster P, Raw G, Thomson N et al.** A modular longitudinal approach to the investigation of sick building syndrome. *Procs of CIB Conference, Research and Healthy Buildings, 1990*, 14–20

[126] **Smith J T and Webb B C.** Cooling technology. A fair cop? *Building Services (The CIBSE Journal)*, September 1993 **15** (9) 26–7

[127] **European Commission.** Code of good practice for the reduction of emissions of chlorofluorocarbons (CFCs) R 11 and R 12 in refrigeration and air conditioning applications. EUR 9509 EN. Brussels, CEC, 1984

Further reading

Brundrett G W. *Legionella and building services.* Oxford, Butterworth-Heinemann, 1992

Chartered Institution of Building Services Engineers. Minimising the risk of Legionnaires' disease. *CIBSE Technical Memorandum* TM 13. London, CIBSE, 1991

Chapter 2.5

[128] **W & R Chambers Ltd and Cambridge University Press.** *Chambers science and technology dictionary.* Edinburgh, W & R Chambers Ltd, 1988

[129] **British Standards Institution.** Specification for rating and performance of air source heat pumps with electrically driven compressors. *British Standard* BS 6901: 1987. London BSI, 1987

[130] **BRE.** *Annual Report 1977.* London, The Stationery Office, 1978

[131] **Energy Efficiency Best Practice programme.** All electric, air-conditioned office uses heat pump technology. GIL 24. London, DETR, 1994

[132] **Grigg P F.** An assessment of the cost-effectiveness and potential of heat pumps for domestic hot water heating. *BRE Information Paper* IP 8/85 Garston, Construction Research Communications Ltd, 1985

[133] **McCall M J and Grigg P F.** Domestic heat pumps. *Building Services (The CIBSE Journal)* (1986) **8** (5) p71

[134] **Butler D J G.** The cost-effectiveness of heat pumps in highly insulated dwellings: an assessment. *BRE Information Paper* 7/85. Garston, BRE, 1985

Further reading
Grigg P and McCall M. *Domestic heat pumps performance and economics.* BRE Report. Garston, Construction Research Communications Ltd, 1988

Chapter 2.6
[135] **Energy Efficiency Best Practice programme.** Using solar energy in schools. GIL 16. London, DETR, 1995
[136] **Department of the Environment and The Welsh Office.** *Sunlight and daylight: planning criteria for the design of buildings.* London, The Stationery Office, 1971
[137] **BRE.** Climate and site development: general climate of the UK (Part 1); influence of microclimate (Part 2); improving microclimate through design (Part 3). *BRE Digest* 350. Garston, Construction Research Communications Ltd, 1990
[138] **Littlefair P J.** *Solar shading of buildings.* BRE Report. Garston, Construction Research Communications Ltd, 1999
[139] **European Union of Agrément (UEAtc).** Directives for the assessment of solar collectors with liquid heat transfer. *Method of Assessment and Test* No 40. Garston, British Board of Agrément, 1986
[140] **European Union of Agrément (UEAtc).** Directives for the assessment of air solar collectors. *Method of Assessment and Test* No 41. Garston, British Board of Agrément, 1986
[141] **Munro D.** Power of the sun. *Building Services (The CIBSE Journal)*, October 1994
[142] **Wozniak S.** Solar water heating at a Tunbridge Wells school. *Heating and Ventilating Engineer*, July/August 1978

Further reading
Littlefair P J. Daylighting design for display-screen equipment. *BRE Information Paper* IP 10/95. Garston, Construction Research Communications Ltd, 1995
Energy Efficiency Best Practice programme. Using solar energy in schools. GIL 16. London, DETR, 1995
British Blind and Shutter Association. *Specifiers' guide and directory.* Rickmansworth, British Blind and Shutter Association, 1998
Littlefair P J. Solar dazzle reflected from sloping glazed façades. *BRE Information Paper* IP 3/87. Garston, Construction Research Communications Ltd, 1987

Chapter 3
[143] **Department of the Environment and The Welsh Office.** *The Building Regulations 1991 Approved Document F: Ventilation* (1995 edition). London, The Stationery Office, 1994
[144] **Raw G J and Fox.T A.** Condensation, heating and ventilation in small homes. *Procs of CIB W67 International Symposium, Moisture and Climate in Buildings, Rotterdam, 1990*
[145] **The Scottish Office.** *The Building Standards (Scotland) Regulations 1990.* Edinburgh, The Stationery Office, 1990
[146] **Department of the Environment (Northern Ireland).** *The Building Regulations (Northern Ireland) 1994.* Belfast, The Stationery Office, 1994

Chapter 3.1
[147] **Perera M D A E S and Kolokotroni M.** Natural ventilation in office-type buildings. *Procs of ASHRAE Indoor Air Quality and Energy Conference, 1998*
[148] **Perera M D A E S, Gilham A V and Croome D J.** Design issues for natural ventilation in the UK: commercial and public buildings. *Procs of CIBSE National Conference, Brighton, 2–4 October 1994, 83–9*
[149] **Stephen R K.** *Airtightness in UK dwellings: BRE's test results and their significance.* BRE Report. Garston, Construction Research Communications Ltd, 1998
[150] **Aizlewood C E, Oseland N A and Raw G J.** The new comfort equation: where are we now? *Procs of CIBSE National Conference, Brighton, 2–4 October 1994*
[151] **Walker R R.** Single-sided natural ventilation in a deep office space. *Building Services (The CIBSE Journal)* (1992) **14** (4)
[152] **Chartered Institution of Building Services Engineers.** *CIBSE Guide*, Volume A1. London, CIBSE, 1999
[153] **British Standards Institution.** Code of practice for ventilation principles and designing for natural ventilation. *British Standard* BS 5925:1991. London, BSI, 1991.
[154] **Walker R R and White M K.** Single-sided natural ventilation: how deep an office? *Building Services Engineering Research and Technology* (1992) **13** (4) 231–6
[155] **White M K, Kolokotroni M and Perera M D A E S.** Trickle ventilators in offices. *BRE Information Paper* IP 12/98. Garston. Construction Research Communications,1998
[156] **Perera M D A E S and Kolokotroni M.** Natural ventilation in office-type buildings. *Procs of ASHRAE Indoor Air Quality and Energy Conference, 1998*
[157] **British Standards Institution.** Performance of windows. Classification for weathertightness. *British Standard* BS 6375-1: 1989. London, BSI, 1989

[158] **Edwards M, Linden P and Walker R R.** Theory and practice: natural ventilation modelling. *Procs of CIBSE National Conference, Brighton, 2–4 October 1994*
[159] **National House Building Council.** Thermal insulation and ventilation. London, NHBC
[160] **Uglow C E.** *Background ventilation of dwellings: a review.* BRE Report. Garston, Construction Research Communications Ltd, 1989
[161] **Stephen R K, Parkins L M and Wooliscroft M.** Passive stack ventilation systems: design and installation. *BRE Information Paper* IP 13/94. Garston, Construction Research Communications Ltd, 1994
[162] **Chartered Institution of Building Services Engineers.** *CIBSE Guide*, Volume A. London, CIBSE, 1999
[163] **Chartered Institution of Building Services Engineers.** *CIBSE Guide*, Volume B. London, CIBSE, 1999
[164] **Perera M D A E S, Kaleem R, Penwarden A D and Tull R G.** Proximity effects: air infiltration and ventilation heat loss of a low-rise office block near a tall slab building. *Procs of 14th AIVC Conference, Energy Impact of Ventilation and Air Infiltration, 1993*
[165] **Kolokotroni M.** Night ventilation for cooling office buildings. *BRE Information Paper* IP 4/98, Garston, Construction Research Communications Ltd, 1998
[166] **Webb B and Kolokotroni M.** Night cooling a 1950s office. *Architects Journal*, 13 June 1996, 54–5
[167] **Kolokotroni M, Tindale A W and Irving S J.** Office night ventilation predesign tool. *Procs of 18th AIVC Conference, Optimum Ventilation and Air flow Control in Buildings, Greece, September 1997*
[168] **Chartered Institution of Building Services Engineers.** Natural ventilation in non-domestic buildings. *CIBSE Applications Manual* AM 10. London, CIBSE, 1997
[169] **Brown V M, Crump D R and Gardiner D.** Sources and concentrations of formaldehyde in household dusts. *Procs of Intl Conference, Indoor and Ambient Air Quality, 1988*
[170] **Kukadia V.** Ventilation and air pollution strategies for buildings located in urban areas. *Air Infiltration Review* (June 1997) **18** (3) 1–4
[171] **Raw G J and Coward S K D.** Exposure to nitrogen dioxide in homes in the UK. *Procs of Conference, Unhealthy Housing, Univ of Warwick, 1991*
[172] **BRE.** *Radon: guidance on protective measures for new dwellings.* BRE Report. Garston, Construction Research Communications Ltd, 1991 (1992)

[173] **Scivyer C R.** *Surveying dwellings with high indoor radon levels: a BRE guide to radon remedial measures in existing dwellings.* BRE Report. Garston, Construction Research Communications Ltd, 1993

[174] **Scivyer C R and Gregory T J.** *Radon in the workplace.* BRE Report. Garston, Construction Research Communications Ltd, 1995

[175] **Health and Safety Executive.** Ventilation of the workplace. EH22 (Rev). Sudbury, HSE Books, 1988

[176] *Control of Substances Hazardous to Health Regulations 1988* (COSHH). London, The Stationery Office, 1988

[177] **Department of the Environment.** *Smoking in public places: guidance for owners and managers of places visited by the public.* London, DOE, 1991

[178] **Perera M D A E S, Walker R R, Hathaway M B, Oglesby O D and Warren P R.** *Natural ventilation in large and multicelled buildings: theory, measurement and prediction.* BRE Report. Garston, BRE, 1984. (Also published by CEC as EUR-10552, 1986)

[179] **Perera M D A E S and Webb B C.** Minimising air leakage in commercial premises. *Building Services (The CIBSE Journal)*, February 1997, 45–6

[180] **BRE.** Continuous mechanical ventilation in dwellings: design, installation and operation. *BRE Digest* 398. Garston, Construction Research Communications Ltd, 1994

[181] **Perera M D A E S, Powell G, Walker R R and Jones P J.** Using pressurisation measurements to predict ventilation performance and heating energy requirements of a large industrial building. *Procs of 11th AIVC Conference, Ventilation System Performance, Belgirate, Italy, September 1990*

[182] **Walker R R and Smith M G.** A passive solution. *Building Services (The CIBSE Journal)* (1993) **15** (10) p57

Further reading

Perera M D A E S. Use of tracer gas techniques in the measurement of natural ventilation in buildings. *Procs of Professional Development Session, 2nd Intl Symposium, Ventilation for Contaminant Control, September 1988*

BRE. Buildings and radon. *BRE Good Building Guide* GBG 25. Garston, Construction Research Communications Ltd, 1996

Rennie D and Parand F. *Environmental design guide for naturally ventilated and daylit offices.* BRE Report. Garston, Construction Research Communications Ltd, 1998

Stephen R. Airtightness in UK dwellings. *BRE Information Paper* IP 1/00. Garston, Construction Research Communications Ltd, 2000

Stephen R. Humidistat-controlled extract fans: performance in dwellings. *BRE Information Paper* IP 5/99. Garston, Construction Research Communications Ltd, 2000

White M, McCann G, Stephen R and Chandler M. Ventilators: ventilation and acoustic effectiveness. *BRE Information Paper* IP 4/99. Garston, Construction Research Communications Ltd, 1999

Chapter 3.2

[183] **Stephen R K, Parkins L M and Wooliscroft M.** Passive stack ventilation systems: design and installation. *BRE Information Paper* IP 13/94 Garston, Construction Research Communications Ltd, 1994

[184] **Chartered Institution of Building Services Engineers.** Natural ventilation in non-domestic buildings. *CIBSE Applications Manual AM 10.* London, CIBSE, 1997

[185] **Parkins L M.** Case studies of passive stack ventilation systems in occupied dwellings. *Procs of 15th AIV Conference, Buxton, 1994*

Chapter 3.3

[186] **Stephen R K.** *Positive pressurisation: a BRE guide to radon remedial measures.* BRE Report. Garston, Construction Research Communications Ltd, 1995

[187] **Stephen R K.** Domestic mechanical ventilation: guidelines for designers and installers. *BRE Information Paper* IP 18/88. Garston, Construction Research Communications Ltd, 1988

[188] **Evans B.** Summer cooling: using thermal capacity. *Architects Journal* (1992) **196** (7) 38–41

[189] **British Standards Institution.** Specification for installation in domestic premises of gas-fired ducted-air heaters of rated input not exceeding 60 kW. *British Standard* BS 5864:1989. London, BSI, 1989

[190] **Raatschen W and Walker R R.** Measuring air exchange efficiency in a mechanically ventilated industrial hall. *ASHRAE Transactions* (1991) **97** (2) 1112–8

[191] **Fargus R and Hepworth S.** Taking control with neural networks. *Building Services (The CIBSE Journal)* (1994) **16** (6) 51–2

[192] **BRE.** *BREEZE software package for estimating ventilation rates.* BRE Software Package. Garston, Construction Research Communications Ltd, 1991

[193] **BRE.** Minimising noise from domestic fan systems. *BRE Good Building Guide* GBG 26. Garston, Construction Research Communications Ltd, 1996

[194] **Ling M K.** Ventilation and acoustics. *Air Infiltration and Ventilation Centre Technical Note.* Coventry, AIVC, 2000

[195] BRE cautions on conflict with radon membrane advice. *AJ Focus*, 7 May 1994

[196] **Perera M D A E S and Tull R G.** Assessing internal air pollution. Wind tunnel method to assess intake contamination from building exhaust. *Building Services (The CIBSE Journal)* (1992) **14** (2) p47

[197] **Shepherd T A.** Spillage of flue gases from open-flued combustion appliances. *BRE Information Paper* IP 21/92. Garston, Construction Research Communications Ltd, 1992

[198] **Nickel Development Institute.** *Stainless steel in swimming pool buildings. A guide to selection and use.* Birmingham, Nickel Development Institute, 1995

Further reading

Morgan H P and Gardner J P. *Design principles for smoke ventilation in enclosed shopping centres.* BRE Report. Garston, Construction Research Communications Ltd, 1990

BRE. Continuous mechanical ventilation in dwellings: design, installation and operation. *BRE Digest* 398 Garston, Construction Research Communications Ltd, 1994

British Standards Institution. Fire precautions in the design, construction and use of buildings. Code of practice for ventilation and air conditioning ductwork. *British Standard* BS 5588-9:1989. London, BSI, 1989

Chapter 4.1

[199] **Ministry of Agriculture, Fisheries and Food.** Farm Survey (MAF 32 series). Kew, Public Record Office, 1940

[200] **BRE.** Durability of metals in natural waters. *BRE Digest* 98. Garston, Construction Research Communications Ltd, 1977

[201] *Digest of environmental protection and water statistics.* London, HMSO, 1992

[202] **Shouler M C, Griggs J C and Hall J.** Water conservation. *BRE Information Paper* IP 15/98. Garston, Construction Research Communications Ltd, 1998

[203] **Rump M E.** Potential water economy measures in dwellings: their feasibility and economics. *BRE Current Paper* CP 65/78. Garston, Construction Research Communications Ltd, 1978

[204] **British Standards Institution.** Specification for design, installation, testing and maintenance of services supplying water for domestic use within buildings and their curtilages. *British Standard* BS 6700:1987. London, BSI, 1987

[205] **BRE.** Pipes and fittings for domestic water supply. *BRE Digest* 15 (Second series). Garston, Construction Research Communications Ltd, 1977

[206] **BRE.** Trussed rafter roofs: tank supports – specification. *BRE Defect Action Sheet (Design)* DAS 43. Garston, Construction Research Communications Ltd, 1984

[207] **The Technical Committee on Backsiphonage in Water Installations, Department of the Environment.** *Report of the committee on backsiphonage in water installations.* London, The Stationery Office, 1974

[208] **Ball E F.** Noise in water systems. *Plumbing,* January 1974

[209] **British Standards Institution.** Thermoplastics pipes and associated fittings for hot and cold water for domestic purposes and heating installations in buildings. General requirements. *British Standard* BS 7291-1: 1990. London, BSI, 1990

[210] **British Standards Institution.** Thermoplastics pipes and associated fittings for hot and cold water for domestic purposes and heating installations in buildings. Specification for polybutylene (PB) pipes and associated fittings. *British Standard* BS 7291-2:1990. London, BSI, 1990

[211] **British Standards Institution.** Thermoplastics pipes and associated fittings for hot and cold water for domestic purposes and heating installations in buildings. Specification for crosslinked polyethylene (PE-X) pipes and fittings. *British Standard* BS 7291-3:1990. London, BSI, 1990

[212] **British Standards Institution.** Thermoplastics pipes and associated fittings for hot and cold water for domestic purposes and heating installations in buildings. Specification for chlorinated polyvinyl chloride (PVC-C) pipes and associated fittings and solvent cement. *British Standard* BS 7291-4: 1990. London, BSI, 1990

[213] **Building Research Station.** Thawing frozen water pipes by electricity. *BRS Miscellaneous Paper* 6. Garston, Construction Research Communications Ltd, 1964. (Originally published in *Illustrated Carpenter and Builder*, 18th December 1964, **153**)

[214] **British Standards Institution.** Workmanship on building sites. Code of practice for hot and cold water services (domestic scale). *British Standard* BS 8000-15:1990. London, BSI, 1990

Further reading

Pitts N J, Griggs J C and Hall J. Water conservation: low-flow showers and flow restrictors. *BRE Information Paper* IP 2/00. Garston, Construction Research Communications Ltd, 2000

British Standards Institution. Chemicals used for treatment of water intended for human consumption. Iron (III) chloride. *British Standard* BS EN 888:1999. London, BSI, 1999

British Standards Institution. Float operated valves. Specification for piston type float operated valves (copper alloy body) (excluding floats). *British Standard* BS 1212-1: 1990. London, BSI, 1990

British Standards Institution. Float operated valves. Specification for diaphragm type (copper alloy body) (excluding floats). *British Standard* BS 1212-2:1990. London, BSI, 1990

British Standards Institution. Float operated valves. Specification for diaphragm type (plastics bodied) for cold water services only (excluding floats). *British Standard* BS 1212-3:1990. London, BSI, 1990

British Standards Institution. Specification for floats for ballvalves (copper). *British Standard* BS 1968:1953. London, BSI, 1953

British Standards Institution. Specification for floats (plastics) for float operated valves for cold water services. *British Standard* BS 2456:1990. London, BSI, 1990

British Standards Institution. Specification for cold water storage and combined feed and expansion cisterns (polyolefin or olefin copolymers) up to 500 L capacity used for domestic purposes. *British Standard* BS 4213:1991. London, BSI, 1991

British Standards Institution. Specification for servicing valves (copper alloy) for water services. *British Standard* BS 6675:1986. London, BSI, 1986

British Standards Institution. Specification for storage cisterns up to 500 L actual capacity for water supply for domestic purposes. *British Standard* BS 7181:1989. London, BSI, 1989

British Standards Institution. Capillary and compression tube fittings of copper and copper alloy. Specification for capillary and compression fittings for copper tubes. *British Standard* BS 864-2:1983. London, BSI, 1983

British Standards Institution. Capillary and compression tube fittings of copper and copper alloy. Specification for compression fittings for polyethylene pipes with outside diameters to BS 5556. *British Standard* BS 864-3:1990. London, BSI, 1990

British Standards Institution. Specification for wrought steel pipe fittings (screwed BS 21 R-series thread). Metric units. *British Standard* BS 1740-1:1971. London, BSI, 1971

British Standards Institution. Specification for elastomeric seals for joints in pipework and pipelines. *British Standard* BS 2494: 1990. London, BSI, 1990

British Standards Institution. Specification for copper and copper alloys. Tubes. Copper tubes for water, gas and sanitation. *British Standard* BS 2871-1:1971. London, BSI, 1971

British Standards Institution. Specification for bitumen-based coatings for cold application, suitable for use in contact with potable water. *British Standard* BS 3416: 1991. London, BSI, 1991

British Standards Institution. Specification for unplasticized polyvinyl chloride (PVC-U) pressure pipes for cold potable water. *British Standard* BS 3505:1986. London, BSI, 1986

British Standards Institution. Specification for light gauge stainless steel tubes, primarily for water applications. *British Standard* BS 4127:1994. London, BSI, 1994

British Standards Institution. Joints and fittings for use with unplasticized PVC pressure pipes. Injection moulded unplasticized PVC fittings for solvent welding for use with pressure pipes, including potable water supply. *British Standard* BS 4346-1: 1969. London, BSI, 1969

British Standards Institution. Joints and fittings for use with unplasticized PVC pressure pipes. Mechanical joints and fittings principally unplasticized PVC. *British Standard* BS 4346-2:1970. London, BSI, 1970

British Standards Institution. Specification for ductile iron pipes and fittings. *British Standard* BS 4772:1988. London, BSI, 1988

British Standards Institution. Specification for copper alloy globe, globe stop and check, check and gate valves. *British Standard* BS 5154:1991. London, BSI, 1991

British Standards Institution. Plastics pipework (thermoplastics materials). Specification for the installation of thermoplastics pipes and fittings for use in domestic hot and cold water services and heating systems. *British Standard* BS 5955-8: 1990. London, BSI, 1990

British Standards Institution. Devices with moving parts for the prevention of contamination of water by backflow. Specification for check valves of nominal size up to including DN 54. *British Standard* BS 6282-1:1982. London, BSI, 1982

British Standards Institution. Devices with moving parts for the prevention of contamination of water by backflow. Specification for terminal anti-vacuum valves of nominal size up to and including DN 54. *British Standard* BS 6282-2:1982. London, BSI, 1982

British Standards Institution. Devices with moving parts for the prevention of contamination of water by backflow. Specification for in-line anti-vacuum valves of nominal size up to and including DN 42. *British Standard* BS 6282-3:1982. London, BSI, 1982

British Standards Institution. Devices with moving parts for the prevention of contamination of water by backflow. Specification for combined check and anti-vacuum valves of nominal size up to and including DN 42. *British Standard* BS 6282-4: 1982. London, BSI, 1982

BRE Water Centre. Protecting pipes from freezing. *BRE Good Building Guide* GBG 40. Garston, Construction Research Communications Ltd, 2000

Chapter 4.2

[215] **BRE.** Unvented domestic hot water systems. *BRE Digest* 308. Garston, Construction Research Communications Ltd, 1986

[216] **British Standards Institution.** Specification for vessels for use in heating systems. Calorifiers and storage vessels for central heating and hot water supply. *British Standard* BS 853-1:1996. London, BSI, 1996

[217] **European Union of Agrément (UEAtc).** Assessment of unvented hot water storage systems and the approval and surveillance of installers. *Method of Assessment and Test* No 38. Garston, British Board of Agrément, 1986

[218] **British Standards Institution.** Specification for draw-off taps and stop valves for water services (screw-down pattern). Draw-off taps and above ground stop valves. *British Standard* BS 1010-2:1973. London, BSI, 1973

[219] **British Standards Institution.** Laboratory tests on noise emission from appliances and equipment intended for use in water supply installations. Method for measurement. *British Standard* BS 6864-1: 1987. London, BSI, 1987

[220] **British Standards Institution.** Acoustics. Laboratory tests on noise emission from appliances and equipment used in water supply installations. Mounting and operating conditions for draw-off taps and mixing valves. *British Standard* BS EN 1 SO 3822-2:1996. London, BSI, 1996

[221] **Hall J.** Safety devices for water heating equipment: temperature or pressure? *Procs CIB W62, Intl Symposium Water Supply and Drainage, Brighton, 26–29 September 1994*

[222] **Griggs J C and Hall J.** Hard water scale in hot water storage cylinders. *BRE Information Paper* IP 13/93. Garston, Construction Research Communications Ltd, 1993

[223] **Hall J.** Hardware testing at BRE. *Procs of BRE/BBA Seminar, Unvented domestic hot-water systems, 19–20 March 1986*

[224] **British Standards Institution.** Specification for expansion vessels using an internal diaphragm for unvented hot water supply systems. *British Standard* BS 6144: 1990. London, BSI, 1990

[225] **British Standards Institution.** Suitability of non-metallic products for use in contact with water intended for human consumption with regard to their effect on the quality of the water. Growth of aquatic micro-organisms test. *British Standard* BS 6920-2.4: 1996. London, BSI, 1996

[226] **Shouler M C and Hall J.** Unvented hot water storage systems: microbial growth in expansion vessels. *BRE Information Paper* IP 1/97. Garston, Construction Research Communications Ltd, 1997

Further reading

British Standards Institution. Specification for black polyethylene pipes up to nominal size 63 for above ground use for cold potable water. *British Standard* BS 6730:1986. London, BSI, 1986

British Standards Institution. Thermoplastics pipes and associated fittings for hot and cold water for domestic purposes and heating installations in buildings. General requirements. *British Standard* BS 7291-1: 1990. London, BSI, 1990

British Standards Institution. Thermoplastics pipes and associated fittings for hot and cold water for domestic purposes and heating installations in buildings. Specification for polybutylene (PB) pipes and associated fittings. *British Standard* BS 7291-2:1990. London, BSI, 1990

British Standards Institution. Thermoplastics pipes and associated fittings for hot and cold water for domestic purposes and heating installations in buildings. Specification for crosslinked polyethylene (PE-X) pipes and fittings. *British Standard* BS 7291-3:1990. London, BSI, 1990

British Standards Institution. Thermoplastics pipes and associated fittings for hot and cold water for domestic purposes and heating installations in buildings. Specification for chlorinated polyvinyl (PVC-C) pipes and fittings and solvent cement. *British Standard* BS 7291-4:1990. London, BSI, 1990

Chapter 4.3

[227] **British Standards Institution.** Sanitary installations. Code of practice for space requirements for sanitary appliances. *British Standard* BS 6465-2:1996. London, BSI, 1996

[228] **Comité Européan de Normalisation.** Specification for finished baths for domestic purposes made of acrylic material. EN 198. (See **British Standards Institution.** Baths for domestic purposes made of acrylic material. Specification for finished baths. *British Standard* BS 4305-1:1989. London, BSI, 1989)

[229] **British Standards Institution.** Baths. Connecting dimensions. *British Standard* BS EN 232:1992. London, BSI, 1992

[230] **Comité Européan de Normalisation.** Specification for cast acrylic sheet for baths and shower trays for domestic purposes. EN 263. (See **British Standards Institution.** Specification for cast acrylic sheet for baths and shower trays for domestic purposes *British Standard* BS 7015:1989. London, BSI, 1989)

[231] **Comité Européan de Normalisation.** Specification for shower trays for domestic purposes made of acrylic materials. EN 249

[232] **British Standards Institution.** Shower trays. Connecting dimensions. *British Standard* BS EN 251:1992. London, BSI, 1992

[233] **Comité Européan de Normalisation.** Specification for cast acrylic sheet for baths and shower trays for domestic purposes. EN 263

[234] **Comité Européan de Normalisation.** Sanitary tapware – Waste fittings for shower trays – General technical specifications. EN 329

[235] **Comité Européan de Normalisation.** Pedestal wash basins – Connecting dimensions. EN 31:1998. (See **British Standards Institution.** Specification for wash basins. Pedestal wash basins. Connecting dimensions. *British Standard* BS 5506-1:1977. London, BSI, 1977)

[236] **Comité Européan de Normalisation.** Wall hung wash basins – Connecting dimensions. EN 32:1998. (See **British Standards Institution.** Specification for wash basins. Wall hung wash basins. Connecting dimensions. *British Standard* BS 5606-2:1977. London, BSI, 1977)

[237] **Comité Européan de Normalisation.** Wall-hung hand rinse basins – Connecting dimensions. EN 37:1998

[238] **Comité Européan de Normalisation.** Wall-hung hand rinse basins – Connecting dimensions. EN 111. (See **British Standards Institution.** Specification for wall hung hand rinse basins. Connecting dimensions. *British Standard* BS 6731:1988. London, BSI, 1988)

[239] **British Standards Institution.** Fireclay sinks: dimensions and workmanship. *British Standard* BS 1206:1974. London, BSI, 1974

[240] **British Standards Institution.** Metal sinks for domestic purposes. Imperial units with metric equivalents *British Standard* BS 1244-1:1956. London, BSI, 1956; Metal sinks for domestic purposes. Specification for sit-on and inset sinks. *British Standard* BS 1244-2:1988. London, BSI, 1988

[241] **Comité Européan de Normalisation.** Sanitary tapware – Waste fittings for sinks – General technical specifications. EN 411

[242] **Comité Européan de Normalisation.** Kitchen sinks – Connecting dimensions. EN 695

[243] **British Standards Institution.** Safety of electrical appliances, particular requirements for toilets. *British Standard* BS EN 60335-2.84:1999. London, BSI, 1999

[244] **Comité Européan de Normalisation.** Pedestal WC pans with close-coupled flushing cistem – Connecting dimensions. EN 33, 1998

[245] **Comité Européan de Normalisation.** Wall hung WC pan with close coupled cistern – Connecting dimensions. EN 34. (See **British Standards Institution.** Specification for wall hung WC pans. Wall hung WC pan with close coupled cistern. Connecting dimensions. *British Standard* BS 5504-1:1977. London, BSI, 1977)

[246] **Comité Européan de Normalisation.** Pedestal WC pans with independent water supply – Connecting dimensions. EN 37

[247] British Standards Institution. Incinerators. *British Standard* BS 3316:1987. London, BSI, 1987

[248] Comité Européan de Normalisation. Pedestal bidets over-rim supply only – Connecting dimensions. EN 35. (See **British Standards Institution.** Specification for bidets. Pedestal bidets, over rim supply only. Connecting dimensions. *British Standard* BS 5505-1:1977. London, BSI, 1977)

[249] Comité Européan de Normalisation. Wall-hung bidets with over-rim supply – Connecting dimensions. EN 36:1998. (See **British Standards Institution.** Specification for bidets. Wall-hung bidets with over rim supply. Connecting dimensions. *British Standard* BS 5505-2:1977. London, BSI, 1977)

[250] British Standards Institution. Urinals. Stainless steel slab urinals. *British Standard* BS 4880-1:1973. London, BSI, 1973

[251] British Standards Institution. Specification for vitreous china bowl urinals. Rimless type. *British Standard* BS 5520:1977. London, BSI, 1977

[252] Comité Européan de Normalisation. Wall-hung urinals – Connecting dimensions. EN 80

[253] British Standards Institution. Kitchen furniture. Co-ordinating sizes for kitchen furniture and kitchen appliances. *British Standard* BS EN 1116:1996. London, BSI, 1996

[254] British Standards Institution. Code of practice for sanitary pipework. *British Standard* BS 5572:1994. London, BSI, 1994

[255] Comité Européan de Normalisation. Gravity drainage systems inside buildings – Waste water systems – Layout and calculation. EN 12056-2:1995

[256] British Standards Institution. Baths for domestic purposes made of acrylic material. Specification for finished baths. *British Standard* BS 4305-1:1989. London, BSI, 1989

[257] British Standards Institution. Specification for baths made from porcelain enamelled cast iron. *British Standard* BS 1189:1986. London, BSI. 1986

[258] British Standards Institution. Specification for baths made from vitreous enamelled sheet steel. *British Standard* BS 1390:1990. London, BSI. 1990

[259] British Standards Institution. Shower units. Guide on choice of shower units and their components for use in private dwellings *British Standard* BS 6340-1:1983. London, BSI, 1983

[260] British Standards Institution. Specification for cast acrylic sheet for baths and shower trays for domestic purposes. *British Standard* BS 7015:1989. London, BSI, 1989

[261] British Standards Institution. Shower units. Prefabricated shower trays made from acrylic material. *British Standard* BS 6340-5:1983. London, BSI, 1983

[262] British Standards Institution. Shower units. Specification for prefabricated shower trays made from porcelain enamelled cast iron. *British Standard* BS 6340-6:1983. London, BSI, 1983

[263] British Standards Institution. Shower units. Specification for prefabricated shower trays made from vitreous enamelled sheet steel. *British Standard* BS 6340-7:1983. London, BSI, 1983

[264] British Standards Institution. Shower units. Specification for prefabricated shower trays made from glazed ceramic. *British Standard* BS 6340-8:1985. London, BSI, 1985

[265] British Standards Institution. Specification for wash basins. Wash basins (one or three tap holes). Materials, quality, design and construction. *British Standard* BS 5506-3:1977. London, BSI, 1977

[266] British Standards Institution. Fireclay sinks. Dimensions and workmanship. *British Standard* BS 1206:1974. London, BSI, 1974

[267] British Standards Institution. Metal sinks for domestic purposes. Imperial units with metric equivalents. *British Standard* BS 1244-1:1956. London, BSI, 1956

[268] British Standards Institution. Specification for vitreous china washdown WC pans with horizontal outlet. *British Standard* BS 5503:1977–90. London, BSI, 1977–90

[269] British Standards Institution. Wall hung WC pan. *British Standard* BS 5504: 1977–90. London, BSI, 1977–90

[270] Griggs J C, Pitts N J, Hall J and Shouler M C. Water conservation: a guide for design of lpw-flush WCs (Part 1); a guide for installation and maintenance of low-flush WCs (Part 2). *BRE Information Paper* IP 8/97. Garston, Construction Research Communications Ltd, 1997

[271] British Standards Institution. Urinals. Stainless steel slab urinals. *British Standard* BS 4880:1973. London, BSI, 1973

[272] British Standards Institution. Specification for vitreous china bowl urinals. Rimless type. *British Standard* BS 5520:1977. London, BSI, 1977

[273] British Standards Institution. Specification for bidets. *British Standard* BS 5505:1977. London, BSI, 1977

[274] British Standards Institution. Code of practice for sanitary pipework. *British Standard* BS 5572:1994. London, BSI, 1994

[275] British Standards Institution. Workmanship on building sites. Code of practice for above ground drainage and sanitary appliances. *British Standard* BS 8000-13:1989. London, BSI, 1989

Chapter 4.4

[276] BRE. Hospital sanitary services: some design and maintenance problems. *BRE Digest* 81. Garston, Construction Research Communications Ltd, 1967

[277] Wise A F E. *Drainage pipework in dwellings.* London, The Stationery Office, 1957

[278] Building Services Research and Information Association. *Space allowances for building services distribution systems: detail design stage.* Bracknell, BSRIA, 1992

[279] Comité Européan de Normalisation. Ventilating pipework – Air admittance valve systems (AVS). prEN 12380

[280] Comité Européan de Normalisation. Gravity drainage systems inside buildings. Waste water systems – Layout and calculation. EN 12056-2

[281] British Standards Institution. Fire tests on building materials and structures. Test methods and criteria for the fire resistance of elements of building construction. *British Standard* BS 476-8: 1972. London, BSI, 1972

[282] Curtis M. Fire spread and plastics pipes. *BRE Current Paper* CP38/77. Garston, Construction Research Communications Ltd, 1977

[283] British Standards Institution. Discharge and ventilating pipes and fittings, sand cast or spun in cast iron. *British Standard* BS 416:1973–90. London, BSI, 1973–90

[284] British Standards Institution. Capillary and compression tube fittings of copper and copper alloy. Specification for capillary and compression fittings for copper tubes. *British Standard* BS 864-2:1983. London, BSI, 1983

[285] British Standards Institution. Copper and copper alloys tubes. Copper tubes for water, gas and sanitation. *British Standard* BS 2871-1:1971. London, BSI, 1971

[286] British Standards Institution. Specification for prefabricated drainage stack units. Galvanised steel. *British Standard* BS 3868:1973. London, BSI, 1973

[287] British Standards Institution. Specification for unplasticised PVC soil and ventilating pipes, fittings and accessories. *British Standard* BS 4514:1983. London, BSI, 1983

[288] British Standards Institution. Specification for polypropylene waste pipe and fittings (external diameter 34.6 mm , 41.0 mm and 54.1 mm). *British Standard* BS 5254:1976. London, BSI, 1976

[289] British Standards Institution. Themoplastics waste pipe and fittings. *British Standard* BS 5255:1989. London, BSI, 1989

[290] British Standards Institution. Specification, copper and copper alloy traps. *British Standard* BS 1184:1976. London, BSI, 1976

[291] **British Standards Institution.**
Specification for plastics waste traps. *British Standard* BS 3943:1979. London, BSI, 1979

[292] **Comité Européan de Normalisation.**
Cast iron pipes and fittings, their joints and accessories for the evacuation of water from buildings – Requirements, test methods and quality assurance. prEN 877

[293] **Comité Européan de Normalisation.**
Pipes and fittings of longitudinally welded hot-dip galvanized steel pipes with spigot and socket for waste water systems – Requirements, testing, quality control. EN 1123-1

[294] **Comité Européan de Normalisation.**
Pipes and fittings of longitudinally welded stainless steel pipes with spigot and socket for waste water systems – Requirements, testing, quality control. EN 1124-1

[295] **British Standards Institution.**
Sanitary tapware. Waste fittings for basins, bidets and baths. General technical specifications. *British Standard* BS EN 274: 1993. London, BSI, 1993

[296] **Comité Européan de Normalisation.**
Fibre-cement pipes and fittings for discharge systems for buildings – Dimensions, technical terms of delivery. prEN 12763

[297] **Comité Européan de Normalisation.**
Plastics piping systems for soil and waste discharge (low and high temperature) within the building structure – Polypropylene (PP) – Requirements for pipes, fittings and the system. EN 1451-1

[298] **Comité Européan de Normalisation.**
Plastics piping systems for soil and waste discharge (low and high temperature) within the building structure - Acqonitrile-butadiene-styrene (ABS) – Requirements for pipes, fittings and the system. EN 1455-1

[299] **Comité Européan de Normalisation.**
Plastics piping systems for soil and waste discharge (low and high temperature) within the building structure – Polyethylene (PE) – Requirements for pipes, fittings and the system. EN 1519-1

[300] **Comité Européan de Normalisation.**
Plastics piping systems for soil and waste discharge (low and high temperature) within the building structure – Styrene-Copolyrner-Blends (SAN + PVC) – Requirements for pipes, fittings and the system. EN 1565-1

[301] **Comité Européan de Normalisation.**
Plastics piping systems for soil and waste discharge (low and high temperature) within the building structure – Chlorinated poly(vinyl chloride) (PVC-C) – Requirements for pipes, fittings and the system. EN 1566-1

[302] **Comité Européan de Normalisation.**
Plastics piping systems for soil and waste discharge (low and high temperature) within the building structure – Unplasticized poly(vinyl chloride) (PVC-U) – Requirements for pipes, fittings and the system. prEN 1329-1

[303] **Comité Européan de Normalisation.**
Plastics piping systems with structured wall pipes for soil and waste discharge (low and high temperature) within the building structure – Unplasticized poly(vinyl chloride) (PVC-U) – Requirements for pipes, fittings and the system. prEN 1453-1

[304] **Comité Européan de Normalisation.**
Plastics piping systems for underground drainage and sewerage under pressure – Unplasticized poly(vinyl chloride) (PVC-U) – General. prEN 1456-1

[305] **BRE.** Plastics sanitary pipework: specifying for outdoor use. *BRE Defect Action Sheet (Design)* DAS 101. Garston, Construction Research Communications Ltd, 1987

[306] **British Standards Institution.**
Workmanship on building sites. Code of practice for above ground drainage and sanitary appliances. *British Standard* BS 8000-13:1989. London, BSI, 1989

Chapter 4.5

[307] **Malhotra H L.** *Fire safety in buildings.* BRE Report. Garston, Construction Research Communications Ltd, 1986

[308] **Hinkley P L.** The effects of venting on the opening of the first sprinkler. *Fire Safety Journal* (1987) **11** (3)

[309] **BRE.** Sprinkler protection of storages. *BRE Digest* 385. Garston, Construction Research Communications Ltd, 1993

[310] **British Standards Institution.**
Components for smoke and heat control systems. Specification for natural smoke and heat exhaust ventilators. *British Standard* BS 7346-1:1990. London, BSI, 1990

[311] **British Standards Institution.**
Components for smoke and heat control systems. Specification for powered smoke and heat exhaust ventilators. *British Standard* BS 7346-2:1990. London, BSI, 1990

[312] **British Standards Institution.**
Specification for smoke curtains. *British Standard* BS 7346-3:1990. London, BSI, 1990

[313] **British Standards Institution.** Fire extinguishing installations and equipment on premises. Specification for sprinkler systems. *British Standard* BS 5306-2:1990. London, BSI, 1990

[314] **British Standards Institution.** Fire precautions in the design, construction and use of buildings. Code of practice for smoke control in protected escape routes using pressure differentials. *British Standard* BS 5588-4:1998. London, BSI, 1998

[315] **BRE.** Smoke control in buildings. *BRE Digest* 396. Garston, Construction Research Communications Ltd, 1994

[316] **Hansell G O and Morgan H P.** *Design approaches for smoke control in atrium buildings.* BRE Report. Garston, Construction Resarch Communications,1994

[317] **Gustaffson N-E.** Smoke ventilation and sprinklers. A sprinkler specialist's view. Paper presented at seminar at the Fire Research Station, Borehamwood, 11 May 1992

[318] **The Loss Prevention Certification Board.** List of approved fire and security products and services. Borehamwood, LPCB, 1999

Further reading

British Standards Institution. Fixed fire fighting systems. Components for sprinkler and water spray systems. *British Standard* BS EN 12259-1:1999. London, BSI, 1999.

Chapter 4.6

[319] **British Standards Institution.**
Installation of low pressure gas pipework of up to 28 mm diameter in domestic premises. *British Standard* BS 6891:1998. London, BSI, 1998

[320] **Institution of Gas Engineers.** Gas Services, TD 4. London, IGE.

[321] **Mainstone R J, Nicholson H G and Alexander S J.** *Structural damage in buildings caused by gaseous explosions and other accidental loadings 1971–1977.* London, The Stationery Office, 1978

Further reading

Ellis B R and Crowhurst D. On the elastic response of panels to gas explosions. *Procs of Conference, Structure under Shock and Impact, Cambridge, Massachusetts, 11–13 July 1989*

British Standards Institution. Domestic butane and propane gas-burning installations. Specification for installations at permanent dwellings. *British Standard* BS 5482-1:1994. London, BSI, 1994

Chapter 4.7

[322] **Department of the Environment, The Scottish Office and The Welsh Office.** Environmental Protection Act 1990. London, The Stationery Office, 1990

[323] **Department of the Environment.** *A survey of English local authority recycling plans.* London, The Stationery Office, 1993

[324] **Shouler M G.** *Household waste: storage provision and recycling.* BRE Report. Garston, Construction Research Communications Ltd, 1998

[325] **British Standards Institution.**
Specification for mild steel dustbins. *British Standard* BS 792:1973. London, BSI, 1973

[326] **Department of the Environment.** *This common inheritance.* London, The Stationery Office, 1990

[327] **Department of the Environment and the Welsh Office.** *Making waste work. A strategy for sustainable waste management in England and Wales.* London, The Stationery Office, 1995

[328] **British Standards Institution.** Specification for plastics and rubber dustbin lids. *British Standard* BS 3735:1987. London, BSI, 1987

[329] **British Standards Institution.** Specification for moulded thermoplastics dustbins excluding lids. *British Standard* BS 4998:1985. London, BSI, 1985

[330] **Smith J T.** *An appraisal of the pneumatic refuse collection system at Lisson Green, Westminster.* Garston, BRE, 1976

[331] **Department of the Environment and The Welsh Office.** *The Building Regulations 1991 Approved Document H: Drainage and waste disposal.* London, The Stationery Office, 1991 (amended 1992)

[332] **British Standards Institution.** Code of practice for storage and on-site treatment of solid waste from buildings. *British Standard* BS 5906:1980. London, BSI, 1980

Chapter 5.1

[333] **British Standards Institution.** Medical electrical equipment. *British Standard* BS EN 60601 (various parts):1993–96. London, BSI, 1993–96

[334] **Brister A.** Body-building. *Building Services (The CIBSE Journal)* (1993) **15** (4) 14–8

[335] **Rayment R.** Energy from the wind. *Building Services Engineer* (1976) **44** (3). (Reprinted as BRE CP 59/76)

[336] **British Standards Institution.** Switches for household and similar fixed electrical installations, particular requirements, time-delay switches. *British Standard* BS EN 60669-2–3:1999. London, BSI, 1999

[337] **Ministry of Housing and Local Government.** *Homes for today and tomorrow.* (Parker Morris Report). London, The Stationery Office, 1961

[338] **Energy Efficiency Best Practice programme.** Energy efficient lighting in factories. BAT (UK & Export) Ltd, Millbrook, Southampton. *BRECSU Good Practice Case Study* 169. London, DETR, 1994

[339] **Energy Efficiency Best Practice programme.** Energy efficient lighting in factories. Tulip International (UK) Bacon Division Ltd, Thetford, Norfolk. *BRECSU Good Practice Case Study* 174. London, DETR, 1994

[340] **Energy Efficiency Best Practice programme.** Energy efficient lighting in industrial buildings. *BRECSU Good Practice Case Study* 258. London, DETR, 1995

[341] **Energy Efficiency Best Practice programme.** Energy efficient lighting in factories. *BRECSU Good Practice Case Study* 158. London, DETR, 1993

[342] **BRE.** Electrical services: avoiding cable overheating. *BRE Defect Action Sheet (Design)* DAS 62. Garston, Construction Research Communications Ltd, 1985

[343] **BRE.** Electric cables – stripping plastics sheathing. *BRE Defect Action Sheet (Site)* DAS 48. Garston, Construction Research Communications Ltd, 1984

[344] **Bromley A K R.** The Electromagnetic Compatibility Directive and its implications for equipment installers. *Electrical Contractor*, July 1990, 28–30

[345] **BRE.** Installing BMS to meet electromagnetic compatibility requirements. *BRE Digest* 424, Garston, Construction Research Communications Ltd, 1997

[346] **British Standards Institution.** Lifts and service lifts. *British Standard* BS 5655 (various parts):1979–95. London, BSI, 1979–95

[347] **Aldersey-Williams A G.** Electricity supply and distribution. *Building Research Station Factory Building Study* 10. London, The Stationery Office, 1961

[348] **Comité Européan de Normalisation.** Qualifications of electrical installation contractors. EN 59004:1998

[349] **Institution of Electrical Engineers.** Inspection and testing. *IEE Guidance Note* No 3. London, IEE, 1997

Further reading

British Standards Institution. Electrical accessories – residual current monitors for household and similar uses (RCMs). *British Standard* BS EN 62020:1999. London, BSI, 1999

British Standards Institution. Earthing and protection against electric shock. *British Standard* BS IEC 60050-195:1998. London, BSI, 1998

Institution of Electrical Engineers. *On-site guide to BS 7671:1992.* London, IEE, 1992 (revised 1996)

Lawrence M. *The Which? book of wiring and lighting.* London, Which? Ltd, 1995

British Standards Institution. Common test methods for cables under fire conditions – Test for resistance to vertical flame propagation for a single insulated conductor or cable. *British Standard* BS EN 50265:1999. London, BSI, 1999

British Standards Institution. Common test methods for cables under fire conditions – Tests on gases evolved during combustion of materials from cables. *British Standard* BS EN 50267:1999. London, BSI, 1999

Chapter 5.2

[350] **British Standards Institution.** Code of practice for installation of apparatus intended for connection to certain telecommunication systems. *British Standard* BS 6701-1:1994. London, BSI, 1994

[351] **Bromley A K R.** The electromagnetic compatibility of building cables. *Building Services (The CIBSE Journal)* (1994) **16** (2) p43

[352] **BRE.** Electrical interference in buildings. *BRE Digest* 335. Garston, Construction Research Communications Ltd, 1988

[353] **Contract Flooring Association.** *The CFA guide to contract flooring.* Rickmansworth, Phebruary Publications Ltd, 1991

[354] **Fardell P J, Rogers S, Colwell R and Chitty R.** Cable fires in concealed spaces: a full scale test facility for standards development. *Procs of Conference, Interflam '96*

[355] **Hoover J, Caudill E I, Chapin T and Clarke F.** Full-scale fire tests on concealed space communication cables. *Procs of Conference, Interflam '96*

[356] **Health and Safety Executive.** *Health and Safety (Display Screen Equipment) Regulations 1992.* London, The Stationery Office, 1992

[357] **Telecommunications Industry Association.** Transmission performance specifications for field testing of unshielded twisted-pair cabling systems. TIA publication TSB67. Arlington Va, TIA, 1995

Further reading

British Standards Institution. Information technology. Generic cabling systems. *British Standard* BS EN 50173:1996. London, BSI, 1996

Bromley A K R and Wylds S. Magnetic fields and building services. *BRE Information Paper* IP 2/97. Garston, Construction Research Communications Ltd, 1997

Atkey D A. Cabling installations: user friendly guide. *Electrical Research Association Report* 98-0668. Leatherhead, ERA Technology Ltd, 1998

Chapter 5.3

[358] **British Standards Institution.** Fire detection and alarm systems for buildings. Code of practice for system design, installation and servicing. *British Standard* BS 5839-1:1988. London, BSI, 1988

[359] **British Standards Institution.** Intruder alarms systems. Specification for installed systems with local audible and/or remote signalling. *British Standard* BS 4737-1:1986. London, BSI, 1986

[360] **British Standards Institution.** Code of practice for remote centres for alarm systems. *British Standard* BS 5979:1993. London, BSI, 1993

[361] **British Standards Institution.** Specification for intruder alarm systems for consumer installation. *British Standard* BS 6707:1986. London, BSI, 1986

[362] **British Standards Institution.** Code of practice for wire-free intruder alarm systems. *British Standard* BS 6799:1986. London, BSI, 1986

[363] **British Standards Institution.** Specification for home and personal security devices. *British Standard* BS 6800:1986. London, BSI, 1986

[364] British Standards Institution. Specification for high security intruder alarm systems in buildings. *British Standard* BS 7042:1988. London, BSI, 1988

[365] British Standards Institution. Code of practice for intruder alarm systems with mains wiring communication. *British Standard* BS 7150:1989. London, BSI, 1989

[366] British Standards Institution. Fire detection and alarm systems for buildings: Specification for manual call points. *British Standard* BS 5839-2:1983. London, BSI, 1983

[367] British Standards Institution. Fire detection and alarm systems for buildings. Specification for optical beam smoke detectors. *British Standard* BS 5839-5:1988. London, BSI, 1988

[368] British Standards Institution. Specification for carbon monoxide detectors (electrical) for domestic use. *British Standard* BS 7860:1996 London, BSI, 1996

[369] Ross D. Carbon monoxide detectors. *BRE Good Building Guide* GBG 30. Garston, Construction Research Communications Ltd, 1999

[370] British Standards Institution. Fire detection and alarm systems for buildings. Specification for automatic release mechanisms for certain fire protection equipment. *British Standard* BS 5839-3:1988. London, BSI, 1988

[371] British Standards Institution. Luminaires. Particular requirements. Specification for luminaires for emergency lighting. *British Standard* BS 4533-102.22: 1990. London, BSI, 1990

[372] BRE. Human behaviour in fire. *BRE Digest* 388, Garston, Construction Research Communications Ltd, 1993

[373] Bromley A K R. Electromagnetic screening of offices for protection against electrical interference and security against electronic eavesdropping. *Procs of CIBSE 1991 National Conference, University of Kent, Canterbury, 7–9 April 1991*, p430

[374] Bromley A K R and Wylds S. Magnetic fields and building services. *BRE Information Paper* IP 2/97. Garston, Construction Research Communications Ltd, 1997

[375] The Loss Prevention Certification Board. *List of approved fire and security products and services.* Borehamwood, LPCB, 1999

[376] National Approval Council for Security Systems. *Official list of NACOSS recognised firms.* Maidenhead, NACOSS, 1999

Further reading

Pascoe T. Domestic burglaries: the burglar's view. *BRE Information Paper* IP 19/93. Garston, Construction Research Communications Ltd, 1993

Pascoe T. Domestic burglaries: the police view. *BRE Information Paper* IP 20/93. Garston, Construction Research Communications Ltd, 1993

Chapter 5.4

[377] British Standards Institution. Sound insulation and noise reduction for buildings. Code of practice. *British Standard* BS 8233: 1999. London, BSI, 1999

Further reading

British Standards Institution. Sound systems for emergency purposes. *British Standard* BS EN 60849:1998. London, BSI, 1998

British Standards Institution. Fire detection and alarm systems for buildings. Code of practice for the design, installation and servicing of voice alarm systems. *British Standard* BS 5839-8:1998. London, BSI, 1998

British Standards Institution. Code of practice for the design, planning, testing and maintenance of sound systems. *British Standard* BS 6259:1997. London, BSI, 1997

Chapter 6

[378] British Standards Institution. Fire precautions in the design, construction and use of buildings. *British Standard* BS 5588 (various parts):1978–99. London, BSI, 1978–99

Chapter 6.1

[379] British Standards Institution. Specification for lifts, escalators, passenger conveyors and paternosters. *British Standard* BS 2655 (various parts):1970–72. London, BSI, 1970–72

[380] British Standards Institution. Safety rules for the construction and installation of lifts. *British Standard* BS EN 81:1998. London, BSI, 1998

[381] Energy Efficiency Best Practice programme. Electricity savings in a large acute hospital. *BRECSU Good Practice Case Study* 196. London, DETR, 1994

[382] Baker P, Barrick J and Wilson R. *Building sight.* London, The Stationery Office and Royal National Institute for the Blind, 1995

[383] Chartered Institution of Building Services Engineers. *Code of practice for interior lighting.* London, CIBSE, 1994

[384] Health and Safety Executive. Lifts: thorough examination and testing. *HSE Guidance Note* PM7. Sudbury, HSE Books, 1982

[385] BRE. *BRE Annual Report 1971.* The Stationery Office, London, 1971

[386] British Standards Institution. Code of practice for safe working on lifts. *British Standard* BS 7255:1989. London, BSI, 1989

[387] British Standards Institution. Code of practice for safe working on escalators. *British Standard* BS 7801:1995. London, BSI, 1995

[388] British Standards Institution. Specification for powered domestic lifts with partially enclosed cars and no lift-well. *British Standard* BS 5900:1999. London, BSI, 1999

Further reading

Lifts Regulations. Statutory Instrument 831. London, The Stationery Office, 1997

Building Services Research and Information Association. Lifts. *BSRIA Bibliography* LB 2/96. Bracknell, BSRIA, 1996

Thorpe S. Wheelchair stair lifts and platform lifts. *Handbook* 2. London, Centre for Accessible Environments, 1993

Cooke G M E. Assisted means of escape of disabled people from fires in tall buildings. *BRE Information Paper* IP 16/91. Garston, Construction Research Communications Ltd, 1991

Chapter 6.2

[389] British Standards Institution. Vehicle lifts. Specification for fixed lifts. *British Standard* BS AU 161-1b:1983. London, BSI, 1983

[390] British Standards Institution. Lifts and service lifts. Specification for hand-powered service lifts and platform hoists. *British Standard* BS 5655-14:1995. London, BSI, 1995

[391] British Standards Institution. Specification for power driven overhead travelling cranes, semi-goliath and goliath cranes for general use. *British Standard* BS 466:1984, London, BSI, 1984

[392] Bolt J. Getting automated. *Department of the Environment Construction Publication* 56. Croydon, DOE, 1986

[393] British Standards Institution. Permanently installed suspended access equipment. *British Standard* BS 6037:1990. London, BSI, 1990

[394] BRE. Sprinkler protection of storages. *BRE Digest* 385. Garston, Construction Research Communications Ltd, 1993

Further reading

British Standards Institution. Code of practice for safe use of cranes. *British Standard* BS 7121:1989–97. London, BSI, 1989–97

Index

Building services is systematically structured to enable the reader seeking particular information to identify quickly the parts of the book relevant to his or her search. The broad structure of chapter and sub-chapter titles will be seen in the Contents list on page iii.

Chapter 0, the introduction to *Building services*, describes the development of these services since late Victorian times, and typical aspects and problems that have arisen, and continue to arise.

Chapter 1 and its sub-chapters explain the habitability of whole buildings – such as thermal comfort and lighting.

Chapters 2–6 describe the main types of building services. (Some, such as services for hospitals (eg medical gases) and hydraulic mains (eg in London), are too specialised to include.)

In the other books in the BRE Building Elements series, the authors have used standard headings (or adaptations of the standard headings) to help readers identify particular areas of interest.

With *Building services*, the subject has necessarily been treated differently with headings reflecting the variety of technical criteria.

Contents of sub-chapters in Chapters 2–6
Characteristic details
Contents (and therefore section headings) describe what building services consist of, including:
- how buildings are supplied with water, electricity, gas etc
- what the services need in terms of features (pipes, cables, appliances etc) and the spaces that these features occupy

Main performance requirements and defects (not necessarily in the order shown)
The following broadly drawn sections describe what building services do (or fail to do), and the consequences they have for building users:
- outputs required or provided
- strength and stability
- choice of solution
- testing
- unwanted side effects
- health and safety
- durability
- maintenance

Work on site
The main section headings are:
- Workmanship
- Supervision of critical features
- Inspection (in panel)

Using the Index
The Index excludes words and expressions that are already presented in the list of Contents or the list of section headings. Therefore the reader should undertake his or her search in the following order.
1 list of Contents (pages iii and iv)
2 list of standard section headings (previous column)
3 Index (starts next page)

Where words and expressions which appear in the Contents and/or list of section headings also appear in other contexts, they may be shown in the Index.

Page references to captions to illustrations are shown in bold.